MICRONEEDLES

MICRONEEDLES

KEVIN ITA
Professor of Pharmaceutics, Department of Biological and Pharmaceutical Sciences, College of Pharmacy, Touro University California, Mare Island-Vallejo, CA, United States

ELSEVIER

ACADEMIC PRESS
An imprint of Elsevier

Academic Press is an imprint of Elsevier
125 London Wall, London EC2Y 5AS, United Kingdom
525 B Street, Suite 1650, San Diego, CA 92101, United States
50 Hampshire Street, 5th Floor, Cambridge, MA 02139, United States
The Boulevard, Langford Lane, Kidlington, Oxford OX5 1GB, United Kingdom

Library of Congress Cataloging-in-Publication Data
A catalog record for this book is available from the Library of Congress

British Library Cataloguing-in-Publication Data
A catalogue record for this book is available from the British Library

ISBN: 978-0-323-91811-4

For information on all Academic Press publications visit our website at https://www.elsevier.com/books-and-journals

Publisher: Stacy Masucci
Acquisitions Editor: Andre G. Wolff
Editorial Project Manager: Barbara Makinster
Production Project Manager: Selvaraj Raviraj
Cover Designer: Mark Rogers

Typeset by TNQ Technologies

I dedicate this book to my wife, Stella, and my children, Precious-Mary and Treasure-Daniel. I am grateful for their love and support.

Contents

Acknowledgments

I am grateful to Touro University California, USA, (Dean James Scott, former Dean Rae Matsumoto, former interim Dean Debbie Sasaki-Hill, Associate Dean for Academic Affairs and Research Nathalie Bergeron, and the Department of Biological and Pharmaceutical Sciences Chair Professor David Evans) for giving me the opportunity to teach Pharmaceutics and conduct research on microneedles. I thank my former students in the Master of Science in Medical Health Sciences program for the time spent on microneedle research. These students include Monika Kaur, Kathy Nguyen, Mylien Hoang, Emmy Luu, Julia Nguyen, Iqra Ahmad, Ifeoma Anne Abiandu, and Karna Ghimirey.

CHAPTER 1

Introduction

Background

Microneedle technology is promising from therapeutic, cosmetic, and diagnostic standpoints (Bhatnagar et al., 2017; Elahpour et al., 2021; Sadeghi et al., 2021). Microneedles (MNs) can be used to extract biological samples from the skin. These devices can also be used for localized or systemic administration of drugs, vaccines and other types of biologics (Bhatnagar et al., 2017). These systems are currently being studied as platforms for the delivery of COVID-19 vaccines (Ita, 2021; Kim, Erdos, et al., 2020; Kim, Yang et al., 2020). Microneedles (MNs) refer to micron-sized needles measuring 50–2000 μm long, 10–50 μm wide, possessing a tip diameter of 1–100 μm with a base width of 25–500 μm (Liu et al., 2020; Shakya et al., 2019). The microscopic needles are typically formed on a small patch and can be used for drug delivery (Ghimirey & Ita, 2020; Hao et al., 2017; Lee, Goudie et al., 2020; Lee, Jeong et al., 2020; Lee, Lee et al., 2020; Liu et al., 2020; Turner et al., 2021; Waghule et al., 2019). It is common to use a roller, patch, applicator for the administration of MNs (Rzhevskiy et al., 2018). A syringe can also be utilized for the application of hollow MNs via electric force or manually (Rzhevskiy et al., 2018). Five types of microneedles have been identified (Ita, 2020): solid, dissolving, coated, hollow, and hydrogel-forming (Fig. 1.1).

The idea of using MNs for drug delivery was first documented on May 17, 1971, in a United States patent filed by Gerstel and Place (ALZA Corporation) (Bhatnagar et al., 2017). The patent was granted on June 22, 1976 (Bhatnagar et al., 2017). In the scientific literature, the first use of microneedles for percutaneous drug delivery is ascribed to Henry and co-workers who fabricated silicon MNs and used them to enhance the transdermal transport of calcein by 4 orders of magnitude (Henry et al., 1998). The fabrication of MNs for percutaneous drug transport is usually linked to the development of high precision microstructure technology and microfabrication tools (Henry et al., 1998; Sabri et al., 2020). In earlier

Microneedles
ISBN 978-0-323-91811-4
https://doi.org/10.1016/B978-0-323-91811-4.00003-1

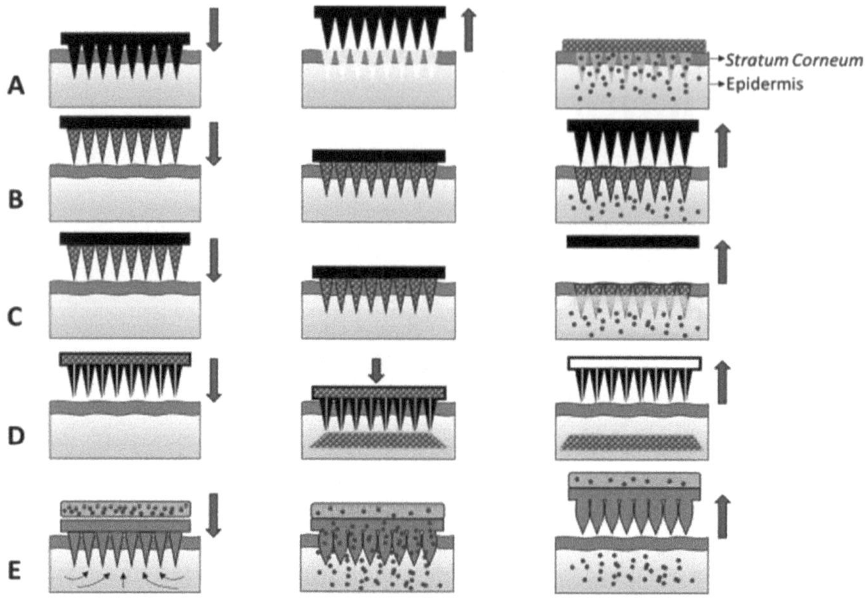

Fig. 1.1 Five types of microneedles. (A) Solid MNs for increasing the permeability of a drug formulation by creating micro-holes across the skin. (B) Coated MNs for rapid dissolution of the coated drug into the skin. (C) Dissolvable MNs for rapid or controlled release of the drug incorporated within the microneedles. (D) Hollow MNs used to puncture the skin and enable release of a liquid drug following active infusion or diffusion of the formulation through the needle bores. (E) Hydrogel-forming MNs take up interstitial fluids from the tissue, inducing diffusion of the drug located in a patch through the swollen microprojections. *(From Larrañeta, E., Lutton, R. E. M., Woolfson, A. D. & Donnelly, R. F. (2016). Microneedle arrays as transdermal and intradermal drug delivery systems: Materials science, manufacture and commercial development.* Materials Science and Engineering: R: Reports, 104, *1–32. https://doi.org/10.1016/j.mser.2016.03.001.)*

research studies, microneedles were used to enhance drug transport across the stratum corneum, an almost impermeable barrier for hydrophilic and large molecules (Rzhevskiy et al., 2018). MNs disorder the skin transiently forming micro-channels thus facilitating the transport of drugs across the skin into systemic circulation (Ghimirey & Ita, 2020).

Microneedles can be inserted into human skin either manually or with a mechanical insertion system. A spring-actuated insertion device that places microneedles into skin tissue at a velocity of 2 m/s was recently used (Wu et al., 2021). During insertion, MNs may fail due to buckling and the bending (Radhika & Gnanavel, 2021). Microneedle buckling takes place when the applied load is more than the critical buckling load. Radhika and

Gnanavel reported a critical buckling load of 0.6 N for hollow polymer MNs (Radhika & Gnanavel, 2021). This load is safe and adequate for insertion (Radhika & Gnanavel, 2021). The mechanical strength of microneedles is usually evaluated by compressing the entire MN array against a flat surface, after which the rupture force of a single microneedle is computed by dividing the total rupture force by the number of needles (Du et al., 2021). Micromanipulation is another approach used for mechanical strength investigation. It entails the measurement of the bursting force of MNs (Du et al., 2021). Force-displacement curves plotted from the compression of single objects can be utilized for the calculation of mechanical parameters such as displacement at rupture and rupture force (Du et al., 2021). Young's modulus and rupture stress of the MNs are then derived (Du et al., 2021). Single microneedles were compressed between a stage and a glass probe with a compression speed of 2 μm/s (Du et al., 2021). The authors documented the applied force on single microneedles as a function of their displacement (Du et al., 2021). MNs pierce the skin barrier thus forming tiny holes (Fig. 1.2) which allow direct permeation of drug molecules into the epidermis (Ronnander et al., 2020).

Although microneedle-based drug delivery systems were conceptualized more than three decades ago, research toward clinical application has seen tremendous growth in the past 15 years (Moreira et al., 2019). Injections delivered from hypodermic needles cause considerable pain (Ita, 2021; Kim, Erdos, et al., 2020; Kim, Yang et al., 2020). Furthermore, utilization of hypodermic needles can result in bruising, discomfort, hemorrhage and

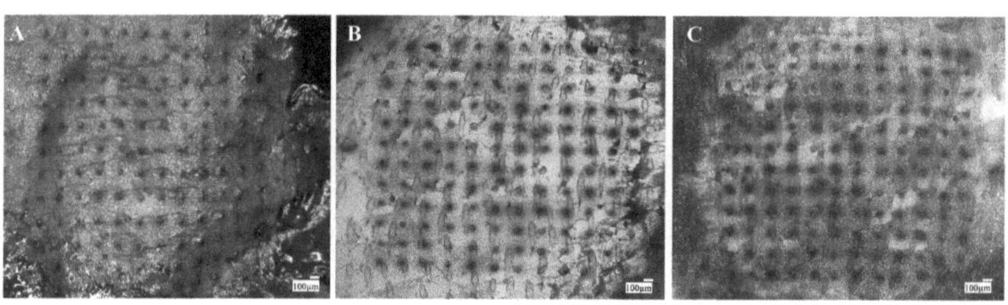

Fig. 1.2 Digital images of methylene blue (MB) deposited holes on neonatal porcine skin after insertion of MB (2% w/w) loaded pullulan (20% w/w) dissolving microneedles. Dissolving microneedles at different time points: (A) 5 min, (B) 6 h and (C) 24 h at 37°C. *(From Vora, L. K., Courtenay, A. J., Tekko, I. A., Larrañeta, E. & Donnelly, R. F. (2020). Pullulan-based dissolving microneedle arrays for enhanced transdermal delivery of small and large biomolecules.* International Journal of Biological Macromolecules, 146, *290—298. https://doi.org/10.1016/j.ijbiomac.2019.12.184.)*

hypersensitivity at the injection site (Ita, 2015). There are other challenges with hypodermic needles such as the necessity to train medical staff and accidental needle stick injury (Ita, 2015). MNs can pierce the skin without causing bleeding, pain, or infection (Ghimirey & Ita, 2020; Hao et al., 2017; Lee, Goudie et al., 2020; Lee, Jeong et al., 2020; Lee, Lee et al., 2020; Turner et al., 2021; Waghule et al., 2019; Nagarkar et al., 2020). These devices are useful because they cause little tissue damage, as they only penetrate a portion of epidermis that is not innervated as well as parts of the dermis that are free of nociceptors (Ita & Ashong, 2020). Additionally, these devices possess the advantages of hypodermic needles (drug delivery to the bloodstream) without the associated adverse event (pain). Similar to traditional transdermal patches, transdermal microneedles are convenient resulting in improved patient compliance (Lopez-Ramirez et al., 2020; Nguyen et al., 2019). It has been shown that the use of MNs can result in improved patient compliance as they do not stimulate nociceptors and patients with needle phobia are more likely to use these devices (Ita, 2017). The use of microneedles has generated significant interest in scientific circles due to the benefits of this drug delivery system. Besides painlessness and convenience, the possibility of fabricating a variety of MNs is another advantage. The diversity of features and different mechanisms for increasing percutaneous drug transport imbue MNs with a high degree of applicability (Rzhevskiy et al., 2018). A faster rate of healing at injection sites has also been documented for microneedles in comparison with hypodermic needles (Bariya et al., 2012). Microneedles are used to transiently breach the stratum corneum and skin integrity should return to physiological levels after microneedle-assisted drug delivery to the skin (Shah & Choudhury, 2017).

Microneedle-assisted delivery of therapeutic agents

Microneedle fractional radiofrequency (MNRF) is a method that uses MNs to convey radiofrequency (RF) energy into the integument (Hansen et al., 2019). A research group administered a 7 × 7 array of 1500 μm microneedles on two contiguous regions of the right hip of five healthy subjects (Hansen et al., 2019). MNRF using high RF energy was delivered to one region while the second area was subjected to MNRF with low RF energy. The structure of the micropores was studied quantitatively and qualitatively using optical coherence tomography (OCT) and reflectance confocal microscopy (RCM) (Hansen et al., 2019). One participant was administered punch biopsy in two areas (Hansen et al., 2019). Content, shape as well as heat-induced coagulation zone (CZ) of the micropores was observed with RCM (Hansen et al., 2019). When high RF energy was used, the obtained microconduits contained hyperreflective granules, possessed concentric shape, and the tissue was coagulated from epidermis to the dermo-epidermal junction (Hansen et al., 2019). On the other hand, microconduits created using low RF energy possessed a stellate shape, had no content and the CZs were observed only in the epidermis (Hansen et al., 2019). The use of high RF energy led to more easily identifiable and deeper (150 μm) micropores in contrast to shallow microconduits (70 μm) obtained with low RF energy (Hansen et al., 2019). Tissue coagulation was observed even at a depth of 1500 μm at high RF energy, but at low RF energy, perturbation was only observed in the epidermis (Hansen et al., 2019). The use of MNRF resulted in the formation of micropores with distinct features depending on RF energy as confirmed with OCT and RCM (Hansen et al., 2019). In vivo imaging techniques permit qualitative, and quantitative evaluation of MNRF effect on the skin following the application of MNs (Hansen et al., 2019).

Gupta carried out human experiments that demonstrated skin resealing following microneedle insertion (Gupta et al., 2011). The authors reported that occlusion markedly impeded skin barrier resealing following the application of microneedles. The research also showed that enhanced skin permeability did not depend on the length, number, or cross-sectional area of the microneedles (Gupta et al., 2011). In another study, it was shown that rat skin retained its complete integrity within 6 h of inserting microneedles prepared from polyvinylpyrrolidone (Shah & Choudhury, 2017). Microneedles are usually fabricated from metals, polymers, glass, and silicon (Liu et al., 2020; Shakya et al., 2019). Researchers are able to fabricate MNs possessing different shapes (snake-fang like, cylindrical, pyramidal conical, spiky, etc) (Liu et al., 2020). Another advantage of MNs is that these devices can encapsulate biomacromolecules in the dry state, which increases drug stability and leads to a reduction in the utilization of cold chain (Yang et al., 2020). Some materials used for the fabrication of dissolving MNs also have solubilizing properties. As an example, polyvinylpyrrolidone (PVP) possesses excellent solubilizing ability and can be useful for the transdermal delivery of poorly soluble drugs such as rapamycin (Mao et al., 2020).

Polyvinylpyrrolidone microneedles are mechanically robust and thus capable of penetrating the stratum corneum (Ghimirey & Ita, 2020; Hao et al., 2017; Lee, Goudie et al., 2020; Lee, Jeong et al., 2020; Lee, Lee et al.,

2020; Turner et al., 2021; Waghule et al., 2019). MNs can be used to deliver both low molecular weight (e.g. doxorubicin) and macromolecular therapeutic agents such as vaccines, genetic materials (e.g. plasmid DNA and siRNA), proteins (e.g. ovalbumin) or even nanomedicines (Moreira et al., 2019). MNs provide a convenient means of administering inactivated poliovirus vaccine (IPV) (Donadei et al., 2019). MN technology is convenient and eliminates hazardous waste leading to improved immunization programs (Donadei et al., 2019). Furthermore, vaccines delivered from MNs have exhibited enhanced storage stability in comparison to traditional formulations (Jeong et al., 2019). As research in MNs has widened over the last few decades, many scientists have produced evidence showing that MNs can be used to target the immune system within the skin (Henry et al., 1998; Sabri et al., 2020). The epidermal layer of the skin contains keratinocytes, melanocytes, T lymphocytes, Langerhans cells (LCs) (Das et al., 2019). The dermis contains gamma delta (γδ), natural killer (NK) cells, dendritic cells (DCs), T-cells, macrophages and innate lymphoid cells (ILCs) (Das et al., 2019). Lymph and blood vessels can also be targeted.

Lymphotropic delivery of drugs and vaccines is useful for the management of conditions such as cancer and human immunodeficiency viral infection (Sabri et al., 2020). The use of MNs for vaccine delivery is advantageous from an economic standpoint (Rouphael et al., 2017): avoidance of costs linked to the disposal of sharps, reduced cold chain needs due to enhanced thermostability, low cost of production reduced transport, storage, and disposal costs (Rouphael et al., 2017). It is also challenging and unsafe for untrained people to use hypodermic needles and syringes for self-administration at home (Liu et al., 2020). One of the key considerations when designing MNs is their length; the needles must be long enough to pierce the skin while also minimizing injury to the tissue or nerves (Ghimirey & Ita, 2020; Hao et al., 2017; Lee, Jeong, et al., 2020; Turner et al., 2021; Waghule et al., 2019). Despite the aforementioned benefits, MNs also possess disadvantages. Due to the small volume of MNs, the quantity of medication that can be encapsulated in a conventional MN is limited (Yang et al., 2020). Secondly, MNs can cause allergy or skin irritation (Ghimirey & Ita, 2020; Hao et al., 2017; Lee, Jeong, et al., 2020; Turner et al., 2021; Waghule et al., 2019). Thirdly, microneedle tip breakage can occur because the needle size is microscopic and thinner in comparison to hair thickness (Waghule et al., 2019). Several challenges have to be dealt with for MNs to find extensive clinical use (Ita, 2015). These concerns include irritation as well as contamination by microbes (Ita, 2015).

In addition, therapeutically relevant concentrations of medications may not be delivered via certain MNs (Ita, 2015). There may also be poor modulation of drug delivery, inadequate drug loading, the lack of suitable biomaterials for fabrication (Ita, 2015). MNs, in some cases, may not be mechanically robust (Ita, 2015).

Microneedles have been investigated for the prevention, diagnosis and monitoring of diseases (Ghimirey & Ita, 2020; Hao et al., 2017; Lee, Jeong, et al., 2020; Turner et al., 2021; Waghule et al., 2019). The use of novel technologies (e.g. genomics and bio-sensing) has led to a quicker and more precise diagnosis as well as the monitoring of pathological disorders with a small quantity of body fluid (Lee, Jeong, et al., 2020). However, traditional sampling techniques, such as those using syringes, needles, and lancets, have not matched detection innovations (Lee, Jeong, et al., 2020). MNs can be used to solve some of these problems. MNs made from metals, silicon or glass are typically used to form microchannels in the skin through which drugs or vaccines can be transported into the bloodstream (Vora et al., 2020). The drawbacks MNs are: the restricted quantity of drug that can be loaded into the MNs, and the potential risk of MNs breaking (Vora et al., 2020). In addition, microneedles typically have straight and symmetrical shapes making it difficult for them to bind to the skin tightly (Zhang et al., 2021). To overcome this limitation and using eagle's claws as a bioinspired model, Zhang et al. developed a special microneedle system that incorporates liquid metal (Zhang et al., 2021). Liquid metals possess outstanding thermal and electrical conductivity, tunable mechanical properties, high plasticity, and low vapor pressure, (Sun et al., 2020). In addition, they are biocompatible and fluidic as the name indicates (Sun et al., 2020).

Five categories of MNs are currently investigated (Ita, 2020). They include coated, dissolving, solid, hollow, and hydrogel-forming microneedles (Fig. 1.1). MN arrays are usually made from dissolving or biodegradable polymers. Drug molecules are loaded into a dissolvable polymer and the drug is released into the skin following polymeric dissolution (Vora et al., 2020). The "poke-and-patch" principle guides the use of solid MNs for drug transport into the skin (He et al., 2019). These devices permeate the skin and form temporary micro-conduits. Subsequently, a transdermal patch loaded with the medication is administered onto the skin from which it diffuses into the bloodstream (He et al., 2019). With the "coat and poke" approach, microneedles coated with the medication are injected into the skin (Prausnitz, 2004). There is usually no drug reservoir thus the entire medication to be administered is on the needle itself (Prausnitz, 2004). In an interesting study, carboxymethyl chitosan (CMC) nanoparticles loaded with 5-fluorouracil were used to coat stainless steel microneedles. Following administration of the MNs, the nanoparticles dissolved and diffused into a pig skin (Park & Kim, 2021). The MNs can be self-administered and effectively delivered the drug thus inhibiting human keloid fibroblasts (Park & Kim, 2021). The authors concluded that a system that combines 5-FU-loaded CMC nanoparticles with MNs can be used for the management of keloid scars (Park & Kim, 2021). In another study, microneedles coated with rhodamine B was injected into pig skin (Park & Kim, 2021). The depth of penetration of rhodamine B was enhanced 4.5-fold for long and 2-fold for short microneedles respectively compared to the control formulation (Park & Kim, 2021). Crosslinked poly (ethylene glycol) and Gantrez®S-97 were used to produce single 4.5 mm and 2 mm polymeric microneedles (Park & Kim, 2021). The compound from different skin sections was quantified and it was shown that shorter needles transported 10 μg of rhodamine B into porcine skin at a depth of 1.5—2.0 mm while long needles (4.5 mm) delivered 5 μg of the compound into deeper layers (2.0—3.0 mm) thus providing evidence that coated polymeric needles can be utilized for deep and rapid intradermal drug administration (Park & Kim, 2021).

Different coating formulations and methods can be used to preserve vaccines and to facilitate vaccine delivery into the skin (Witting et al., 2015). A feasibility study was carried out to determine a suitable needle length for protein administration into the epidermis. The authors also prepared a protein-stabilizing coating formulation (Witting et al., 2015). Following the use of coated MNs, 90% of asparaginase bioactivity was preserved for 3 months and the obtained protein concentration was between 10 and 23 μg (Witting et al., 2015). Following a single administration, an intraepidermal deposition of 68.0% ± 11.7% of asparaginase was achieved (Witting et al., 2015). However, the authors reported a slightly greater interleukin 8 concentration following MN insertion and this was attributed to minor skin irritation (Witting et al., 2015). It was shown by the authors that microneedle-mediated intraepidermal deposition of proteins or peptides' with MNs measuring 300 μm feasible (Witting et al., 2015).

Microneedles (MNs) can also be utilized for the intradermal delivery of autoantigens (Arikat et al., 2020). To date, only single peptide autoantigens have been investigated and this can restrict immunotherapy of patients who express specific human leukocyte antigen (HLA) molecules

(Arikat et al., 2020). Arikat et al. developed MNs that were used to target proinsulin, a massive multi-epitope protein that can induce tolerance in a heterogeneous group of type 1 diabetes patients (belonging to various HLA subpopulations) (Arikat et al., 2020). An optimized coating formulation comprising a surfactant, diluent and proinsulin resulted in a reproducible and uniform coating of more than 30 μg of the proinsulin on a stainless steel MN array made up of 30 needles (500 μm long) (Arikat et al., 2020). The MNs coated with proinsulin efficiently punctured the skin following administration to a murine model. A significant portion of the proinsulin (86%) was reproducibly transported into the local tissue using an insertion time of 150 s (Arikat et al., 2020). When coated MNs were utilized for the localized administration of proinsulin to non-obese mice with diabetes (NODs), the spread of adoptively transferred $CD8^+$ T cells was observed in the axillary lymph nodes in the MN-treated batch of rodents but not in the untreated batches or those treated with traditional intradermal injections (Arikat et al., 2020). This study shows that targeted delivery of the multi-epitope proinsulin antigen to antigen-specific T cells can be achieved (Arikat et al., 2020).

In another study, a frozen spray-coating technique was used to prepare double-layered microneedles (MNs) (Ning et al., 2020). Hydrophilic (FITC, FITC-Dextran, Insulin) and hydrophobic molecules (Nile red, Cy5) were used to coat swellable methacrylated hyaluronic acid (MeHA)-based MNs (Ning et al., 2020). MeHA double-layered MNs coated with insulin was used to deliver the drug into the bloodstream of mice resulting in the control of plasma glucose concentration (Ning et al., 2020). The investigators also extracted interstitial fluid for biomarker (glucose) analysis (Ning et al., 2020).

Dissolving MNs can be fabricated out of biodegradable or water-soluble polymer that incorporates the medication within the MN matrix (Ita, 2021; Kim, Erdos, et al., 2020). The MNs are capable of fully dissolving or degrading in the skin and releasing the encapsulated drug leaving behind no biohazardous waste (sharps) (Kim et al., 2012). Chi and coworkers developed a microneedle patch from chitosan and incorporated a smart responsive system for wound healing (Chi et al., 2020). Chitosan has excellent properties such as antibacterial activity and has been widely utilized for wound healing. Another important attribute of this system is that loaded medications can be effectively delivered into the target area thus circumventing undue adhesion between the patch and the skin (Chi et al., 2020). The authors used a thermally sensitive gel to incorporate vascular endothelial growth factor (VEGF) into the micro-conduits of the microneedle via the use of temperature-sensitive chitosan hydrogel (Chi et al., 2020). The smart release of the drugs was thus regulated through increase in temperature caused by an inflammatory response at the wound site (Chi et al., 2020). The microneedle array can inhibit collagen deposition, inflammation, and angiogenesis. In addition, the tissue can regenerate leading to wound closure (Chi et al., 2020).

Fonseca et al. used pullulan (PL), a water-soluble polysaccharide, for the fabrication of insulin-loaded microneedles (Fonseca et al., 2020). PL MNs patches micromolding was used for the production of the PL MNs. The microneedles possessed good mechanical properties (>0.15 N per needle) and thermal stability (Tdmax = 294°C). The MNs penetrated human abdominal skin by up to 381 μm as shown by porcine skin experiments (Fonseca et al., 2020). Following administration in vitro, the microneedle array dissolved within 120 min releasing up to 87% of the drug (Fonseca et al., 2020). Circular dichroism spectropolarimetry was used to prove insulin stability following storage at 4, 20 and 40°C for 4 weeks (Fonseca et al., 2020). Under these conditions, the secondary structure of the peptide did not change (Fonseca et al., 2020). The microneedles did not display cytotoxicity toward human keratinocytes and can be used for the transdermal delivery of insulin thus circumventing the subcutaneous mode of administration (Fonseca et al., 2020).

Dissolving microneedles (DMNs) can be used for the percutaneous delivery of therapeutic agents with the aim of managing a variety of diseases (Kim, Yang et al., 2020). One constraint is that most DMN production techniques entail drug reconstitution, which can lead to limited loading or drug degradation (Kim, Yang et al., 2020). Kim et al. fabricated powder-carrying microneedles (PCMs) with no need for reconstitution (Kim, Yang et al., 2020). These microneedles place insulin powder directly into the skin avoiding the need to administer a sticky patch (Kim, Yang et al., 2020). PCMs were used to deliver insulin with less degradation and solubility issues in comparison with DMNs of the same geometries (Kim, Yang et al., 2020). Furthermore, PCMs displayed a more sustained release kinetics and greater stability (Kim, Yang et al., 2020).

Dutasteride is normally used as an oral capsule for the management of symptomatic benign prostatic hyperplasia (Giffen et al., 2020). Giffen et al. assessed an alternative route of administering the drug with microneedles (Giffen et al., 2020). The poor water solubility, low oral dose and slow rate of elimination of the drug motivated the use of a once-weekly microneedle approach (Giffen et al., 2020). The concept of sustained release was initially proven in mini-pigs whereby simple intradermal administration of a nanomilled dutasteride suspension (0.12 mg/kg) was associated with an exposure period of at least 1 month (Giffen et al., 2020). Dissolvable microarrays were successfully manufactured using the suspension and were administered to rats at doses up to 0.32 mg/kg. In these studies, serum dutasteride was quantifiable for approximately 2 weeks after a single application (Giffen et al., 2020). In silico modeling of the rat data using a two-compartment intradermal model was conducted and it was predicted that, in humans, a once-weekly dose of 2 mg, given as a microarray, could deliver cumulative and therapeutically relevant levels of dutasteride in a manner which is comparable to that observed with the current oral regimen (Giffen et al., 2020).

Shu et al. used a casting technique to prepare sinomenine hydrochloride-loaded dissolving microneedles from chondroitin sulfate and polyvinylpyrrolidone (Shu et al., 2020). The authors carried out in vitro transdermal drug delivery experiments and showed that the cumulative permeation and transdermal flux rate of the drug from microneedles (SH-MN) were 5.31 and 5.06 times greater compared to that of the sinomenine hydrochloride gel (Shu et al., 2020). Pharmacokinetic experiments determined that the values of the area under the curve in the skin and blood following the application of SH-MN were 1.43- and 1.63-fold higher respectively compared to the gel (SH-G) (Shu et al., 2020). It was shown that 4 h after drug administration, the tip of the microneedle was almost totally dissolved, and the drug was delivered transdermally to a depth of 200 μm. These results provided evidence that dissolving microneedles can be used effectively for transporting sinomenine hydrochloride across rat skin (Shu et al., 2020).

Lee et al. studied dissolving microneedles loaded with lidocaine (Li-DMN) (Ghimirey & Ita, 2020; Hao et al., 2017; Lee, Jeong, et al., 2020; Turner et al., 2021; Waghule et al., 2019). The authors used centrifugal lithography to prepare Li-DMN from carboxymethyl cellulose (Lee, Lee, et al., 2020). The authors used fracture force analyzer and optical microscopy to evaluate the physical properties of the microneedles. Histology section in vitro and ear thickness experiments in vivo were used to investigate biocompatibility (Lee, Lee, et al., 2020). The efficacy of the MNs in vitro was evaluated via electrophysiological recordings in primary cultured sensory neurons (Lee, Lee, et al., 2020). In addition, von Frey test was conducted on rats' hind paws in vivo. The microneedles were robust enough for to pierce the skin (Lee, Lee, et al., 2020). The authors incorporated 331.20 ± 6.30 μg of lidocaine into the patch. It was shown that microneedle cytotoxicity to neuronal cells was insignificant for when an effective dose of lidocaine was used (Lee, Lee, et al., 2020). Electrophysiological parameters confirmed the inhibitory action of the voltage-gated sodium channel current following the use of the lidocaine microneedle arrays (Lee, Lee, et al., 2020). The histology of the rats' ears following the administration of the dissolving microneedles loaded with lidocaine was insignificant (Lee, Lee, et al., 2020). Skin reaction to the edema test was negative. Administration of the microneedle patch also decreased the nocifensive reaction of the rats (Lee, Lee, et al., 2020).

Li and coworkers investigated the mechanical properties of polylactic acid MNs and the influence of drug concentration, administration time, MN dimensions, and the viscosity of the formulation on drug delivery into pretreated porcine skin (Li et al., 2017). The authors showed that 600 μm microneedles had good mechanical stability while MNs with the density of 256 needles per cm^2 were the most efficient in enhancing transdermal drug delivery (Li et al., 2017). Furthermore, increasing drug concentration enhanced the cumulative amount of drug delivered but not the drug permeation rate. There was an inverse relationship between drug viscosity and the amount of the drug delivered (Li et al., 2017). The amount of the model compound (sulforhodamine B) delivered was constant after 1 h (Li et al., 2017). Significantly, the MNs were used to deliver insulin in vivo (Li et al., 2017). Blood glucose levels of diabetic mice decreased to 29% of initial level at 5 h because of enhanced transdermal delivery of insulin following the application of MNs (Li et al., 2017). It was shown that biodegradable polymer solid microneedles can be used for the painless drug delivery (Li et al., 2017).

Hollow microneedles (HMNs) possess a basic structure that is similar to the traditional hypodermic needle, i. e, the presence of a lumen through which a solution flows, but HMNs are shorter (normally less than 1000 μm) (Ita, 2018). When an HMN is used, the drug solution is transported either actively utilizing a pressure-driven flow through the bore of the needle or passively by diffusion (Ita, 2018). When medications are administered via HMNs, the flow rate of the drug solution or other compounds can be controlled via traditional flow-control devices, such as micropumps and syringes (Ita, 2018). Chen et al. used a hollow-structured microneedles to facilitate the transdermal delivery of cold atmospheric plasma (CAP) for the lysis of tumor cells (Chen et al., 2020). The maturation of dendritic cells in the tumor-draining lymph nodes is usually induced by the release of tumor-associated antigens (Chen et al., 2020). This cascade eventually leads to the initiation of T cell-mediated immunity (Chen et al., 2020). An immune checkpoint inhibitor (anti-programmed death-ligand 1 antibody (aPDL1)) was delivered from the MNs to boost the immune response directed against tumor cells (Chen et al., 2020). The authors found out that these hollow microneedles can be used to immune checkpoint blockade with cold atmospheric plasma thus inhibiting tumor growth and extending the rodents (Chen et al., 2020).

Several researchers have used hydrogel-forming microneedles (HFMs) for the percutaneous administration of medications (Courtenay et al., 2020; Tekko et al., 2020). These devices absorb interstitial skin fluid, swell and create micropores thus facilitating drugs transport into blood vessels in the skin (Courtenay et al., 2020). Typically, there are no medications in the matrices of HFMs but these systems swell in the skin thus enhancing drug transport from an attached reservoir (Migdadi et al., 2018). Subsequently, the drug diffuses from the reservoir layer through the dermal microcirculation into the bloodstream (Migdadi et al., 2018). Once administered to the skin, hydrogel-forming MNs can be removed intact, with no polymer remaining (Migdadi et al., 2018). This is more beneficial compared to dissolving MNs (Migdadi et al., 2018).

The management of treatment resistant depression with conventional therapeutic procedures is challenging (Courtenay et al., 2020). However, esketamine can be used as a therapeutic alternative in patients suffering from depressive crisis states (Courtenay et al., 2020). It is important to emphasize that intravenous drug administration can lead to several adverse events. Researchers have suggested that advanced delivery methods can be used to administer esketamine parenterally (Courtenay et al., 2020). Courtenay et al. formulated esketamine-loaded hydrogel-forming microneedles (Courtenay et al., 2020). The microneedles were fully characterized and skin insertion studies were carried out (Courtenay et al., 2020). Lyophilized reservoirs and polymeric films loaded with esketamine were evaluated (Courtenay et al., 2020). The authors also conducted content recovery and dissolution experiments (Courtenay et al., 2020). Transdermal transport experiments were conducted using 350 μm thick neonatal porcine skin in modified Franz cell apparatus (Courtenay et al., 2020). The authors chose lead reservoir candidates by determining physicochemical properties with in vivo studies conducted in female Sprague-Dawley rats (Courtenay et al., 2020). Blood samples were assayed for esketamine with reverse phase high performance liquid chromatography (Courtenay et al., 2020). Lyophilized reservoirs and polymeric films were used to achieve drug blood levels that were higher than the therapeutically relevant concentration of 0.15−0.3 μg/mL over 24 h (Courtenay et al., 2020). Mean plasma concentrations in rats, 24 h following the administration of the microneedles with some drug reservoirs were 0.260 and 0.498 μg/mL (Courtenay et al., 2020).

Inspired by endoparasites that swell their proboscis to anchor to host's intestines, Jeon et al. fabricated a hydrogel-forming double-layered adhesive microneedle (MN) patch made up of a non-swellable silk fibroin (SF)-based core and a swellable mussel adhesive protein (MAP)-based shell (Jeon et al., 2019). The MN device has tissue insertion capability that is 7-times higher than the force required for pig skin penetration (Jeon et al., 2019). Due to selective swelling-assisted physical association and surface adhesion engendered by the use of mussel adhesive protein, the use of these MNs led to excellent wound sealing ex vivo against luminal leaks (139.7 ± 14.1 mmHg) (Jeon et al., 2019). The effect was similar to that obtained with suture (151.0 ± 23.3 mmHg) (Jeon et al., 2019). The authors also showed that in vivo the performance against wet and/or dynamic external and internal tissues was superior (Jeon et al., 2019). The double-layered adhesive microneedle MN patch can be utilized for the management of gastrointestinal and vascular wound healing as well as for the percutaneous transport of anti-inflammatory or proregenerative compounds (Jeon et al., 2019).

Hydrogel-forming microneedles (HFMs) can also be used to deliver water-insoluble drugs (Zhou, 2020). With this class of MNs, direct medication loading into the hydrophilic matrices is challenging (Zhou, 2020). Zhou used β-cyclodextrin (GelMA-β-CD) and gelatin methacryloyl derivatives to fabricate biodegradable MNs (Zhou, 2020). The authors loaded curcumin, a water-insoluble and an unstable anticancer drug into the MNs. The use of these microneedles boosted both stability and solubility of the drug. This was ascribed to the creation of an inclusion complex (Zhou, 2020). The polymer-drug complex (GelMA-β-CD/CUR) was loaded into MNs with the needles possessing adequate mechanical strength to pierce the skin and provide tunable drug release (Zhou, 2020). Anticancer effect of the drug was evaluated in three-dimensional B16F10 melanoma cells (Zhou, 2020). GelMA-β-CD/CUR MN displayed relatively greater therapeutic effect in comparison with a non-transdermal patch (Zhou, 2020). In vivo experiments also confirmed degradability and biocompatibility of the GelMA-β-CD MN MNs (Zhou, 2020).

MNs can also be divided into two types (out-of-plane and in-plane) based on the structure (He et al., 2019). In out-of-plane MNs, the length of MNs is perpendicular to the substrate while the length of MNs is parallel to the substrate in the case of in-plane MNs (He et al., 2019). Because the lengths of the microneedles protrude out of the substrate surface in out-of-plane MNs, it is less challenging to make this category of MNs in 1D or 2D arrays (Ashraf et al., 2011). Nevertheless, it is difficult to produce out-of-plane MNs with high aspect ratio structure (Ashraf et al., 2011). The principal benefit of in-plane MNs is that microneedle length can be accurately and easily achieved during the manufacturing process (Ashraf et al., 2011). The disadvantage of in-plane MNs is that it is very challenging to make these this class of MN arrays with two-dimensional geometry (Ashraf et al., 2011). Another shortcoming of in-plane MNs is the limited density (Ashraf et al., 2011). Even though microneedles are being investigated for a wide range of applications, several technologies are used for the fabrication of these devices. In Chapter two, we will examine these technologies.

References

Arikat, F., Hanna, S. J., Singh, R. K., Vilela, L., Wong, F. S., Dayan, C. M., Coulman, S. A., & Birchall, J. C. (2020). Targeting proinsulin to local immune cells using an intradermal microneedle delivery system; a potential antigen-specific immunotherapy for type 1 diabetes. *Journal of Controlled Release, 322*, 593–601. https://doi.org/10.1016/j.jconrel.2020.02.031

Ashraf, M. W., Tayyaba, S., & Afzulpurkar, N. (2011). Micro Electromechanical Systems (MEMS) based microfluidic devices for biomedical applications. *International Journal of Molecular Sciences, 12*(6), 3648–3704. https://doi.org/10.3390/ijms12063648

Bariya, S. H., Gohel, M. C., Mehta, T. A., & Sharma, O. P. (2012). Microneedles: An emerging transdermal drug delivery system. *Journal of Pharmacy and Pharmacology, 64*(1), 11–29. https://doi.org/10.1111/j.2042-7158.2011.01369.x

Bhatnagar, S., Dave, K., & Venuganti, V. V. K. (2017). Microneedles in the clinic. *Journal of Controlled Release, 260*, 164–182. https://doi.org/10.1016/j.jconrel.2017.05.029

Chen, G., Chen, Z., Wen, D., Wang, Z., Li, H., Zeng, Y., Dotti, G., Wirz, R. E., & Gu, Z. (2020). Transdermal cold atmospheric plasma-mediated immune checkpoint blockade therapy. *Proceedings of the National Academy of Sciences of the United States of America, 117*(7), 3687–3692. https://doi.org/10.1073/pnas.1917891117

Chi, J., Zhang, X., Chen, C., Shao, C., Zhao, Y., & Wang, Y. (2020). Antibacterial and angiogenic chitosan microneedle array patch for promoting wound healing. *Bioactive Materials, 5*(2), 253–259. https://doi.org/10.1016/j.bioactmat.2020.02.004

Courtenay, A. J., McAlister, E., McCrudden, M. T. C., Vora, L., Steiner, L., Levin, G., Levy-Nissenbaum, E., Shterman, N., Kearney, M. C., McCarthy, H. O., &

Donnelly, R. F. (2020). Hydrogel-forming microneedle arrays as a therapeutic option for transdermal esketamine delivery. *Journal of Controlled Release, 322*, 177–186. https://doi.org/10.1016/j.jconrel.2020.03.026

Das, D., Akhtar, S., Kurra, S., Gupta, S., & Sharma, A. (2019). Emerging role of immune cell network in autoimmune skin disorders: An update on pemphigus, vitiligo and psoriasis. *Cytokine and Growth Factor Reviews, 45*, 35–44. https://doi.org/10.1016/j.cytogfr.2019.01.001

Donadei, A., Kraan, H., Ophorst, O., Flynn, O., O'Mahony, C., Soema, P. C., & Moore, A. C. (2019). Skin delivery of trivalent Sabin inactivated poliovirus vaccine using dissolvable microneedle patches induces neutralizing antibodies. *Journal of Controlled Release, 311–312*, 96–103. https://doi.org/10.1016/j.jconrel.2019.08.039

Du, G., Zhang, Z., He, P., Zhang, Z., & Sun, X. (2021). Determination of the mechanical properties of polymeric microneedles by micromanipulation. *Journal of the Mechanical Behavior of Biomedical Materials, 117*, 104384. https://doi.org/10.1016/j.jmbbm.2021.104384

Elahpour, N., Pahlevanzadeh, F., Kharaziha, M., Bakhsheshi-Rad, H. R., Ramakrishna, S., & Berto, F. (2021). 3D printed microneedles for transdermal drug delivery: A brief review of two decades. *International Journal of Pharmaceutics, 597*, 120301. https://doi.org/10.1016/j.ijpharm.2021.120301

Fonseca, D. F. S., Costa, P. C., Almeida, I. F., Dias-Pereira, P., Correia-Sá, I., Bastos, V., Oliveira, H., Duarte-Araújo, M., Morato, M., Vilela, C., Silvestre, A. J. D., & Freire, C. S. R. (2020). Pullulan microneedle patches for the efficient transdermal administration of insulin envisioning diabetes treatment. *Carbohydrate Polymers, 241*, 116314. https://doi.org/10.1016/j.carbpol.2020.116314

Ghimirey, K. B., & Ita, K. (2020). Microneedle-assisted percutaneous transport of magnesium sulfate. *Current Drug Delivery, 17*(2), 140–147. https://doi.org/10.2174/1567201817666191217093936

Giffen, P. S., Bhuiya, R., Brackenborough, K., Hobbs, M. J., Qian, L., & Burke, M. D. (2020). Controlled delivery of dutasteride using dissolvable microarrays: Initial formulation and in vivo evaluation. *Journal of Pharmaceutical Sciences, 109*(3), 1303–1311. https://doi.org/10.1016/j.xphs.2019.11.012

Gupta, J., Gill, H. S., Andrews, S. N., & Prausnitz, M. R. (2011). Kinetics of skin resealing after insertion of microneedles in human subjects. *Journal of Controlled Release, 154*(2), 148–155. https://doi.org/10.1016/j.jconrel.2011.05.021

Hansen, F. S., Wenande, E., Haedersdal, M., & Fuchs, C. S. K. (2019). Microneedle fractional radiofrequency-induced micropores evaluated by in vivo reflectance confocal microscopy, optical coherence tomography, and histology. *Skin Research and Technology, 25*(4), 482–488. https://doi.org/10.1111/srt.12676

Hao, Y., Li, W., Zhou, X. L., Yang, F., & Qian, Z. Y. (2017). Microneedles-based transdermal drug delivery systems: A review. *Journal of Biomedical Nanotechnology, 13*(12), 1581–1597. https://doi.org/10.1166/jbn.2017.2474

Henry, S., McAllister, D. V., Allen, M. G., & Prausnitz, M. R. (1998). Microfabricated microneedles: A novel approach to transdermal drug delivery. *Journal of Pharmaceutical Sciences, 87*(8), 922–925. https://doi.org/10.1021/js980042+

He, X., Sun, J., Zhuang, J., Xu, H., Liu, Y., & Wu, D. (2019). Microneedle system for transdermal drug and vaccine delivery: Devices, safety, and prospects. *Dose-Response, 17*(4). https://doi.org/10.1177/1559325819878585, 155932581987858.

Ita, K. (2015). Transdermal delivery of drugs with microneedles—potential and challenges. *Pharmaceutics*, 7(3), 90–105. https://doi.org/10.3390/pharmaceutics7030090

Ita, K. (2017). Dissolving microneedles for transdermal drug delivery: Advances and challenges. *Biomedicine and Pharmacotherapy, 93*, 1116–1127. https://doi.org/10.1016/j.biopha.2017.07.019

Ita, K. (2018). Ceramic microneedles and hollow microneedles for transdermal drug delivery: Two decades of research. *Journal of Drug Delivery Science and Technology, 44*, 314–322. https://doi.org/10.1016/j.jddst.2018.01.004

Ita, Kevin (2020). Chapter 8 - microneedles. In *Transdermal drug delivery: Concepts and application* (pp. 143–181). Academic Press. https://doi.org/10.1016/B978-0-12-822550-9.00008-9

Ita, Kevin (2021). Coronavirus disease (COVID-19): Current status and prospects for drug and vaccine development. *Archives of Medical Research, 52*(1), 15–24. https://doi.org/10.1016/j.arcmed.2020.09.010

Ita, K., & Ashong, S. (2020). Percutaneous delivery of antihypertensive agents: Advances and challenges. *AAPS PharmSciTech, 21*(2). https://doi.org/10.1208/s12249-019-1583-9

Jeong, H. R., Park, S., Park, J. H., Bae, J. Y., Kim, G.y., Baek, S. K., Park, M. S., & Park, J. H. (2019). Preparation of H1N1 microneedles by a low-temperature process without a stabilizer. *European Journal of Pharmaceutics and Biopharmaceutics, 143*, 1–7. https://doi.org/10.1016/j.ejpb.2019.08.005

Jeon, E. Y., Lee, J., Kim, B. J., Joo, K. I., Kim, K. H., Lim, G., & Cha, H. J. (2019). Bio-inspired swellable hydrogel-forming double-layered adhesive microneedle protein patch for regenerative internal/external surgical closure. *Biomaterials, 222*. https://doi.org/10.1016/j.biomaterials.2019.119439

Kim, E., Erdos, G., Huang, S., Kenniston, T. W., Balmert, S. C., Carey, C. D., Raj, V. S., Epperly, M. W., Klimstra, W. B., Haagmans, B. L., Korkmaz, E., Falo, L. D., & Gambotto, A. (2020). Microneedle array delivered recombinant coronavirus vaccines: Immunogenicity and rapid translational development. *EBioMedicine, 55*, 102743. https://doi.org/10.1016/j.ebiom.2020.102743

Kim, Y. C., Park, J. H., & Prausnitz, M. R. (2012). Microneedles for drug and vaccine delivery. *Advanced Drug Delivery Reviews, 64*(14), 1547–1568. https://doi.org/10.1016/j.addr.2012.04.005

Kim, S., Yang, H., Eum, J., Ma, Y., Fakhraei Lahiji, S., & Jung, H. (2020). Implantable powder-carrying microneedles for transdermal delivery of high-dose insulin with enhanced activity. *Biomaterials, 232*, 119733. https://doi.org/10.1016/j.biomaterials.2019.119733

Lee, K. J., Goudie, M. J., Tebon, P., Sun, W., Luo, Z., Lee, J., Zhang, S., Fetah, K., Kim, H. J., Xue, Y., Darabi, M. A., Ahadian, S., Sarikhani, E., Ryu, W. H., Gu, Z., Weiss, P. S., Dokmeci, M. R., Ashammakhi, N., & Khademhosseini, A. (2020). Non-transdermal microneedles for advanced drug delivery. *Advanced Drug Delivery Reviews, 165–166*, 41–59. https://doi.org/10.1016/j.addr.2019.11.010

Lee, K. J., Jeong, S. S., Roh, D. H., Kim, D. Y., Choi, H. K., & Lee, E. H. (2020). A practical guide to the development of microneedle systems – in clinical trials or on the market. *International Journal of Pharmaceutics, 573*. https://doi.org/10.1016/j.ijpharm.2019.118778

Lee, B. M., Lee, C., Lahiji, S. F., Jung, U. W., Chung, G., & Jung, H. (2020). Dissolving microneedles for rapid and painless local anesthesia. *Pharmaceutics, 12*(4). https://doi.org/10.3390/pharmaceutics12040366

Liu, G.-S., Kong, Y., Wang, Y., Luo, Y., Fan, X., Xie, X., Yang, B.-R., & Wu, M. X. (2020). Microneedles for transdermal diagnostics: Recent advances and new horizons. *Biomaterials, 232*, 119740. https://doi.org/10.1016/j.biomaterials.2019.119740

Li, Q. Y., Zhang, J. N., Chen, B. Z., Wang, Q. L., & Guo, X. D. (2017). A solid polymer microneedle patch pretreatment enhances the permeation of drug molecules into the skin. *RSC Advances*, 7(25), 15408–15415. https://doi.org/10.1039/c6ra26759a

Lopez-Ramirez, M. A., Soto, F., Wang, C., Rueda, R., Shukla, S., Silva-Lopez, C., Kupor, D., McBride, D. A., Pokorski, J. K., Nourhani, A., Steinmetz, N. F., Shah, N. J., & Wang, J. (2020). Built-in active microneedle patch with enhanced autonomous drug delivery. *Advanced Materials, 32*(1). https://doi.org/10.1002/adma.201905740

Mao, J., Wang, H., Xie, Y., Fu, Y., Li, Y., Liu, P., Du, H., Zhu, J., Dong, L., Hussain, M., Li, Y., Zhang, L., Zhu, J., & Tao, J. (2020). Transdermal delivery of rapamycin with poor water-solubility by dissolving polymeric microneedles for anti-angiogenesis. *Journal of Materials Chemistry B, 8*(5), 928–934. https://doi.org/10.1039/c9tb00912d

Migdadi, E. M., Courtenay, A. J., Tekko, I. A., McCrudden, M. T. C., Kearney, M. C., McAlister, E., McCarthy, H. O., & Donnelly, R. F. (2018). Hydrogel-forming microneedles enhance transdermal delivery of metformin hydrochloride. *Journal of Controlled Release, 285*, 142–151. https://doi.org/10.1016/j.jconrel.2018.07.009

Moreira, A. F., Rodrigues, C. F., Jacinto, T. A., Miguel, S. P., Costa, E. C., & Correia, I. J. (2019). Microneedle-based delivery devices for cancer therapy: A review. *Pharmacological Research, 148*, 104438. https://doi.org/10.1016/j.phrs.2019.104438

Nagarkar, R., Singh, M., Nguyen, H. X., & Jonnalagadda, S. (2020). A review of recent advances in microneedle technology for transdermal drug delivery. *Journal of Drug Delivery Science and Technology, 59*. https://doi.org/10.1016/j.jddst.2020.101923

Nguyen, T. T., Choi, J. A., Kim, J. S., Park, H., Yang, E., Lee, W. J., Baek, S. K., Song, M., & Park, J. H. (2019). Skin immunization with third-generation hepatitis B surface antigen using microneedles. *Vaccine, 37*(40), 5954–5961. https://doi.org/10.1016/j.vaccine.2019.08.036

Ning, X., Wiraja, C., Lio, D. C. S., & Xu, C. (2020). A double-layered microneedle platform fabricated through frozen spray-coating. *Advanced Healthcare Materials, 9*(10). https://doi.org/10.1002/adhm.202000147

Park, J., & Kim, Y. C. (2021). Topical delivery of 5-fluorouracil-loaded carboxymethyl chitosan nanoparticles using microneedles for keloid treatment. *Drug Delivery and Translational Research, 11*(1), 205–213. https://doi.org/10.1007/s13346-020-00781-w

Prausnitz, M. R. (2004). Microneedles for transdermal drug delivery. *Advanced Drug Delivery Reviews, 56*(5), 581–587. https://doi.org/10.1016/j.addr.2003.10.023

Radhika, C., & Gnanavel, B. K. (2021). Buckling analysis of polymer microneedle for transdermal drug delivery. *Materials Today: Proceedings*. https://doi.org/10.1016/j.matpr.2020.12.397

Ronnander, P., Simon, L., & Koch, A. (2020). Experimental and mathematical study of the transdermal delivery of sumatriptan succinate from polyvinylpyrrolidone-based microneedles. *European Journal of Pharmaceutics and Biopharmaceutics, 146*, 32–40. https://doi.org/10.1016/j.ejpb.2019.11.007

Rouphael, N. G., Paine, M., Mosley, R., Henry, S., McAllister, D. V., Kalluri, H., Pewin, W., Frew, P. M., Yu, T., Thornburg, N. J., Kabbani, S., Lai, L., Vassilieva, E. V., Skountzou, I., Compans, R. W., Mulligan, M. J., Prausnitz, M. R., Beck, A., Edupuganti, S., … Nesheim, W. (2017). The safety, immunogenicity, and acceptability of inactivated influenza vaccine delivered by microneedle patch (TIV-MNP 2015): A randomised, partly blinded, placebo-controlled, phase 1 trial. *The Lancet, 390*(10095), 649–658. https://doi.org/10.1016/S0140-6736(17)30575-5

Rzhevskiy, A. S., Singh, T. R. R., Donnelly, R. F., & Anissimov, Y. G. (2018). Microneedles as the technique of drug delivery enhancement in diverse organs and tissues. *Journal of Controlled Release, 270*, 184–202. https://doi.org/10.1016/j.jconrel.2017.11.048

Sabri, A. H., Kim, Y., Marlow, M., Scurr, D. J., Segal, J., Banga, A. K., Kagan, L., & Lee, J. B. (2020). Intradermal and transdermal drug delivery using microneedles – fabrication, performance evaluation and application to lymphatic delivery. *Advanced Drug Delivery Reviews*, 195–215. https://doi.org/10.1016/j.addr.2019.10.004

Sadeghi, I., Byrne, J., Shakur, R., & Langer, R. (2021). Engineered drug delivery devices to address global health challenges. *Journal of Controlled Release, 331*, 503–514. https://doi.org/10.1016/j.jconrel.2021.01.035

Shah, V., & Choudhury, B. K. (2017). Fabrication, physicochemical characterization, and performance evaluation of biodegradable polymeric microneedle patch system for enhanced transcutaneous flux of high molecular weight therapeutics. *AAPS PharmSciTech, 18*(8), 2936–2948. https://doi.org/10.1208/s12249-017-0774-5

Shakya, A. K., Ingrole, R. S. J., Joshi, G., Uddin, M. J., Anvari, S., Davis, C. M., & Gill, H. S. (2019). Microneedles coated with peanut allergen enable desensitization of peanut sensitized mice. *Journal of Controlled Release, 314*, 38–47. https://doi.org/10.1016/j.jconrel.2019.09.022

Shu, Z., Cao, Y., Tao, Y., Liang, X., Wang, F., Li, Z., Li, Z., & Gui, S. (2020). Polyvinylpyrrolidone microneedles for localized delivery of sinomenine hydrochloride: Preparation, release behavior of in vitro & in vivo, and penetration mechanism. *Drug Delivery, 27*(1), 642–651. https://doi.org/10.1080/10717544.2020.1754524

Sun, X., Yuan, B., Sheng, L., Rao, W., & Liu, J. (2020). Liquid metal enabled injectable biomedical technologies and applications. *Applied Materials Today, 20*, 100722. https://doi.org/10.1016/j.apmt.2020.100722

Tekko, I. A., Chen, G., Domínguez-Robles, J., Thakur, R. R. S., Hamdan, I. M. N., Vora, L., Larrañeta, E., McElnay, J. C., McCarthy, H. O., Rooney, M., & Donnelly, R. F. (2020). Development and characterisation of novel poly (vinyl alcohol)/poly (vinyl pyrrolidone)-based hydrogel-forming microneedle arrays for enhanced and sustained transdermal delivery of methotrexate. *International Journal of Pharmaceutics, 586*. https://doi.org/10.1016/j.ijpharm.2020.119580

Turner, J. G., White, L. R., Estrela, P., & Leese, H. S. (2021). Hydrogel-forming microneedles: Current advancements and future trends. *Macromolecular Bioscience, 21*(2). https://doi.org/10.1002/mabi.202000307

Vora, L. K., Courtenay, A. J., Tekko, I. A., Larrañeta, E., & Donnelly, R. F. (2020). Pullulan-based dissolving microneedle arrays for enhanced transdermal delivery of small and large biomolecules. *International Journal of Biological Macromolecules, 146*, 290–298. https://doi.org/10.1016/j.ijbiomac.2019.12.184

Waghule, T., Singhvi, G., Dubey, S. K., Pandey, M. M., Gupta, G., Singh, M., & Dua, K. (2019). Microneedles: A smart approach and increasing potential for transdermal drug delivery system. *Biomedicine and Pharmacotherapy, 109*, 1249–1258. https://doi.org/10.1016/j.biopha.2018.10.078

Witting, M., Obst, K., Pietzsch, M., Friess, W., & Hedtrich, S. (2015). Feasibility study for intraepidermal delivery of proteins using a solid microneedle array. *International Journal of Pharmaceutics, 486*(1–2), 52–58. https://doi.org/10.1016/j.ijpharm.2015.03.046

Wu, L., Shrestha, P., Iapichino, M., Cai, Y., Kim, B., & Stoeber, B. (2021). Characterization method for calculating diffusion coefficient of drug from polylactic acid (PLA) microneedles into the skin. *Journal of Drug Delivery Science and Technology, 61*, 102192. https://doi.org/10.1016/j.jddst.2020.102192

Yang, P., Lu, C., Qin, W., Chen, M., Quan, G., Liu, H., Wang, L., Bai, X., Pan, X., & Wu, C. (2020). Construction of a core-shell microneedle system to achieve targeted codelivery of checkpoint inhibitors for melanoma immunotherapy. *Acta Biomaterialia, 104*, 147–157. https://doi.org/10.1016/j.actbio.2019.12.037

Zhang, X., Chen, G., Sun, L., Ye, F., Shen, X., & Zhao, Y. (2021). Claw-inspired microneedle patches with liquid metal encapsulation for accelerating incisional wound healing. *Chemical Engineering Journal, 406*, 126741. https://doi.org/10.1016/j.cej.2020.126741

Zhou, X., et al. (2020). Biodegradable β-cyclodextrin conjugated gelatin methacryloyl microneedle for delivery of water-insoluble drug. *Advanced Healthcare Materials, 9*(11), e2000527.

CHAPTER 2
Fabrication of microneedles

Introduction

The dramatic rise in the use of novel technologies by the microfabrication industry has resulted in the development of several micron-sized devices (Sabri et al., 2020). As a consequence, there are numerous technologies available for the fabrication of microneedles (MNs). It has been mentioned in the previous chapter that there are five categories of MNs: dissolving, solid, hollow, coated and hydrogel forming. Several techniques are available for the fabrication of MNs (Evens et al., 2020; Xenikakis et al., 2021). These include laser micromachining, electro-drawing, thermal drawing, photolithography, magnetorheological drawing lithography, micromolding, micro-injection compression molding, droplet-born air blowing, drawing lithography, 3D printing and layer-by-layer (LbL) assembly (Demuth et al., 2012; Indermun et al., 2014; Kathuria et al., 2020; Kim et al., 2013; Lee & Jung, 2012; Martanto et al., 2004; Nguyen et al., 2018; Ruggiero et al., 2018; Yao et al., 2020). Wire electrical discharge or infrared laser machining can also be used to fabricate metal MNs from metal sheets by cutting needle structures from stainless steel sheets with subsequent bending of the needle at 90° out of the plane of the metal sheet (Chen, Ren et al., 2018; Chen, Wang et al., 2018). Ablation or micromilling has been utilized to obtain required structures from the slab of a material (Indermun et al., 2014). MN arrays can also be produced by cutting needles from stainless steel sheets (SS 304, 75-μm thick) via an infrared laser (Martanto et al., 2004). In one study, the orientation and shape of the MN arrays were first drafted in a CAD file and subsequently the metal sheet was ablated and the MNs produced in the plane of the sheet (Martanto et al., 2004). The laser was operated at 1000 Hz with an energy density of 20 J/cm^2 with the investigators needing approximately 4 min to cut the array of solid MNs (Martanto et al., 2004). The microneedles were washed with hot water and rinsed with de-ionized water. Each needle was subsequently bent out of the plane of the metal sheet at 90° (Martanto et al., 2004).

Microneedles
ISBN 978-0-323-91811-4
https://doi.org/10.1016/B978-0-323-91811-4.00004-3

Lithography is a commonly used microfabrication operation (Ita, 2015; Larrañeta et al., 2016; Wang et al., 2021). During photolithography (Fig. 2.1), a light-sensitive polymer (photo-resist) is subjected to ultraviolet (UV) radiation (Ita, 2015). A thin layer of photo-resist is coated onto the oxide surface of a silicon wafer via a spin coating approach (Larrañeta et al., 2016). The resist-coated wafers are then subjected to UV radiation (150 and 500 nm) via a mask (Larrañeta et al., 2016). Photomask is a nontransparent plate that permits radiation only in a certain pattern (Ita, 2015). When subjected to radiation, photo-resist degrades and becomes more soluble in a developer. The photo-resist is detached via de-scumming to obtain a defined photo-resist pattern (Ita, 2015; Larrañeta et al., 2016) and finally the wafers are post-baked (Larrañeta et al., 2016). Planar pattern-to-cross-section technology (PCT) technique via LIGA (Photolithography, Galvanogormung, Abformung) technology (comparable to an X-ray mask

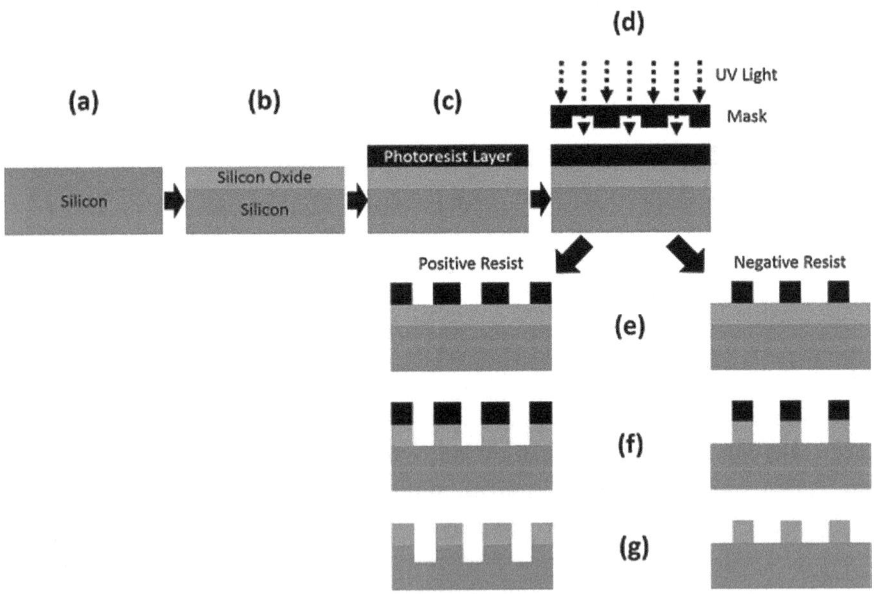

Figure 2.1 Sequential processes in the transfer of a pattern to the substrate surface. (A) Si wafer; (B) Si wafer with oxide coating; (C) spin-coated photoresistive material; (D) mask guided UV light exposure on the photoresistive material; (E) development process to remove the soluble resist material; (F) etching of SiO_2 film; and (G) photoresist removal. *(From Larrañeta, E., Lutton, R. E. M., Woolfson, A. D., & Donnelly, R. F. (2016). Microneedle arrays as transdermal and intradermal drug delivery systems: Materials science, manufacture and commercial development.* Materials Science and Engineering: R: Reports, 104, *1–32. https://doi.org/10.1016/j.mser.2016.03.001).*

pattern) has also been used to obtain a three-dimensional structure (Wang et al., 2021). The authors combined this method with electroplating and polydimethylsiloxane (PDMS) transfer technology to fabricate metal microneedles (Wang et al., 2021).

Recent advances in the fabrication of microneedles

A low-cost process called Cross-Over Lines (COLs) laser engraving has also been utilized for microneedle fabrication (Sadeqi et al., 2018). This method employs carbon dioxide laser-cutter to form MN molds on acrylic sheet. Polydimethylsiloxane (PDMS) is then cast on the acrylic sheet mold to create base PDMS MNs (Sadeqi et al., 2018). The PDMS templates are subsequently deployed for the formation of polymeric needles (Sadeqi et al., 2018). The authors cast and cured chitosan solution on the PDMS micromolds which were conveniently detached leading to the creation of hollow chitosan MNs (Sadeqi et al., 2018). MNs can also be fabricated via centrifugal casting. Traditional ceramic microneedles are usually dense devices with restricted free space for the incorporation of bioactive compounds (Gholami et al., 2019). It is desirable that an uncomplicated approach be used for the preparation of ceramic microneedles with intertwined pores (Gholami et al., 2019). Gholami et al. applied a facile technique of centrifugal casting to alumina suspensions to produce microporous microneedles (Gholami et al., 2019). The slurry were cast into micromolds under centrifugal force and subsequently sintered (Gholami et al., 2019). The authors studied the influence of solvent type, particle size, sintering temperature, binder concentration and resin quantity on the mechanical properties and microstructure of MNs. By carefully choosing the best process parameters, highly porous (up to 60%) MNs with interwoven microconduits (of diameter $\sim 1-1.5$ μm) were fabricated (Gholami et al., 2019). The fabricated MNs were robust and biocompatible. The authors utilized the MNs for the accurate and speedy percutaneous extraction of glucose from a rat skin. Furthermore, a swift release and efficient loading of insulin was demonstrated (Gholami et al., 2019).

Electro-drawing entails the self-assembly of droplets into 3D MNs with the aid of electrohydrodynamic force (EHD) (Chen, Wang, et al., 2018). MNs are shaped in one step beginning with the application of EHD for the deformation of immobile drops of polymer solution (Ruggiero et al., 2018). Electro-drawing depends on quick solidification and surface drawing (Kathuria et al., 2020). This fabrication method is fast but less flexible in comparison to molding (Kathuria et al., 2020). The current trend is to produce MNs via stamp-less methods. Electro-drawing is an option that entails the deployment of fast and low temperature approach in comparison to stamp-based technologies (Ruggiero et al., 2018). ED relies on the combined use of titanium micro-heaters with pyroelectric substrate which is the electric field generator that facilitates electro-drawing (Ruggiero et al., 2018). In one study, the configuration of micro-heaters was optimized via numerical simulations to provide consistent heating on large areas in contrast to the conventional local heating provided by a hot tip (Ruggiero et al., 2018). A uniform electric field was also produced and this is a fundamental condition for in-parallel production of MNs (Ruggiero et al., 2018).

In another study, an electro-drawing system employing a pyroelectric cartridge as a substitute for syringe piston was reported (Vecchione et al., 2014). The cartridge uses electro-drawing to produce biodegradable, drug-loaded, polymer MN arrays (Vecchione et al., 2014). The MNs were fabricated on a flexible substrate/holder via direct drawing from reservoirs, utilizing a tightly regulated electro-hydrodynamic force (Vecchione et al., 2014). This contact-free and mold-free technique allows the formation of ready-to-use dissolving MNs that deliver medications following a pre-defined chrono-program (Vecchione et al., 2014). A polar dielectric crystal (lithium tantalate) with a polymer reservoir drop was used for electro-drawing. The approach is based on the application of pyroelectric effect on a lithium tantalate crystal that functions as a driving plate (Vecchione et al., 2014). The polymer drops are then deposited onto a poly (dimethyl-siloxane) (PDMS) strip. A thermal stimulus is used to create an electric field that causes an EHD force that subsequently draws the polymer into an MN shape (Vecchione et al., 2014). Some authors have combined thermal drawing and mold-based technology leading to the replication of free-form style MNs without the need to rely on complicated and costly master mold fabrication (Lee et al., 2015).

Lee et al. prepared MNs by using a spatially-discrete thermal drawing (SDTD) system to make polylactic acid-*co*-glycolic acid MN masters with various shapes (normal, lender and bullet type MNs) (Lee et al., 2015). Polylactic-*co*-glycolic acid (PLGA) film was heated up to 200°C and then stainless-steel micro-pillars that were heated up to 130°C were dropped into the heated PLGA film (Lee et al., 2015). The micro-pillars measured 270 μm in diameter. Subsequently, the micro-pillars were used to draw up the heated PLGA. The authors stabilized device structure by heating both the micro-pillars and the PLGA film at 160°C (Lee et al., 2015). Subsequently, PLGA was drawn up to 600 μm (Lee et al., 2015). The authors reduced the temperature of the PLGA substrate to 60°C and heated the micro-pillars up to 200°C to create sharp MN tips (Lee et al., 2015). Next, PDMS resin was poured on the PLGA MN masters and cured at room temperature for one day to fabricate negative molds for the MNs (Lee et al., 2015). Lastly, the authors poured an aqueous SF solution on the negative PDMS molds and conducted degassing several times via vacuum with subsequent drying at room temperature for 24 h (Lee et al., 2015).

The fabrication of microneedles is challenging because of the high aspect ratio as well as the three-dimensional (3D) conical geometry of these devices (Nejad et al., 2018). MNs can be fabricated via molding technologies. The molds are usually produced via complex microfabrication steps in a cleanroom (Nejad et al., 2018). Fabrication operations include Lithographie (lithography) Galvanoformung (electroplating) Abformung (molding) collectively known as LIGA (Demuth et al., 2012; Indermun et al., 2014; Kathuria et al., 2020; Kim et al., 2013; Lee & Jung, 2012; Martanto et al., 2004; Nguyen et al., 2018; Ruggiero et al., 2018; Yao et al., 2020). Ultraviolet radiation can be used in conjunction with LIGA to fabricate devices with relatively low aspect ratios (Kim et al., 2012). On the other hand, X-Ray LIGA utilizes X-rays generated by a synchrotron to fabricate high-aspect ratio devices (Kim et al., 2012). In one study, poly (methyl methacrylate) was used as a material for MN fabrication in conjunction with vertical X-ray exposure and inclined deep X-ray treatment (Kim et al., 2012).

SU-8 is an epoxy based, negative photoresist used for micromachining (Wouters & Puers, 2009). The UV-light absorptivity of SU-8 is less than that of other resins making it possible for high-aspect ratio devices to be fabricated from this compound (Chen et al., 2010). SU-8 is usually supplied as a photoinitiator and a resin. When exposed to near-ultraviolet radiation, the photoinitiator induces SU-8 crosslinking and curing (Chen et al., 2010). Ultraviolet radiation measuring 193–436 nm in wavelength is passed through a photomask (Ita, 2015). Photomask refers to an opaque plate which permits radiation to cross only in a predetermined pattern (Ita, 2015). In the exposed region, the photo-resist is degraded and becomes more soluble in a developer solution (Ita, 2015). Park and colleagues coated a silicon wafer with SU-8 and lithographically patterned the wafer into cylinders (100 μm diameter) thus defining the shape of the MNs (Park et al., 2005). Kathuria et al. prepared an MN patch via an uncomplicated and efficient method under mild conditions, using a combination of temperature and photopolymerization (Kathuria et al., 2020). The authors combined heat and pre-polymerization of vinylpyrrolidone followed by photolithography to fabricate the MN patch (Kathuria et al., 2020). When heat was used, less UV exposure time was needed for the N-vinylpyrrolidone solution, thus decreasing the total fabrication time (Kathuria et al., 2020). In practical terms, 2 mL of N-vinylpyrrolidone with 1% w/v of 4,4′-azobis (4-cyanovaleric acid) was first pre-polymerized using heat (for 2 min at 90°C) and then the solution was allowed to cool down for 1 min at room temperature (Kathuria et al., 2020). Subsequently, the pre-polymerized mixture was poured into a PDMS micromold, vacuumed at -95 kPa for 10—20 min to fill the micro cavities and then treated with UV light at 100 W for 8 min (Kathuria et al., 2020). The authors showed that the proportion of MN that permeated human cadaver skin was more than 33.9% (Kathuria et al., 2020). Sugiyama and colleagues also developed a technique for fabricating three-dimensional devices via deep x-ray lithography (Sugiyama et al., 2004). Curved surfaces and sloped sidewalls were prepared by subjecting a moving x-ray resist to synchrotron radiation (Sugiyama et al., 2004). The method (plain-pattern to cross-section transfer) is an extension of the traditional 2.5-dimension (2.5D) lithography (Sugiyama et al., 2004).

Porous silicon, silicon oxide and silicon nitride are frequently used for MN preparation (Dardano et al., 2015). But silicon is fragile and it is not biocompatible because the inflammatory properties of this material can induce silicosis (Dardano et al., 2015). The first set of MNs were fabricated from silicon via utilization of traditional microfabrication procedures, including thin film deposition, photolithography and etching (Dardano et al., 2015). Thin-film deposition can be accomplished via chemical or physical vapor deposition (Ita, 2015). Physical vapor deposition (PVD) entails film formation via the direct movement of atoms from the source to the substrate via a gas phase (Ita, 2015). PVD can be conducted as either sputtering or evaporation. When chemical vapor deposition (CVD) is used for deposition, film formation is through chemical reaction on the surface of substrate (Ita, 2015). Using deep reactive ion etching, Henry et al. deposited a chromium mask on silicon wafers followed by patterning of dots comparable in diameter to the base of the proposed MNs (Henry et al., 1998). The wafers were subsequently subjected to reactive ion etching and then put through a meticulously controlled fluorine/oxygen plasma to etch high aspect ratio and very deep troughs into the wafers (Henry et al., 1998).

The areas covered by the metal mask remained and formed the MNs (Henry et al., 1998). Dry etchants sulfur hexafluoride (SF_6) and oxygen (O_2) were utilized by the authors for vertical silicon etching, while tetrafluoromethane (CF_4) was employed (Bosch method) to avoid lateral undercutting leading to the fabrication of high aspect ratio (AR) MNs (Henry et al., 1998).

In another set of studies, a single-crystal silicon was used to create MNs (Li et al., 2010). The first procedure entailed the deposition of nitride using low pressure chemical vapor (Li et al., 2010). Following deposition and patterning of square oxide patterns on a silicon substrate, isotropic-reactive ion etching was used to etch away the region of the Si wafers that was not covered by these patterns (Li et al., 2010). When a certain tip shape is created, the MN body is then fabricated via anisotropic Bosch method (Li et al., 2010). Silicon MNs can also be fabricated via anisotropic wet etching (Lee et al., 2020). The rate of wet etching is a function of silicon crystal plane orientation in contrast to dry etching. When dry etching is used, the etch rate does not depend on crystal planes (Lee et al., 2020). A square-patterned mask can be utilized for the fabrication of molds (Lee et al., 2020). A (100) wafer with a square mask pattern is usually etched with potassium hydroxide solution to create pyramidal MN cavities, tapered at 54.72°(Lee et al., 2020).

Deep-reactive ion etching can be used to produce highly anisotropic structures with high-selectivity relative to photoresist, making it possible to create devices with high aspect ratios i.e., the ratio of depth to width (Li et al., 2010). Research scientists have employed deep reactive ion etching (DRIE) and mono-crystalline $<100>$ silicon wafers for the fabrication of MNs (Coffey et al., 2016). Coffey et al. varied MN height between about 40 and 190 μm by tightly regulating the loop number in the final vertical Bosch process (Coffey et al., 2016). In one study, MN density between 5000 and 20,408 needles per cm^2 for the conventional 100 μm length was achieved by modifying the spot density of photolithography masks that were used for wafer patterning (Coffey et al., 2016). Silicon has superior mechanical properties (high Young's Modulus, nonductile nature and indentation hardness) compared to metal and polymer thus facilitating drug transport across the skin without the MNs breaking (Li et al., 2019). In an interesting study Li et al. fabricated hollow microneedles from -silicon via DRIE (Li et al., 2019). Anisotropic etching of holes and pillars from silicon was carried out via inductively coupled plasma reactive ion etching (ICP-RIE). The authors etched the pillars by employing a conventional dual-phase Bosch DRIE approach (etching and passivation) (Li et al., 2019). A triple-phase modified Bosch technique (i.e., passivation, depassivation, etching) was then used to etch the holes (Li et al., 2019). MN pattern was made on a silicon wafer via standard photolithography techniques (Li et al., 2019).

Takahashi et al. described a scalable technique of making MNs via rotated/inclined ultraviolet lithography (Takahashi et al., 2016). The researchers used inclined/rotated ultraviolet radiation through circular mask patterns to apply conical-shaped UV rays to a negative photoresist leading to the formation of MNs by the photo-polymerized regions (Takahashi et al., 2016). By regulating the exposure dose and inclination angle, the investigators conveniently modified photo-polymerized areas fabricated from identical mask patterns, thus permitting the facile production of MNs with on-demand geometry (Takahashi et al., 2016). MN arrays with different doses of UV exposure and distinct inclination angles of radiation were fabricated (Takahashi et al., 2016).

Poly (ethylene glycol) diacrylate (PEGDA) is a physiologically inactive polymer that quickly solidifies at room temperature in the presence of a photo-initiator or when it is exposed to ultraviolet (UV) radiation (Dardano et al., 2015). This property has been leveraged for the fabrication of PEGDA MNs via standard photolithography (Dardano et al., 2015). Dardano and coworkers produced polymeric MNs via direct photolithography without etching or molding (Dardano et al., 2015). This approach entailed casting polyethylene glycol into a silicone container and exposing it to ultraviolet radiation (365 nm) using a commercially available photocatalyzer for cross-linking (Dardano et al., 2015). MNs from hundred micrometers up to two millimeters were fabricated by changing the exposure time (5−15 s), the radiation dose, and/or the power density used for photolithography (Dardano et al., 2015). In another study, Evens and coworkers used an injection molding process and laser ablated molds to fabricate solid polymer microneedles (Evens et al., 2020). The authors employed a steel mold, femtosecond laser and a cross-hatching strategy to fabricate cone-shaped microtunnels with low tip radii (Evens et al., 2020). Polypropylene microneedles were also fabricated via an injection molding process (Evens et al., 2020).

The fabrication of MNs from silicon and glass is more complicated since these materials are fragile (Azmana et al., 2020). Furthermore, metals are less appealing materials for MN fabrication due to inadequate drug loading, costly production procedures and possible breakage in the skin

(Azmana et al., 2020). Pharmaceutical scientists have recently focused on alternative fabrication methods to circumvent these limitations (Azmana et al., 2020). Normally, a subtractive microfabrication process is used for the production of MNs and this entails the selective fabrication of MN 3D structure from a 2D substrate (Chen, Ren et al., 2018; Chen, Wang et al., 2018). Lithography with dry or wet etching is typically employed (Chen, Ren et al., 2018; Chen, Wang et al., 2018). But clean rooms and sophisticated equipment are needed. Furthermore, toxic waste is produced making lithography complicated, inconvenient, ecologically harmful, and costly (Chen, Ren et al., 2018; Chen, Wang et al., 2018). Additive manufacturing operations like electro-drawing, thermal drawing, 3D printing and droplet-born air blowing can also be used to fabricate 3D polymer MNs from 2D surfaces or droplets (Chen, Ren et al., 2018; Chen, Wang et al., 2018). Recently, Chen and coworkers developed a magnetorheological drawing lithography (MRDL) technique for the fabrication of bioinspired MNs, MN arrays and MNs (Chen, Ren et al., 2018; Chen, Wang et al., 2018). The authors drew a droplet of curable magnetorheological fluid directly from a substrate under an external magnetic field to produce a 3D microneedle (Chen, Ren et al., 2018; Chen, Wang et al., 2018). Although the MRDL technique is simple, it is capable of producing complex bio-inspired MNs (Chen, Ren et al., 2018; Chen, Wang et al., 2018). In a separate study, drawing lithography was used for the preparation of microneedles directly from 2D planar substrates, thus avoiding the limitations of traditional subtractive manufacturing (Lee & Jung, 2012). This method leverages the viscous property of polymers in the glass transition state for manufacturing (Lee & Jung, 2012). When this technique is used, a thermosetting polymer is directly drawn from a two-dimensional solid surface without using radiation or a mask (Ita, 2017; Lee et al., 2010).

Centrifugal casting was used for the creation of microporous ceramic MNs from alumina suspension (Gholami et al., 2019). Alumina suspension or slurry was poured into micromolds under centrifugal force (Gholami et al., 2019). Subsequently, sintering was carried out. A negative aluminum mold with 5 × 5 conical holes was drilled by a machine linked to a computer that utilizes numerical control (Gholami et al., 2019). The authors then dissolved 2% (w/w) polyvinyl alcohol in water at 80°C until the solution became transparent. After that, they added PEG (a plasticizer) to the solution under magnetic stirring before adding alumina powder and pulverizing the mixture in a planetary mill for 15 h (Gholami et al., 2019).

Traditionally, dissolving MNs (DMN) are prepared from a viscous mixture of a polymer and a compound via a stepwise procedure that includes a shape-forming step and a solidification step for the production of mechanically robust MNs (Huh et al., 2018). Micromolding is widely used for fabricating dissolving microneedles (Fig. 2.2) (Lee et al., 2021). The technique involves several steps: (a) the creation of MN master molds, (b) creation of polydimethylsiloxane molds, and (c) production of dissolving microneedle arrays loaded with the active pharmaceutical ingredient(s) (Balmert et al., 2020). The most common approaches for the fabrication of MN master molds include subtractive methods (laser cutting, photolithography, and micromachining) (Balmert et al., 2020). Although considerable progress has been made in the field of dissolving microneedle fabrication, challenges remain. Master mold are usually fabricated via resource intensive microelectronics industry-derived operations (Tarbox et al., 2018). The dependence on such operations can lead to significant design and material impediments (Tarbox et al., 2018). Moreover, multiphase filling and handling operations linked to micromolding are usually batch processes, which can pose scalability problems (Tarbox et al., 2018).

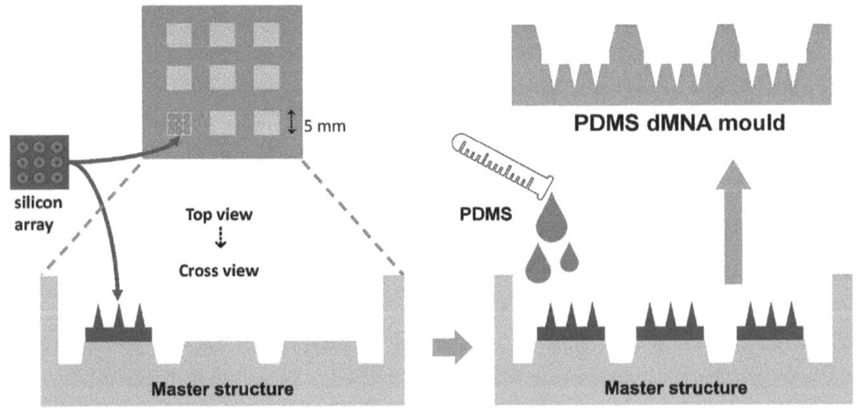

Figure 2.2 Polydimethylsiloxane (PDMS) mold fabrication process. Nine silicon arrays were attached on the pedestals of a polymethylmethacrylate (PMMA) master structure (left). A PDMS solution was poured into the master structure followed by overnight curing (right bottom). The resulting PDMS mold contains nine arrays and each array has nine individual microneedles (right top). *(From Lee, J., van der Maaden, K., Gooris, G., O'Mahony, C., Jiskoot, W., & Bouwstra, J. (2021). Engineering of an automated nano-droplet dispensing system for fabrication of antigen-loaded dissolving microneedle arrays.* International Journal of Pharmaceutics, 600, *120473. https://doi.org/10.1016/j.ijpharm.2021.120473).*

Recently, micro-additive manufacturing (μAM) has gained attention as a substitute approach that is simple with broad geometric capabilities, thus allowing the production of more designs (Balmert et al., 2020). Master molds can be prepared from polydimethylsiloxane (PDMS). Polydimethylsiloxane molds are usually made by preparing a mixture of silicone elastomer base and curing agent in a 10:1 ratio (Kim et al., 2019; Nguyen et al., 2018). The mixture is then poured into a tube for degassing via centrifugation. In one study, Sylgard 184® was utilized to prepare PDMS mold (Pattarabhiran et al., 2019). The mold was used to prepare MNs from polyvinyl alcohol (PVA) and polyvinylpyrrolidone (Pattarabhiran et al., 2019). Twenty percent w/v of PVA was dissolved in hot water (60°C) and then 20% w/v of PVP was added slowly while stirring (Pattarabhiran et al., 2019). The formulation was allowed to settle to remove entrapped air bubbles before being transferred to PDMS mold and dried at room temperature for 24 h (Pattarabhiran et al., 2019). The fabricated MNs were kept in a dryer (Pattarabhiran et al., 2019). In another study, laser engraving technique was used to produce a PDMS mold (Feng, 2020). Sulforhodamine B was incorporated into the formulation and the polymeric matrix vacuum dried to increase drug concentration in the tips of the microneedles (Feng, 2020). The authors dried the MNs in a ventilated location before demolding (Feng, 2020). Tian et al. also fabricated dissolving microneedles from a composite of poly (acrylic acid) (PAA) and graphene (Tian et al., 2019). Each microneedle array was made up of 100 (10×10) pyramidal needles with a height of 550 μm tip-to-tip distance of 500 μm. The authors used graphene and this increased the mechanical robustness of the MNs and aided their ability to pierce the skin (Tian et al., 2019). Drug transport from the graphene/PAA microneedles was enhanced and this was attributed to the enhanced drug solubility induced by poly (acrylic acid) (Tian et al., 2019).

Electrospun fibers refer to mechanically robust material with interconnected pores (Creighton et al., 2020). Creighton and colleagues used a fiber patterning technique to incorporate electrospun fibers (EFs) into DMNs (Creighton et al., 2020). The authors created a two-layer collector in the shape of a MN array and filled the collectors with a polymer dispersion to produce the MNs (Creighton et al., 2020). The MNs consisted of polycaprolactone fibers electrospun onto a collector with conical patterns (Creighton et al., 2020). Following electrospinning, an aqueous dispersion of poly (acrylic acid) was used to form a dissolvable matrix (Creighton et al., 2020). An intertwined system of electrospun fibers dispersed throughout the matrix was formed within the microneedle framework (Creighton et al., 2020). Compressive testing of the MNs proved that integrated fiber MNs (IFMNs) were stiffer than MNs with no fibers produced via by the same method (Creighton et al., 2020). IFMN failure forces were greater than targets for tissue puncture, with a mean failure force per needle of 0.055 N in comparison to the traditional matrix MNs with no fibers which failed at 0.027 N/needle (Creighton et al., 2020). It has been postulated that the microstructure of IFMNs can enhance an efficient movement of compressive force from the poly (acrylic acid) dispersion to the polycaprolactone fibers, leading to better compressive characteristics (Creighton et al., 2020).

Percutaneous transport of water-insoluble medications through hydrogel-based microneedles is important for boosting pharmacological effect (Zhou, 2020). Nevertheless, the direct encapsulation of a water-insoluble medication into hydrophilic polymers is difficult. A biodegradable microneedle patch fabricated from the polymer conjugates of gelatin methacryloyl and cyclodextrin (GelMA–CD) has been documented by Zhou and coworkers (Zhou, 2020). Curcumin, a water-insoluble and an unstable anticancer medication was incorporated into this patch leading to enhanced solubility and stability. The improved properties were attributed to the synthesis of an inclusion complex (Zhou, 2020). A dispersion of GelMA–CD/CUR inclusion complex was poured into a polydimethylsiloxane (PDMS) mold and this was subjected to centrifugation (Zhou, 2020). Following crosslinking via UV irradiation and overnight drying, the MN arrays were conveniently removed from the mold. The authors used fluorescence microscopy to confirm efficient loading of curcumin into the MNs (Zhou, 2020). The mechanical strength of the polymer-drug complex GelMA–CD/CUR MNs was adequate for skin penetration and tunable drug delivery (Zhou, 2020). The antitumor effect of the released drug in B16F10 melanoma models was documented. The GelMA–CD/CUR MN demonstrated relatively greater therapeutic efficacy compared with a control patch for non-transdermal drug delivery (Zhou, 2020). In vivo experiments also confirmed that the GelMA–CD MN patch was degradable and biocompatible (Zhou, 2020).

A novel technique of micro-injection compression molding has been recently suggested for the fabrication of MN array with flexible layer (Chen et al., 2020). The mold was made up of 11 laminates, and the micropores between the laminates features a 6×10 array of conventional pyramid MNs

(Chen et al., 2020). The authors reported that MNs fabricated using this technique were robust enough to penetrate the skin without breaking (Chen et al., 2020). Significantly, the MNs with flexible substrate can overcome the "nail bed" effect of the skin compared to metal MNs (Chen et al., 2020). Another manufacturing method is soft lithography (SL) (Leone et al., 2017). SL entails pairing a polymer film with a mold with micro-cavities and passing them through a heated nip (Leone et al., 2017). Subsequently, the mold is put on a flexible, water-soluble layer and this is put through the heated nip. The DMN patch stays on the substrate following separation from the mold (Leone et al., 2017).

Hydrogel-forming MNs can be made via micromolding. Combinations of PEG 10,000 (5.6%) and hydrolyzed PMVE/MA (11.1%) were utilized for the preparation of MNs (600 μm in height, 19 × 19 arrays, interspacing of 50 μm and a base width of 300 μm) (Eltayib et al., 2016). The authors utilized laser-engineered silicone micromolds (Eltayib et al., 2016). The MNs were crosslinked (esterification reaction) via heating at 80°C for 24 h and the sidewalls were detached with a heated blade (Eltayib et al., 2016). Thermally crosslinked pectin (PMVE/MA-PE) and PMVE/MA have also been used for the fabrication of swellable hydrogel MNs (Demir et al., 2017). A spin-casting method was used for the preparation of the hydrogel MNs. An inverse replicate of 3D printed MNs was made from PDMS and utilized as the female mold. Well-mixed aqueous combinations of pectin and PMVE/MA were introduced into the PDMS micromolds and centrifuged at 3500 rpm for 20 min at 5°C (Demir et al., 2017).

Yun and coworkers used a micromilling technique to prepare a metal master mold (Yun et al., 2020). The template consisted of 282 square pyramids placed on a circular disk (Yun et al., 2020). Using the master template, female PDMS molds were fabricated. Polydimethylsiloxane curing agent and monomer were mixed and placed in the master mold. Each PDMS mold was peeled from the master template and utilized for the production of the MNs (Yun et al., 2020). Jin et al. used replica molding to make a polydimethylsiloxane (PDMS) mold from an aluminum MN master (Jin et al., 2014). The reverse PDMS replica mold was prepared from an aluminum master. β-Chitin powder obtained from squid pen was dissolved in hexafluoroisopropanol (0.5% w/v). The chitin/HFIP solution was subsequently lyophilized to produce a chitin sponge, which was redissolved in hexafluoroisopropanol (HFIP) to form a highly concentrated chitin/HFIP gel (Jin et al., 2014). The authors had fabricated an aluminum master via micromachining. The ensuing PDMS replica mold was positioned in a centrifuge tube into which the chitin/HFIP gel was poured (Jin et al., 2014). The authors dried the sample for 48–60 h and manually detached the MNs from the mold (Jin et al., 2014). In another study, A MN patch of thiolated chitosan loaded with tacrolimus was fabricated via a micromold casting method (Ahmad et al., 2020). PDMS mold was utilized for the preparation of the patch. The molds were filled with the thiolated chitosan solution followed by centrifugation at 5000 rpm for 4 h (Ahmad et al., 2020). Subsequently, the filled mold was dried at room temperature for 2 days, and the patch was carefully removed from the mold with an adhesive backing tape (Ahmad et al., 2020).

Drawing lithography is based on an elastic deformation of polymers with subsequent fabrication of 3D MN structures via extensional deformation (stretching deformation) (Lee et al., 2010; Lee & Jung, 2012). Lee et al. produced sharp dissolving MNs from maltose via stepwise controlled drawing lithography (Lee & Jung, 2012). The later entails the simultaneous curing and creation of MNs without 3D molding (Lee & Jung, 2012). During the fabrication process, the triple-state (in liquid, glassy, solid) and viscosity of maltose was controlled by adjusting the drawing temperature (Lee et al., 2010). Maltose triple-state provides the ideal conditions for drawing lithography. The active pharmaceutical ingredient can be mixed with maltose in the liquid state while MNs can be formed during the glassy state and the solid state ensures that the MNs are robust enough to pierce animal or human skin (Lee et al., 2010).

Several investigators have used drawing lithography for the fabrication of microneedles (Chen, Ren et al., 2018; Chen, Wang et al., 2018; Lee et al., 2010; Lee & Jung, 2012). However, with traditional drawing lithographic procedures, it is challenging to fabricate scalable and dense MNs. Xiang et al. used a novel lithographic method that entails the use of a self-loading mold (Xiang et al., 2015). The authors attached a membrane to an SU-8 coated wafer to create the mold. The SU-8 colloid in this mold prevents MN tips from merging (Xiang et al., 2015). Secondly, the SU-8 colloid imparts uniform length to the MNs. Thirdly, low temperature drawing method prevents sharp tips from bending during solidification. This approach leads to the production of microneedles with a variety of lengths. The fabricated MNs possess mechanical strength and can pierce the *stratum corneum* (Xiang et al., 2015). With the right inter-needle spacing, skin piercing can be optimized. The creation of microconduits was confirmed and enhanced transdermal drug delivery was achieved

(Xiang et al., 2015). The authors utilized the MNs for the transdermal transport of insulin in vivo. Lower blood glucose levels were attained with the fabricated microneedles and they can be potentially used to deliver macromolecules across the skin (Xiang et al., 2015).

Droplet-born air blowing (DAB) technique is based on fast (≤10 min) and gentle (4−25°C) MN fabrication conditions. DAB entails the transformation of polymer droplet into MNs via air blowing (Kim et al., 2013). Since air is blown directly to the polymer droplet for subsequent solidification and MN formation, DAB is considered a gentle production option with neither heat nor ultraviolet radiation (Kim et al., 2013). In an interesting study, Kim et al. placed 20 μL of polymer dispersions (10% carboxymethylcellulose, 25% hyaluronic acid and 35% polyvinylpyrrolidone) into two 6 × 9 base arrays of upper and lower plates (10 μL for each plate) (Kim et al., 2013). The authors used air blow to elongate the viscous dispersion at a rate of 5 mm/min with subsequent polymer droplet solidification (Kim et al., 2013). By using the DAB approach, dissolving MNs can be produced within 10 min leading to a reduction in microfabrication cost with the added benefit of maintaining the pharmacological activity of the medication(s) (Kim et al., 2013). In another set of experiments, micro-milling was used to create the initial microneedle master template from which polylactic acid (PLA) MNs were then fabricated (Kim, Park et al., 2018). The authors used oxygen plasma etching to create polydimethylsiloxane (PDMS) MN molds and then prepared multilayer MNs via sequential spraying of polyvinylpyrrolidone (PVP) and poly (lactic-*co*-glycolic acid) (PLGA) solutions onto the molds (Kim, Park et al., 2018). Tests were carried out and it was shown that the MNs (obelisk in shape) were mechanically more robust compared to pyramid shaped MNs (Kim, Park et al., 2018). Furthermore, the investigators provided evidence that the PLGA and PVP layers adhered strongly to themselves (Kim, Park et al., 2018).

Huh and coworkers studied the influence of two droplet-based manufacturing techniques (centrifugal lithography and droplet-born air blowing) on the pharmacological effect of ascorbic acid and epidermal growth factor encapsulated in DMNs (Huh et al., 2018). With the centrifugal lithography (CL) approach, a plate with plasma treated hydrogel sheet was made with a gap between two parallel plates (Huh et al., 2018). Vertical loading into a centrifuge was then carried out. Droplet centrifugation was conducted at 500 g for 1 min using angular acceleration and vacuum at 4°C for polymer solidification. An isolation step was incorporated into the MN production cycle (Huh et al., 2018). When MNs were fabricated via droplet-born air blowing (DAB), stainless-steel plate and plasma treated hydrogel sheet were used. Elongation was carried out to form a biconcave shape. Then the authors used symmetric air blow at room temperature to solidify the shape (Huh et al., 2018). Lastly, the MNs containing epidermal growth factor were peeled off (Huh et al., 2018). DAB and CL are distinct in terms of their solidification conditions and this may influence an encapsulated antigen's biological activity (Lee et al., 2018).

Micro-nano 3D methods can be used to process polymeric materials into different shapes and these techniques have been utilized for MN fabrication without the need for templates (Yao et al., 2020). Three-dimensional printing entails the use of fused deposition modeling, selective laser sintering and stereolithography (Yao et al., 2020). Additive Manufacturing (AM) or 3-Dimensional Printing simplifies MN production since the entire process can be undertaken in one step (Xenikakis et al., 2019). The required shape and dimensions are first designed with a Computer Aided Design (CAD) software before fabrication takes place. 3D printing is simple and flexible. Alterations to the designs can be undertaken at any time via CAD (Xenikakis et al., 2019). Stereolithography (SLA) can be used as an accurate production technique for 3D printing (Xenikakis et al., 2019). Uddin and coworkers fabricated MNs by selectively photopolymerizing layers of a photopolymer resin via stereolithography (SLA) (Uddin et al., 2020). The MNs were coated with cisplatin formulations via inkjet printing (Uddin et al., 2020).

Piezo dispensing method is based on the "drop-on-demand" inkjet printing technology that can produce minute drops (1–70 pL) (Allen et al., 2016). The dispensing of these drops via piezoelectric force is based on administering electric pulses to a piezoelectric material that encloses a liquid reservoir (Allen et al., 2016). The resultant electric field leads to the deformation of the piezoelectric material leading to a pressure wave within the repository (Allen et al., 2016). Subsequently, drops are dispensed from a nozzle. The number of drops produced is a function of the frequency setting (Allen et al., 2016). Allen and colleagues used this technology to dispense a formulation into PDMS molds and the fabrication of bilayer DMNs (Allen et al., 2016). The effect of frequency, voltage and back pressure as well as excipients (polyvinyl alcohol and trehalose concentration) on drop formation was demonstrated (Allen et al., 2016).

In another study, Yao et al. utilized a high-precision digital light processing device to fabricate MNs (Yao et al., 2020). In another paper, layer-by-layer (LbL) method was employed for MN coating (Demuth et al., 2012). Poly (lactide-coglycolide) MNs were coated with multilayer films via layer-by-layer (LbL) assembly of anionic interbilayer-crosslinked multilamellar lipid vesicles (ICMVs) and a cationic biodegradable poly (β-amino ester) (Demuth et al., 2012). A biocompatible, hydrolytically degradable PBAE (Poly-1) was used for the experiments. The authors coated the MNs with (poly-1/ICMV) and poly-1/multilamellar vesicle multilayers (Poly-1/MVM) by dipping in ICMV/MLV solution (5 min) and Poly-1 (5 min) (Demuth et al., 2012). The dips were separated by rinsing in phosphate buffered saline. The investigators also deposited (Poly-1/ovalbumin) multilayers by dipping the MNs in ovalbumin solutions (0.1 mg/mL, 0.2 M sodium acetate) and Poly-1 (2 mg/mL, 0.2 M sodium acetate) (Demuth et al., 2012). Between the dips, the MNs were rinsed in deionized water (Demuth et al., 2012). In another study, Krieger et al. fabricated MNs via a facile mold fabrication method that entailed using an inexpensive scanning-based stereolithographic three-dimensional desktop printer (Krieger et al., 2019). In contrast to the traditional techniques used for MN fabrication, this approach does not depend upon complex and costly manufacturing operations or expertize in microfabrication (Krieger et al., 2019). Even though the majority of contemporary 3D-printed MNs possess inferior tip sharpness and low aspect ratios, the authors demonstrated that by using a two-stage "Print & Fill" mold preparation method, the production of sharp needles with high-aspect ratio was feasible (Krieger et al., 2019). By optimizing MN height, the investigators were able to fabricate needles with tip radii of 20—40 μm and high-aspect ratios. The sharpness of the MN tips was less than what was specified by the resolution of the printer (Krieger et al., 2019). For that reason, the authors printed sharp and high-aspect ratio MN arrays in containers that were backfilled and cured in a second step resulting in the fabrication of microneedles that were less than one millimeter in length (Krieger et al., 2019). The MN arrays served as male masters. Silicone female master molds were subsequently created. Using the molds, both polylactic acid microneedle arrays and carboxymethyl cellulose MNs loaded with rhodamine B were fabricated (Krieger et al., 2019).

Pramanick et al. combined polymer precursor carbonization with UV photolithography for the fabrication of hollow microneedle array (Pramanick et al., 2016). SU-8 microneedle contraction following pyrolysis was analyzed. Furthermore, the authors used Raman spectroscopy to determine the extent of sp2 versus sp3 carbon hybridization thus determining the properties of the pyrolyzed carbon (Pramanick et al., 2016). The authors also determined the chemical composition of the MNs via energy dispersive X-ray analysis (Pramanick et al., 2016). In another interesting research project, Mansoor and colleagues used a novel technique to fabricate metallic microneedles with predetermined spacing, height and lumen size (Mansoor et al., 2013). The authors utilized coated a mold, with an array of pillars via solvent casting. The conductive coating was then utilized as a seed tier during metal electrodeposition (Mansoor et al., 2013). The authors studied the polymer composite conductivity as a function of various filler concentrations. In addition, the investigators also conducted polymer plasma etching (Mansoor et al., 2013). The thickness of the MN array plate was controlled via electroplating. Compression tests were used to assess the strength of the MNs through a series and the transdermal drug delivery potential was investigated by administering fluorescent microspheres into pig skin (Mansoor et al., 2013).

Industrial fabrication of microneedles

One of the most critical issues for researchers in the field of microneedles (MNs) is the development of suitable large-scale manufacturing techniques that will facilitate the translation of this technology into clinical practice (Rad et al., 2017). Manufacturability and scalability obstruct mass commercialization of MNs (Kim, Park et al., 2018; Kim, Theogarajan et al., 2018). The significant impediment of precisely manufacturing micrometer-sized needles has greatly retarded the clinical application and commercialization of MNs (Yang et al., 2021). Sometimes, the geometry of 3D micro-devices cannot be meticulously guided by classical fabrication methods that are based on etching and photolithography (Liao et al., 2019). Other challenges include insufficient permeation efficiency due to mechanical failure, poor skin adhesion, and low biocompatibility (Yang et al., 2021). Some current methods for manufacturing MNs are not suitable for make-to-order batches and are also sensitive to input variation (Lim et al., 2021). In addition, the costly equipment and complex processes required for microfabrication result in long lead times (on the order of months) and constitutes a formidable manufacturing barrier (Johnson & Procopio, 2019). Furthermore, design problems, such as low available aspect ratios, are frequently encountered due to technical limitations of the manufacturing technologies (Johnson & Procopio, 2019). Traditionally, MN manufacturing processes include micromolding, chemical wet etching, injection molding, two-photon polymerization and reactive ion etching (Rad et al., 2017). Other fabrication methods are direct laser micromachining, electrical discharge machining and micromilling (Krieger et al., 2019). These methods can produce microneedles with excellent microscale features. However, the techniques also require either advanced manufacturing facilities such as clean rooms or expensive specialized equipment (Krieger et al., 2019). High-throughput production of MNs is also hindered by the high precision and number of features that need to be manufactured and the multi-step processing techniques needed for the fabrication of microscopic features (Gülçür et al., 2021). Furthermore, MN production is hampered by the three-dimensional geometry as well as the high aspect ratio structures of this drug delivery platform (the height of microneedles can range from 0.5 to 3 mm) (Nejad et al., 2018). An aspect ratio is a proportional relationship between the microneedle's width and height. Aspect ratios are typically documented as a ratio of width to height. Another obstacle is that MN molds are fabricated via complex procedures that entail the use of costly clean room facilities (Nejad et al., 2018).

Cost-effective techniques for the manufacturing of MNs are needed due to the diverse types of materials used and the different applications of these devices (Krieger et al., 2019). It is important to choose the most suitable materials for the fabrication of MNs based on the following criteria: (a) controlled or immediate drug release, (b) gentle manufacturing production procedures and (c) adequate mechanical strength to enable skin permeation (Tucak et al., 2020). Replica molding is beneficial for the production of MNs because it is reproducible, quick, up-scalable, inexpensive, useable and simple (Krieger et al., 2019). Significantly, diverse thermoset, thermoplastic, metals, ceramics and swellable/hydrogel-forming polymers can be utilized for the industrial fabrication of these devices (Economidou & Douroumis, 2021). Fully metallic micrometer-scale 3D devices can be made through a hybrid additive technique (2PTE) that combines electrochemical metal deposition with multi-photon lithography (Aksit et al., 2021). Two-photon templated electrodeposition (2PTE) makes it easier to produce complex, high aspect ratio, difficult-to-make metallic devices such as MNs (Aksit et al., 2021). Two-photon lithography (TPL) is based on the nonlinear dependency of polymerization rate on irradiating light intensity (Harinarayana & Shin, 2021). This technique has been employed for the production of 3-D devices with feature sizes beyond the limit of diffraction (Harinarayana & Shin, 2021). Printing resolution and material properties are the two critical parameters that have the most influential effect on the outcome of every microneedle printing (Economidou & Douroumis, 2021). Three-dimensional printing has been used by several investigators to obtain higher printing resolution (Economidou & Douroumis, 2021). Extrusion-based 3D printing and photopolymerization are particularly useful approaches (Economidou & Douroumis, 2021). The majority of low-cost 3D-printed microneedles exhibit low aspect ratios and poor tip sharpness (Krieger et al., 2019). Krieger and coworkers showed that by incorporating a two-step "Print & Fill" mold fabrication technique, high-aspect ratio sharp needles can be produced (Krieger et al., 2019). With decreasing input height and an increasingly larger than specified needle base diameter, low aspect ratio MNs were fabricated (Krieger et al., 2019).

Another approach for optimized large-scale manufacturing of MNs entails the production of mass-customizable master molds for the fabrication of sharp-tipped biodegradable dissolving MNs (Lim et al., 2021). The investigators combined chip casting and the pre-drying of an ultrathick photoresist layer with a substrate-less, inclined, and rotational exposure (Lim et al., 2021). The authors were able to reduce solvent uniformly across the photoresist thickness resulting in the formation of sharp tips (Lim et al., 2021). In addition, master molds with easily customizable geometry were made (Lim et al., 2021). Direct writing was combined with micromolding to manufacture sharp-tipped, mass-customizable, biodegradable MNs (Lim et al., 2021). Direct laser writing entails the photo-polymerization of a resin that is loaded with a chromophore (Xu et al., 2015). Active centers created through photon absorption by the chromophore triggers polymerization. Mass production to keep the unit cost of MNs low was achieved by using a micromolding approach as the key manufacturing operation; a master mold for micromolding was directly written to attain fabrication cost reduction and optimal design flexibility (Lim et al., 2021). Importantly, the investigators fabricated poly (lactic-*co*-glycolic) acid (PLGA) MNs that successfully pierced pig skin (Lim et al., 2021). Recently, Gülçür and coworkers described a full manufacturing process for the MNs that combine part design, mold laser micromachining, and micromolding together with real-time monitoring and thorough quality assessment (Gülçür et al., 2021). MNs with minute tip radii (about 2 μm) were fabricated and a competitive total machining time of 40 min for a single mold insert was attained (Gülçür et al., 2021). It is noteworthy that MNs require a skin insertion force of about 0.1−5 N. Solid microneedles with tip radius of 30 μm were shown to pierce the human skin with a comparatively low insertion force of about

1.3 N (Gülçür et al., 2021). On the other hand, pencil-like dissolvable MNs with a radius tip radius of approximately 15 μm pierced ex-vivo rat skin with an insertion force of 0.17 N/needle (Gülçür et al., 2021). To put it in context, a mosquito can painlessly breach the human skin with a significantly reduced force of approximately 18 μN (Yang et al., 2019). This lower force and higher penetration efficiency is ascribed to vibration (Yang et al., 2019). The industrial machinery required to produce MNs with such a high degree of efficiency is not yet available but scientists and industrial establishments are taking steps to enhance MN manufacturing efficiency. The coefficient of variation of 2.66% in attaining the desired cavity depth also shows the suitability and reproducibility of the process in a real manufacturing scheme even though it was also negatively impacted by polydimethylsiloxane demolding operations (Gülçür et al., 2021). To ensure quality assurance, the authors constructed a bespoke telecentric measurement system (Gülçür et al., 2021). A real-time monitoring of the manufacturing cycles combined with robot-assisted pass-fail procedures can circumvent the industrial fabrication of defective parts (Gülçür et al., 2021).

The geometries of MNs have to be beveled, sharp, and controllable (Kim, Park et al., 2018; Kim, Theogarajan et al., 2018). It is challenging to comply with these requirements via microelectromechanical fabrication methods (Kim, Park et al., 2018; Kim, Theogarajan et al., 2018). Kim and colleagues utilized silicon etch chemistries to optimize the production of manufacturable and scalable beveled hollow silicon MNs (Kim, Park et al., 2018; Kim, Theogarajan et al., 2018). The authors demonstrated that hollow silicon microneedles with any dimensions can be produced with 5% deviation between different processes and at least 2% uniformity across wafers (Kim, Park et al., 2018; Kim, Theogarajan et al., 2018). The reason for attaining such a high degree of consistency and uniformity is the utilization of a non-agitated $HF-HNO_3$ bath, silicon nitride masks, and surrounding silicon filler materials with precise dimensions (Kim, Park et al., 2018; Kim, Theogarajan et al., 2018). In another interesting study, Johnson and Procopio used a widely available 3D printer to easily and quickly produce microneedle masters (Johnson & Procopio, 2019). The authors investigated the effect of light aberrations on the fabricated devices, aliasing, and stair-stepping (Johnson & Procopio, 2019). Sometimes, there are aberrations that result in variations between the light distribution that reaches the resin and the image sent to the projector (Johnson & Procopio, 2019). "Aliasing" takes place when each layer is projected as an individual image while "stair-stepping" occurs when a continuous part is stripped into layers in the z direction (Johnson & Procopio, 2019). The influence of grayscale, the number of layers and the layer height on the surface texture, sharpness, and dimensional fidelity of the fabricated devices was also studied (Johnson & Procopio, 2019). The authors successfully used the Autodesk Ember to produce sharp MNs with a tip radius of about 15 μm in less than 30 min per. The authors used the grayscale to enhance the sharpness, surface texture of the MNs. A dimensional fidelity of about ±5% was attained (Johnson & Procopio, 2019).

Even though micro-machining, micromolding, lithography and etching are extensively utilized for the development of MNs, they are characterized by several limitations linked to customizability, scalability, and financial viability (Economidou & Douroumis, 2021). As an illustration, mold fabrication or procurement is expensive while etching, micromachining, and lithography are complex, time and labor intensive (Economidou & Douroumis, 2021). These fabrication techniques also require expensive equipment and scientific expertise (Economidou & Douroumis, 2021). Furthermore, injection molding processes can only be economically viable for production lines that make large unit numbers (Economidou & Douroumis, 2021). Additive manufacturing can produce reliable, fast, and cost-effective means of fabricating MN (Economidou & Douroumis, 2021). High reproducibility, versatility, accuracy and simplicity of 3D printing have caught the attention of many research scientists who investigate microneedles (Economidou & Douroumis, 2021). Recently, Wu et al. employed post stretching and extrusion-based 3D printing to produce an MN patch for glucose-responsive insulin delivery (Wu et al., 2020). The authors studied the printability of various bioinks made from hydroxyapatite and alginate (Wu et al., 2020). After printing the substrate and a cylindrical array of the patch, the authors generated needle-like tips by stretching the top surface of the cylindrical array (Wu et al., 2020).

References

Ahmad, Z., Khan, M. I., Siddique, M. I., Sarwar, H. S., Shahnaz, G., Hussain, S. Z., Bukhari, N. I., Hussain, I., & Sohail, M. F. (2020). Fabrication and characterization of thiolated chitosan microneedle patch for transdermal delivery of tacrolimus. *AAPS PharmSciTech, 21*(2). https://doi.org/10.1208/s12249-019-1611-9

Aksit, A., Lalwani, A. K., Kysar, J. W., & West, A. C. (2021). Simulation assisted design for microneedle manufacturing: Computational modeling of two-photon templated electrodeposition. *Journal of Manufacturing Processes, 66*, 211–219. https://doi.org/10.1016/j.jmapro.2021.04.023

Allen, E. A., O'Mahony, C., Cronin, M., O'Mahony, T., Moore, A. C., & Crean, A. M. (2016). Dissolvable microneedle fabrication using piezoelectric dispensing technology. *International Journal of Pharmaceutics, 500*(1–2), 1–10. https://doi.org/10.1016/j.ijpharm.2015.12.052

Azmana, M., Mahmood, S., Hilles, A. R., Mandal, U. K., Saeed Al-Japairai, K. A., & Raman, S. (2020). Transdermal drug delivery system through polymeric microneedle: A recent update. *Journal of Drug Delivery Science and Technology, 60*, 101877. https://doi.org/10.1016/j.jddst.2020.101877

Balmert, S. C., Carey, C. D., Falo, G. D., Sethi, S. K., Erdos, G., Korkmaz, E., & Falo, L. D. (2020). Dissolving undercut microneedle arrays for multicomponent cutaneous vaccination. *Journal of Controlled Release, 317*, 336–346. https://doi.org/10.1016/j.jconrel.2019.11.023

Chen, Y. T., Hsu, C. C., Tsai, C. H., & Kang, S. W. (2010). Fabrication of microneedles. *Journal of Marine Science and Technology, 18*(2), 243–248.

Chen, Z., Ren, L., Li, J., Yao, L., Chen, Y., Liu, B., & Jiang, L. (2018). Rapid fabrication of microneedles using magnetorheological drawing lithography. *Acta Biomaterialia, 65*, 283–291. https://doi.org/10.1016/j.actbio.2017.10.030

Chen, X., Wang, L., Yu, H., Li, C., Feng, J., Haq, F., Khan, A., & Khan, R. U. (2018). Preparation, properties and challenges of the microneedles-based insulin delivery system. *Journal of Controlled Release, 288*, 173–188. https://doi.org/10.1016/j.jconrel.2018.08.042

Chen, S., Wu, D., Liu, Y., Huang, Y., Xu, H., Gao, W., Zhang, J., Sun, J., & Zhuang, J. (2020). Optimal scaling analysis of polymeric microneedle length and its effect on transdermal insulin delivery. *Journal of Drug Delivery Science and Technology, 56*, 101547. https://doi.org/10.1016/j.jddst.2020.101547

Coffey, J. W., Meliga, S. C., Corrie, S. R., & Kendall, M. A. F. (2016). Dynamic application of microprojection arrays to skin induces circulating protein extravasation for enhanced biomarker capture and detection. *Biomaterials, 84*, 130–143. https://doi.org/10.1016/j.biomaterials.2016.01.015

Creighton, R. L., Phan, J., & Woodrow, K. A. (2020). In situ 3D-patterning of electrospun fibers using two-layer composite materials. *Scientific Reports, 10*(1). https://doi.org/10.1038/s41598-020-64846-z

Dardano, P., Caliò, A., Di Palma, V., Bevilacqua, M. F., Di Matteo, A. D., & Stefano, L. (2015). A photolithographic approach to polymeric microneedles array fabrication. *Materials, 8*(12), 8661–8673. https://doi.org/10.3390/ma8125484

Demir, Y. K., Metin, A.Ü., Satıroğlu, B., Solmaz, M. E., Kayser, V., & Mäder, K. (2017). Poly (methyl vinyl ether-co-maleic acid) – pectin based hydrogel-forming systems: Gel, film, and microneedles. *European Journal of Pharmaceutics and Biopharmaceutics, 117*, 182–194. https://doi.org/10.1016/j.ejpb.2017.04.018

Demuth, P. C., Moon, J. J., Suh, H., Hammond, P. T., & Irvine, D. J. (2012). Releasable layer-by-layer assembly of stabilized lipid nanocapsules on microneedles for enhanced transcutaneous vaccine delivery. *ACS Nano, 6*(9), 8041–8051. https://doi.org/10.1021/nn302639r

Economidou, S. N., & Douroumis, D. (2021). 3D printing as a transformative tool for microneedle systems: Recent advances, manufacturing considerations and market potential. *Advanced Drug Delivery Reviews, 173*, 60–69. https://doi.org/10.1016/j.addr.2021.03.007

Eltayib, E., Brady, A. J., Caffarel-Salvador, E., Gonzalez-Vazquez, P., Zaid Alkilani, A., McCarthy, H. O., McElnay, J. C., & Donnelly, R. F. (2016). Hydrogel-forming microneedle arrays: Potential for use in minimally-invasive lithium monitoring. *European Journal of Pharmaceutics and Biopharmaceutics, 102*, 123–131. https://doi.org/10.1016/j.ejpb.2016.03.009

Evens, T., Malek, O., Castagne, S., Seveno, D., & Van Bael, A. (2020). A novel method for producing solid polymer microneedles using laser ablated moulds in an injection moulding process. *Manufacturing Letters, 24*, 29–32. https://doi.org/10.1016/j.mfglet.2020.03.009

Feng, Y. H. (2020). Multiscale simulations of drug distributions in polymer dissolvable microneedles. *Colloids and Surfaces B: Biointerfaces, 189*, 110844. https://doi.org/10.1016/j.colsurfb.2020.110844

Gholami, S., Mohebi, M. M., Hajizadeh-Saffar, E., Ghanian, M. H., Zarkesh, I., & Baharvand, H. (2019). Fabrication of microporous inorganic microneedles by centrifugal casting method for transdermal extraction and delivery. *International Journal of Pharmaceutics, 558*, 299–310. https://doi.org/10.1016/j.ijpharm.2018.12.089

Gülçür, M., Romano, J. M., Penchev, P., Gough, T., Brown, E., Dimov, S., & Whiteside, B. (2021). A cost-effective process chain for thermoplastic microneedle manufacture combining laser micro-machining and micro-injection moulding. *CIRP Journal of Manufacturing Science and Technology, 32*, 311–321. https://doi.org/10.1016/j.cirpj.2021.01.015

Harinarayana, V., & Shin, Y. C. (2021). Two-photon lithography for three-dimensional fabrication in micro/nanoscale regime: A comprehensive review. *Optics & Laser Technology, 142*, 107180. https://doi.org/10.1016/j.optlastec.2021.107180

Henry, S., McAllister, D. V., Allen, M. G., & Prausnitz, M. R. (1998). Microfabricated microneedles: A novel approach to transdermal drug delivery. *Journal of Pharmaceutical Sciences, 87*(8), 922–925. https://doi.org/10.1021/js980042+

Huh, I., Kim, S., Yang, H., Jang, M., Kang, G., & Jung, H. (2018). Effects of two droplet-based dissolving microneedle manufacturing methods on the activity of encapsulated epidermal growth factor and ascorbic acid. *European Journal of Pharmaceutical Sciences, 114*, 285–292. https://doi.org/10.1016/j.ejps.2017.12.025

Indermun, S., Luttge, R., Choonara, Y. E., Kumar, P., Du Toit, L. C., Modi, G., & Pillay, V. (2014). Current advances in the fabrication of microneedles for transdermal delivery. *Journal of Controlled Release, 185*(1), 130–138. https://doi.org/10.1016/j.jconrel.2014.04.052

Ita, K. (2015). Transdermal delivery of drugs with microneedles: Strategies and outcomes. *Journal of Drug Delivery Science and Technology, 29*, 16–23. https://doi.org/10.1016/j.jddst.2015.05.001

Ita, K. (2017). Dissolving microneedles for transdermal drug delivery: Advances and challenges. *Biomedicine & Pharmacotherapy, 93*, 1116–1127. https://doi.org/10.1016/j.biopha.2017.07.019

Jin, J., Reese, V., Coler, R., Carter, D., & Rolandi, M. (2014). Chitin microneedles for an easy-to-use tuberculosis skin test. *Advanced Healthcare Materials, 3*(3), 349–353. https://doi.org/10.1002/adhm.201300185

Johnson, A. R., & Procopio, A. T. (2019). Low cost additive manufacturing of microneedle masters. *3D Printing in Medicine*. https://doi.org/10.1186/s41205-019-0039-x

Kathuria, H., Kang, K., Cai, J., & Kang, L. (2020). Rapid microneedle fabrication by heating and photolithography. *International Journal of Pharmaceutics, 575*, 118992. https://doi.org/10.1016/j.ijpharm.2019.118992

Kim, Y., Bhattacharjee, S. A., Beck-Broichsitter, M., & Banga, A. K. (2019). Fabrication and characterization of hyaluronic acid microneedles to enhance delivery of magnesium ascorbyl phosphate into skin. *Biomedical Microdevices, 21*(4). https://doi.org/10.1007/s10544-019-0455-0

Kim, J. D., Kim, M., Yang, H., Lee, K., & Jung, H. (2013). Droplet-born air blowing: Novel dissolving microneedle fabrication. *Journal of Controlled Release, 170*(3), 430–436. https://doi.org/10.1016/j.jconrel.2013.05.026

Kim, Y. C., Park, J. H., & Prausnitz, M. R. (2012). Microneedles for drug and vaccine delivery. *Advanced Drug Delivery Reviews, 64*(14), 1547–1568. https://doi.org/10.1016/j.addr.2012.04.005

Kim, M. J., Park, S. C., Rizal, B., Guanes, G., Baek, S. K., Park, J. H., Betz, A. R., & Choi, S. O. (May 2018). Fabrication of circular obelisk-type multilayer microneedles using micro-milling and spray deposition. *Frontiers in Bioengineering and Biotechnology, 6.* https://doi.org/10.3389/fbioe.2018.00054

Kim, H., Theogarajan, L. S., & Pennathur, S. (2018). A repeatable and scalable fabrication method for sharp, hollow silicon microneedles. *Journal of Micromechanics and Microengineering, 28*(3). https://doi.org/10.1088/1361-6439/aaa6a8

Krieger, K. J., Bertollo, N., Dangol, M., Sheridan, J. T., Lowery, M. M., & O'Cearbhaill, E. D. (2019). Simple and customizable method for fabrication of high-aspect ratio microneedle molds using low-cost 3D printing. *Microsystems and Nanoengineering, 5*(1). https://doi.org/10.1038/s41378-019-0088-8

Larrañeta, E., Lutton, R. E. M., Woolfson, A. D., & Donnelly, R. F. (2016). Microneedle arrays as transdermal and intradermal drug delivery systems: Materials science, manufacture and commercial development. *Materials Science and Engineering: R: Reports, 104,* 1–32. https://doi.org/10.1016/j.mser.2016.03.001

Lee, K. J., Goudie, M. J., Tebon, P., Sun, W., Luo, Z., Lee, J., Zhang, S., Fetah, K., Kim, H. J., Xue, Y., Darabi, M. A., Ahadian, S., Sarikhani, E., Ryu, W. H., Gu, Z., Weiss, P. S., Dokmeci, M. R., Ashammakhi, N., & Khademhosseini, A. (2020). Non-transdermal microneedles for advanced drug delivery. *Advanced Drug Delivery Reviews, 165–166,* 41–59. https://doi.org/10.1016/j.addr.2019.11.010

Lee, K., & Jung, H. (2012). Drawing lithography for microneedles: A review of fundamentals and biomedical applications. *Biomaterials, 33*(30), 7309–7326. https://doi.org/10.1016/j.biomaterials.2012.06.065

Lee, C., Kim, H., Kim, S., Lahiji, S. F., Ha, N. Y., Yang, H., Kang, G., Nguyen, H. Y. T., Kim, Y., Choi, M. S., Cho, N. H., & Jung, H. (2018). Comparative study of two droplet-based dissolving microneedle fabrication methods for skin vaccination. *Advanced Healthcare Materials,* 7(11). https://doi.org/10.1002/adhm.201701381

Lee, K., Lee, H. C., Lee, D. S., & Jung, H. (2010). Drawing lithography: Three-dimensional fabrication of an ultrahigh-aspect-ratio microneedle. *Advanced Materials, 22*(4), 483–486. https://doi.org/10.1002/adma.200902418

Lee, J. Y., Park, S. H., Seo, I. H., Lee, K. J., & Ryu, W. H. (2015). Rapid and repeatable fabrication of high A/R silk fibroin microneedles using thermally-drawn micromolds. *European Journal of Pharmaceutics and Biopharmaceutics, 94,* 11–19. https://doi.org/10.1016/j.ejpb.2015.04.024

Lee, J., van der Maaden, K., Gooris, G., O'Mahony, C., Jiskoot, W., & Bouwstra, J. (2021). Engineering of an automated nano-droplet dispensing system for fabrication of antigen-loaded dissolving microneedle arrays. *International Journal of Pharmaceutics, 600,* 120473. https://doi.org/10.1016/j.ijpharm.2021.120473

Leone, M., Mönkäre, J., Bouwstra, J. A., & Kersten, G. (2017). Dissolving microneedle patches for dermal vaccination. *Pharmaceutical Research, 34*(11), 2223–2240. https://doi.org/10.1007/s11095-017-2223-2

Liao, C., Anderson, W., Antaw, F., & Trau, M. (2019). Two-photon nanolithography of tailored hollow three-dimensional microdevices for biosystems. *ACS Omega, 4*(1), 1401–1409. https://doi.org/10.1021/acsomega.8b03164

Li, W.-Z., Huo, M.-R., Zhou, J.-P., Zhou, Y.-Q., Hao, B.-H., Liu, T., & Zhang, Y. (2010). Super-short solid silicon microneedles for transdermal drug delivery applications. *International Journal of Pharmaceutics, 389*(1–2), 122–129. https://doi.org/10.1016/j.ijpharm.2010.01.024

Lim, H., Ha, S., Bae, M., & Yoon, S.-H. (2021). A highly robust approach to fabricate the mass-customizable mold of sharp-tipped biodegradable polymer microneedles for drug delivery. *International Journal of Pharmaceutics, 600*, 120475. https://doi.org/10.1016/j.ijpharm.2021.120475

Li, Y., Zhang, H., Yang, R., Laffitte, Y., Schmill, U., Hu, W., Kaddoura, M., Blondeel, E. J. M., & Cui, B. (2019). Fabrication of sharp silicon hollow microneedles by deep-reactive ion etching towards minimally invasive diagnostics. *Microsystems and Nanoengineering, 5*(1). https://doi.org/10.1038/s41378-019-0077-y

Mansoor, I., Liu, Y., Häfeli, U. O., & Stoeber, B. (2013). Arrays of hollow out-of-plane microneedles made by metal electrodeposition onto solvent cast conductive polymer structures. *Journal of Micromechanics and Microengineering, 23*(8), 085011. https://doi.org/10.1088/0960-1317/23/8/085011

Martanto, W., Davis, S. P., Holiday, N. R., Wang, J., Gill, H. S., & Prausnitz, M. R. (2004). Transdermal delivery of insulin using microneedles in vivo. *Pharmaceutical Research, 21*(6), 947–952. https://doi.org/10.1023/B:PHAM.0000029282.44140.2e

Nejad, H. R., Sadeqi, A., Kiaee, G., & Sonkusale, S. (2018). Low-cost and cleanroom-free fabrication of microneedles. *Microsystems & Nanoengineering, 4*, 17073. https://doi.org/10.1038/micronano.2017.73

Nguyen, H. X., Bozorg, B. D., Kim, Y., Wieber, A., Birk, G., Lubda, D., & Banga, A. K. (2018). Poly (vinyl alcohol) microneedles: Fabrication, characterization, and application for transdermal drug delivery of doxorubicin. *European Journal of Pharmaceutics and Biopharmaceutics, 129*, 88–103. https://doi.org/10.1016/j.ejpb.2018.05.017

Park, J. H., Allen, M. G., & Prausnitz, M. R. (2005). Biodegradable polymer microneedles: Fabrication, mechanics and transdermal drug delivery. *Journal of Controlled Release, 104*(1), 51–66. https://doi.org/10.1016/j.jconrel.2005.02.002

Pattarabhiran, S. P., Saju, A., Sonawane, K. R., Manimaran, R., Bhatnagar, S., Roy, G., Kulkarni, R. B., & Venuganti, V. V. K. (2019). Dissolvable microneedle-mediated transcutaneous delivery of tetanus toxoid elicits effective immune response. *AAPS PharmSciTech, 20*(7). https://doi.org/10.1208/s12249-019-1471-3

Pramanick, B., Martinez-Chapa, S. O., & Madou, M. (2016). Fabrication of biocompatible hollow microneedles using the C-MEMS process for transdermal drug delivery. In *ECS transactions* (Vol. 72, pp. 45–50). Electrochemical Society Inc. https://doi.org/10.1149/07201.0045ecst. Issue 1.

Rad, Z., Nordon, R., Anthony, C., Bilston, L., Prewett, P., Arns, J.-Y., Arns, C., Zhang, L., & Davies, G. (2017). High-fidelity replication of thermoplastic microneedles with open microfluidic channels. *Microsystems & Nanoengineering, 3*, 17034. https://doi.org/10.1038/micronano.2017.34

Ruggiero, F., Vecchione, R., Bhowmick, S., Coppola, G., Coppola, S., Esposito, E., Lettera, V., Ferraro, P., & Netti, P. A. (2018). Electro-drawn polymer microneedle arrays with controlled shape and dimension. *Sensors and Actuators B: Chemical, 255*, 1553–1560. https://doi.org/10.1016/j.snb.2017.08.165

Sabri, A. H., Kim, Y., Marlow, M., Scurr, D. J., Segal, J., Banga, A. K., Kagan, L., & Lee, J. B. (2020). Intradermal and transdermal drug delivery using microneedles — fabrication, performance evaluation and application to lymphatic delivery. *Advanced Drug Delivery Reviews*, 195–215. https://doi.org/10.1016/j.addr.2019.10.004

Sadeqi, A., Nejad, H. R., Kiaee, G., & Sonkusale, S. (2018). Cost-effective fabrication of chitosan microneedles for transdermal drug delivery. In *Conference proceedings: annual international conference of the IEEE engineering in medicine and biology society. IEEE engineering in medicine and biology society. Annual conference, 2018* (pp. 5737–5740). https://doi.org/10.1109/EMBC.2018.8513691

Sugiyama, S., Khumpuang, S., & Kawaguchi, G. (2004). Plain-pattern to cross-section transfer (PCT) technique for deep x-ray lithography and applications. *Journal of Micromechanics and Microengineering, 14*(10), 1399–1404. https://doi.org/10.1088/0960-1317/14/10/015

Takahashi, H., Heo, Y. J., Arakawa, N., Kan, T., Matsumoto, K., Kawano, R., & Shimoyama, I. (2016). Scalable fabrication of microneedle arrays via spatially controlled UV exposure. *Microsystems and Nanoengineering, 2*. https://doi.org/10.1038/micronano.2016.49

Tarbox, T. N., Watts, A. B., Cui, Z., & Williams, R. O. (2018). An update on coating/manufacturing techniques of microneedles. *Drug Delivery and Translational Research, 8*(6), 1828–1843. https://doi.org/10.1007/s13346-017-0466-4

Tian, Z., Cheng, J., Liu, J., & Zhu, Y. (2019). Dissolving graphene/poly(acrylic acid) microneedles for potential transdermal drug delivery and Photothermal therapy. *Journal of Nanoscience and Nanotechnology*, 2453–2459. https://doi.org/10.1166/jnn.2019.15884

Tucak, A., Sirbubalo, M., Hindija, L., Rahić, O., Hadžiabdić, J., Muhamedagić, K., Čekić, A., & Vranić, E. (2020). Microneedles: Characteristics, materials, production methods and commercial development. *Micromachines, 11*(11). https://doi.org/10.3390/mi11110961

Uddin, M. J., Scoutaris, N., Economidou, S. N., Giraud, C., Chowdhry, B. Z., Donnelly, R. F., & Douroumis, D. (2020). 3D printed microneedles for anticancer therapy of skin tumours. *Materials Science and Engineering: C, 107*. https://doi.org/10.1016/j.msec.2019.110248

Vecchione, R., Coppola, S., Esposito, E., Casale, C., Vespini, V., Grilli, S., Ferraro, P., & Netti, P. A. (2014). Electro-drawn drug-loaded biodegradable polymer microneedles as a viable route to hypodermic injection. *Advanced Functional Materials, 24*(23), 3515–3523. https://doi.org/10.1002/adfm.201303679

Wang, J., Wang, H., Lai, L., & Li, Y. (2021). Preparation of microneedle array mold based on mems lithography technology. *Micromachines, 12*(1), 1–13. https://doi.org/10.3390/mi12010023

Wouters, K., & Puers, R. (2009). Accurate measurement of the steady-state swelling behavior of SU-8 negative photo resist. *Procedia Chemistry, 1*(Issue 1), 60–63. https://doi.org/10.1016/j.proche.2009.07.015

Wu, M., Zhang, Y., Huang, H., Li, J., Liu, H., Guo, Z., Xue, L., Liu, S., & Lei, Y. (2020). Assisted 3D printing of microneedle patches for minimally invasive glucose control in diabetes. *Materials Science and Engineering: C, 117*, 111299. https://doi.org/10.1016/j.msec.2020.111299

Xenikakis, I., Tsongas, K., Tzimtzimis, E. K., Zacharis, C. K., Theodoroula, N., Kalogianni, E. P., Demiri, E., Vizirianakis, I. S., Tzetzis, D., & Fatouros, D. G. (2021). Fabrication of hollow microneedles using liquid crystal display (LCD) vat polymerization 3D printing technology for transdermal macromolecular delivery. *International Journal of Pharmaceutics, 597*, 120303. https://doi.org/10.1016/j.ijpharm.2021.120303

Xenikakis, I., Tzimtzimis, M., Tsongas, K., Andreadis, D., Demiri, E., Tzetzis, D., & Fatouros, D. G. (2019). Fabrication and finite element analysis of stereolithographic 3D printed microneedles for transdermal delivery of model dyes across human skin in vitro. *European Journal of Pharmaceutical Sciences, 137*, 104976. https://doi.org/10.1016/j.ejps.2019.104976

Xiang, Z., Wang, H., Murugappan, S. K., Yen, S.-C., Pastorin, G., & Lee, C. (2015). Dense vertical SU-8 microneedles drawn from a heated mold with precisely controlled volume. *Journal of Micromechanics and Microengineering, 25*(2), 025013. https://doi.org/10.1088/0960-1317/25/2/025013

Xu, Q., Lv, Y., Dong, C., Sreeprased, T. S., Tian, A., Zhang, H., Tang, Y., Yu, Z., & Li, N. (2015). Three-dimensional micro/nanoscale architectures: Fabrication and applications. *Nanoscale, 7*(25), 10883–10895. https://doi.org/10.1039/c5nr02048d

Yang, J., Liu, X., Fu, Y., & Song, Y. (2019). Recent advances of microneedles for biomedical applications: Drug delivery and beyond. *Acta Pharmaceutica Sinica B, 9*(3), 469–483. https://doi.org/10.1016/j.apsb.2019.03.007

Yang, Q., Zhong, W., Xu, L., Li, H., Yan, Q., She, Y., & Yang, G. (2021). Recent progress of 3D-printed microneedles for transdermal drug delivery. *International Journal of Pharmaceutics, 593*, 120106. https://doi.org/10.1016/j.ijpharm.2020.120106

Yao, W., Li, D., Zhao, Y., Zhan, Z., Jin, G., Liang, H., & Yang, R. (2020). 3D Printed multi-functional hydrogel microneedles based on high-precision digital light processing. *Micromachines, 11*(1). https://doi.org/10.3390/mi11010017

Yun, S., Lee, S.-J., Giri, S. S., Kim, H. J., Kim, S. G., Kim, S. W., Han, S. J., Kwon, J., Oh, W. T., & Chang Park, S. (2020). Vaccination of fish against Aeromonas hydrophila infections using the novel approach of transcutaneous immunization with dissolving microneedle patches in aquaculture. *Fish & Shellfish Immunology, 97*, 34–40. https://doi.org/10.1016/j.fsi.2019.12.026

Zhou, X. (2020). Biodegradable beta-cyclodextrin conjugated gelatin methacryloyl microneedle for delivery of water-insoluble drug. *Advanced Healthcare Materials, 9*(11), e2000527. http://doi:10.1002/adhm.202000527.

CHAPTER 3
Dissolving microneedles

Introduction

Dissolving microneedles (DMNs) are typically less than 2000 μm in length and are fabricated from water-soluble materials from which a medication can be delivered into the skin or other tissues (Ita, 2017). The needles are mechanically robust to pierce the skin and release the drug after polymer dissolution. DMNs encapsulate medications within the soluble polymeric matrix and can dissolve completely upon insertion into the skin (Ye et al., 2018). Following insertion, the polymer matrix dissolves at a defined rate due to contact with interstitial fluid (Jin et al., 2018). Increasingly, dissolving MNs are being used by investigators for transdermal drug delivery and other applications. This can be explained by the numerous benefits associated with the use of DMNs. Apart from being cost-effective from a manufacturing standpoint, drug delivery rate can be controlled when drug molecules are delivered from DMNs (Sabri et al., 2020). Safety, low cost and the ease of preparation are widely reported benefits (Chen, Han et al., 2020; Chen, Quan et al., 2020; Ita, 2017; Ye et al., 2018) and the polymers used for DMN fabrication are typically biodegradable, tough, biocompatible, strong and optical clear (Sabri et al., 2020). The temporal drug release profile can be regulated from several seconds to several months by choosing the appropriate polymeric matrix (Lee et al., 2018). In addition, the stability of proteins is maintained because DMNs are essentially dry (Leone, Du et al., 2020; Leone, Romeijin et al., 2020). Thirdly, the cost of providing cold chain can be eliminated and the shelf-life of proteins/peptides in vaccines prolonged when DMNs are used (Vora et al., 2020). The biomaterials utilized for DMN fabrication of have significant impact on the pharmacological effect of encapsulated drugs and the manufacturing operations employed during microneedle production (Leone, Du et al., 2020; Leone, Romeijin et al., 2020). Leone et al. investigated the influence of molecular weight of hyaluronan (from 4.8 kDa to 1.8 MDa) on MN dissolution (Leone, Du et al., 2020; Leone, Romeijin et al., 2020).

Microneedles
ISBN 978-0-323-91811-4
https://doi.org/10.1016/B978-0-323-91811-4.00009-2

The effect of the variable (MW) on T-cell activation in vitro and antibody response in mice was also studied (Leone, Du et al., 2020; Leone, Romeijin et al., 2020). HA with the molecular weight of 1.8 MDa was unsuitable for the preparation of dissolving microneedles. The 4.8 kDa hyaluronic acid (HA) was used to create DMNs that were not hard enough (Leone, Du et al., 2020; Leone, Romeijin et al., 2020). The authors chose 20 kDa HA as the optimal polymer for the production of DMNs because the DMNs produced from this material were hard but dissolved rapidly in the skin with no deleterious immunogenic effect (Leone, Du et al., 2020; Leone, Romeijin et al., 2020). It is important to ascertain the ability of the DMNs to penetrate the skin. The polymer must be compatible with the active pharmaceutical ingredient and an acceptable and scalable fabrication method should be utilized (Leone, Du et al., 2020; Leone, Romeijin et al., 2020; Leone et al., 2017).

But there are drawbacks linked to the use of DMNs. Increased temperature or ultraviolet (UV) radiation during MN fabrication can lead to the denaturation of biologics and other forms of drug instabilities (Lee et al., 2018). In addition, the incorporation of active pharmaceutical ingredients into the polymer is also limited because a high amount of drug in the DMN will reduce the mechanical strength of the needles (Lee et al., 2018). Furthermore, water-soluble polymers possess poor mechanical strength in comparison to non-dissolving materials such as silicon. Another important consideration is that drug loading can lead to further reduction in mechanical strength (Vora et al., 2020). Only few polymers can be utilized for the production of MNs (Vora et al., 2020).

Dissolving microneedles for drug delivery

A tremendous amount of work has been carried out by researchers regarding the employment of DMNs for drug delivery into distinct tissues in the body such as skin and eye. These devices can deliver medications and vaccines (Ita & Ashong, 2020; Mao et al., 2020; Sullivan et al., 2010). Recently, interesting projects focused on applying DMN for drug transport into the eye (Huang et al., 2018; Khandan et al., 2012) and other tissues (Rzhevskiy et al., 2018) were reported. Furthermore, DMNs have also been investigated as devices for diagnostics (Liu, Chou et al., 2020; Liu, Kong et al., 2020). Franz diffusion cells were used by Chen et al. to investigate the transdermal delivery of polydatin (PD) from DMN across pig ear skin (Chen, Han et al., 2020; Chen, Quan et al., 2020).

PDMS micromolding was the fabrication procedure used by the authors. Different polymers, including hydroxypropyl methylcellulose (HPMC), polyvinyl alcohol, hyaluronic acid (HA), polyvinylpyrrolidone (PVP) and carboxymethylcellulose sodium (CMC-Na) were utilized as matrices for the fabrication of the dissolving microneedles (Chen, Han et al., 2020; Chen, Quan et al., 2020). The cumulative transport of polydatin across porcine skin increased remarkably over 8 h following the use of PVP DMNs in comparison to a saturated solution of the compound. But the use of DMNs fabricated from CMC-Na, HPMC or HA did not result in enhanced polydatin transport in comparison to a saturated solution of the compound (Chen, Han et al., 2020; Chen, Quan et al., 2020). The authors attributed enhanced drug transport from PVP DMNs to increased wettability as well as an accelerated rate of dissolution (Chen, Han et al., 2020; Chen, Quan et al., 2020).

Balmert et al. used a spin-casting technique and replica master molds to fabricate polydimethylsiloxane (PDMS) molds, which were subsequently used to formulate dissolving undercut MNs tip-loaded with ovalbumin (OV) with or without Poly (I:C) (an adjuvant) (Balmert et al., 2020). The stem regions of these MNs were undercut with filleted bases. The amounts of Poly (I:C) and OV in the DMNs were 24.29 ± 1.60 μg and 10.15 ± 0.87 μg respectively (Balmert et al., 2020). The authors prepared DMNs from trehalose and carboxymethylcellulose. These pharmaceutical excipients are approved by the United States Food and Drug Administration. Carboxymethylcellulose is suitable for DMN fabrication because it is mechanically robust and possesses adequate aqueous solubility (Balmert et al., 2020). Trehalose, on the other hand, is a disaccharide that can enhance protein stability (Balmert et al., 2020). The authors also prepared DMNs from PVP/PVA hydrogel. DMNs loaded with Allura Red R40 dye were administered onto human skin samples. Complete dissolution of the DMNs was achieved (Balmert et al., 2020). MN conduits permeating the epidermis into the dermis were created and the transport of OVA and Poly (I:C) labeled with fluorescent dyes to targeted portions of the skin samples was attained (Balmert et al., 2020).

Yao and colleagues used DMNs for the transport of the poorly water-soluble levonorgestrel (contraceptive medication) (Yao et al., 2017). Micromolding was the manufacturing technique used for the fabrication of the needles. The authors incorporated chitosan and beta-sodium glycerophosphate (β-GP) into the MNs (Yao et al., 2017). It was reported that $69.32\% \pm 4.23\%$ of the MNs pierced porcine skin and dissolved within the first 120 min, which was approximately two times greater than that of traditional MNs (Yao et al., 2017). The DMNs prepared from chitosan and β-GP displayed greater solubilizing property. Furthermore, the DMNs showed greater mechanical strength. The DMNs exhibited enhanced drug loading capability (about 30 fold) (Yao et al., 2017). Furthermore, the T_{max} and C_{max} values were 30 min and 189.27 ± 57.46 ng/mL respectively following the use of these DMNs (Yao et al., 2017). The corresponding parameters when the drug was administered from oral suspension were 30 min and 224.71 ± 55.23 ng/mL respectively (Yao et al., 2017). Even though the C_{max} following DMN application was marginally less than what was achieved via oral administration, the average blood level of LNG following the use of DMNs was greater than that of oral suspension within the first 0.5 h (Yao et al., 2017). There was no marked difference in AUC values following the administration of oral suspension or DMNs (Yao et al., 2017).

In a recent report, Yang et al. utilized a bilayer dissolving microneedle array (Fig. 3.1) for the delivery of triamcinolone and 5-fluorouracil (Yang et al., 2021). Equipped with several micro-sized needle tips, the bilayer microneedles can be self-administered into hypertrophic scar (HS) with less pain and uniform drug distribution (Yang et al., 2021). Both in vitro penetration and in vivo HS retention data showed that triamcinolone and 5-fluorouracil are capable of coexisting in a scar tissue for a sustained period (Yang et al., 2021).

In an interesting study, Donadei et al. prepared a dissolvable DMN patch loaded with trivalent inactivated Sabin poliovirus vaccine (sIPV) (Donadei et al., 2019). Pyramidal-shaped silicon MNs comprising 5 × 5 array of 600 μm MN were utilized as the molding template for the formation of MN cavities in polydimethylsiloxane (PDMS) molds (Donadei et al., 2019). The investigators utilized polyvinyl alcohol and trehalose-based formulation mixed with magnesium chloride hexahydrate, sorbitol and monosodium glutamate to fabricate the DMNs (Donadei et al., 2019). The formulation was poured into molds, and the MNs were dried for 36 h at 23°C (Donadei et al., 2019). The DMNs were stable under standard laboratory conditions for at least 0.5 h and effectively pierced pig skin. Significantly, immunoglobulin G antibodies specific to polio and capable of neutralizing the virus were generated in rats following the administration of these MN patches (Donadei et al., 2019). The neutralizing antibodies induced were similar to type 1 and type 3 poliovirus serotypes obtained following the administration of high intramuscular vaccine doses

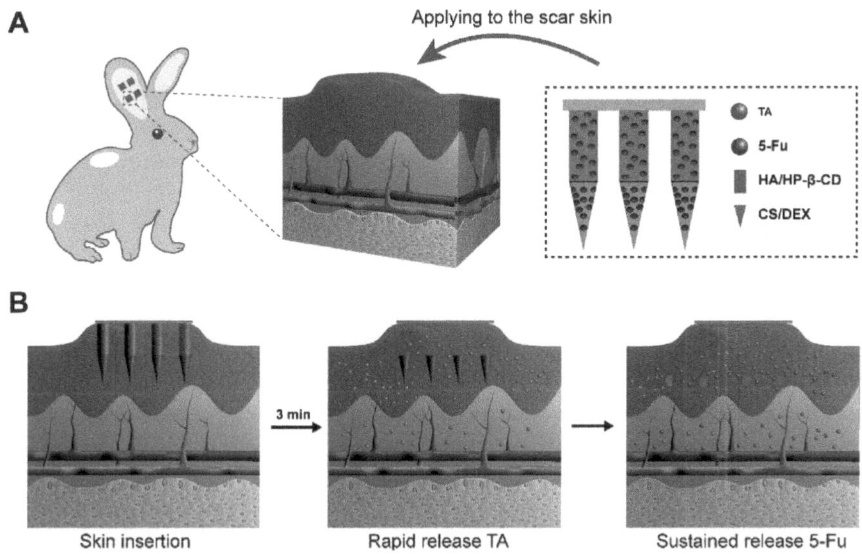

Figure 3.1 Dissolving microneedles for drug delivery to rabbit ear. Bilayer microneedles with 5-fluorouracil in the tip layer and triamcinolone in the tail layer of the needle (A), and drug delivery with biphasic release profile to scar lesion using bilayer microneedles for synergic hypertrophic scar treatment (B). *(From Yang, B., Dong, Y., Shen, Y., Hou, A., Quan, G., Pan, X. & Wu, C. (2021). Bilayer dissolving microneedle array containing 5-fluorouracil and triamcinolone with biphasic release profile for hypertrophic scar therapy.* Bioactive Materials, *6(8), 2400–2411. https://doi.org/10.1016/j.bioactmat.2021.01.014).*

(Donadei et al., 2019). Furthermore, administering the DMNs to the flanks induced a substantially greater antibody response in comparison to ear administration (Donadei et al., 2019).

Most dissolving MNs contain medications that are placed in the tips and the backing layer (Chen et al., 2018). This makes it difficult to control drug loading (Chen et al., 2018). Tip-dissolving MNs (TD MNs) with undissolving backing layer can be used to overcome this issue. Chen and coworkers prepared DMNs from PDMS molds (Chen et al., 2018). The DMNs were made up of two layers: a dissolving tip prepared from a water-soluble polymer (polyvinylpyrrolidone, PVP) loaded with the investigated compounds (methylene blue, meloxicam, silicon dioxide, and tyrosinase) and a backing layer prepared from polycaprolactone (Chen et al., 2018). The undissolving backing layer was fabricated from a water-insoluble polymer (polycaprolactone, PCL) (Chen et al., 2018). A polymer solution (PVP) loaded with meloxicam (MX) was placed in a PDMS mold which was then subjected to high pressure (2 or 3 atm) in an airtight cylinder for 2–3 min (Chen et al., 2018). Next, a pipette was used to remove the solution left on the mold surface. After that, the tip solution in the cavities was dried (Chen et al., 2018). Subsequently, molten PCL was poured into the mold followed by yet another high-pressure processing. Finally, after the PCL had solidified under room temperature, the tip-dissolving MNs were peeled out of the mold (Chen et al., 2018). The low, medium, and high doses of the TD MNs contained 75.98 ± 9.13, 168.33 ± 10.95, and 332.53 ± 13.37 μg of MX in the tips, respectively, were administered on the abdominal skin of rats by hand for 5 min (Chen et al., 2018). The delivery efficiency of various doses of the TD MNs had no significant difference with average delivery efficiency of 79.18%, which was much higher than the common dissolving MNs (49.2%) (Chen et al., 2018). The authors also reported that 24 h after removing the tip-dissolving MNs, rodent skin was devoid of microconduits (Chen et al., 2018). For the tip-dissolving MN group, the authors reported that the peak level, C_{max}, the time when serum MX level reached the peak level, T_{max}, and the area under the curve ($AUC_{0-12\ h}$) were 265.76 ± 40.12 ng/mL, 3 h, and 1762.61 ng h/mL respectively (Chen et al., 2018). The corresponding values in the intragastric group, were 215.89 ± 24.60 ng/mL, 2 h, and 1441.37 ng h/mL, respectively (Chen et al., 2018).

Tumor can circumvent host immunity and T-cell responses via programmed death-1 (PD-1) and PD-1 ligand (PD-L1) (Dong et al., 2017). In the presence of tumor, both PD-1 and PD-L1 are overexpressed leading to T-cell response suppression, immune evasion and tumor growth (Dong et al., 2017). A widely described immune checkpoint is the interplay between PD-1 and its ligand (PD-L1) (Yang et al., 2020). Following the binding of PD-1 to PD-L1, T-cells go through apoptosis and its cytotoxic immune effects are curbed (Yang et al., 2020). Immune checkpoint modulation via monoclonal antibody blockade of the checkpoint ligands, receptors, or both is one of the most successful strategies for the management of cancer (Lewis et al., 2020). Indeed, the use of PD-1 inhibitors, nivolumab and pembrolizumab, has resulted in prolonged survival rates for colorectal and non-small cell lung cancer patients (Lewis et al., 2020). Microneedles can be utilized for tumor immunotherapy but the quantity of drug that can be encapsulated in the conventional microneedle is limited (Yang et al., 2020). Furthermore, traditional MNs cannot concentrate the encapsulated drug in the tips of the MNs and it is challenging achieving complete transdermal transport due to elastic skin deformation (Yang et al., 2020). On the other hand, with a two-layer hybrid structure, a core-shell microneedle (CSMN) system is a more functional device for on-demand drug loading (Yang et al., 2020). The matrix material of each layer can be optimized for drug stability, encapsulation efficiency, and release profile in accordance with the physicochemical properties of the drug (hydrophilicity, charge and hydrogen bond donors/acceptors) (Yang et al., 2020).

Recently, Yang and coworkers fabricated a chore-shell microneedle system comprising a hydrophilic core as well as a charged shell and used it to co-deliver 1-methyl-D,L-tryptophan (1- MT) and an *anti*-PD-1/L1 antibody (aPD-1/aPD-L1) (Yang et al., 2020). The 1-methyl-D,L-tryptophan molecule possesses only one hydrogen bond acceptor (indolyl nitrogen) since its carboxyl and amino groups are ionized (Yang et al., 2020). Furthermore, this molecule has no hydrogen bond donor. The authors utilized different polymers containing multiple hydrogen bond donors (hydroxyl) such as dextran (DEX), polyvinyl alcohol (PVA) and hyaluronic acid (HA) to prepare supersaturated 1-MT solutions (Yang et al., 2020). The use of this system resulted in enhanced drug encapsulation. PVA was superior in comparison to other polymers in keeping 1-MT supersaturated (Yang et al., 2020). The authors used polydimethylsiloxane (PDMS) female molds with micron-sized holes inversely reproduced from a brass template to fabricate the CSMNs (Yang et al., 2020). Centrifugation molding was used for the fabrication process (Yang et al., 2020). Initially, the shell of CSMN was fabricated via PDMS molding from sodium alginate solution or chitosan solution with subsequent centrifugation. To prepare the CSMN core, the authors prepared an aqueous dispersion consisting of 25% (w/v) PVP K30, 15% (w/v) PVA and 15 mg/mL 1-MT. The fabrication procedure entailed centrifugation (Yang et al., 2020). A dispersion of PVP K90 in ethanol was used to create the CSMN base. Following drying at room temperature for 3 days, the CSMN was peeled from the mold and stored in a hermetically sealed container (Yang et al., 2020). The use of CSMN led to higher transdermal drug delivery efficiency and a longer topical retention time of aPD-L1 (2 days) (Yang et al., 2020). In addition, drug-encapsulated CSMNs achieved higher anti-tumor action compared to intra-tumor injection and this was ascribed to the recruitment of more T lymphocytes to the tumor site (Yang et al., 2020).

In an interesting study, Moreira et al. loaded PVP MNs with doxorubicin and gold-core silica shell nanorods via sequential electrospraying of poly (vinyl alcohol) and chitosan (Moreira et al., 2020). The system combined the photothermal therapeutic efficacy of gold nanoparticles with the drug delivery capability of mesoporous silica (Moreira et al., 2020).

Bevel-like tipped MNs were fabricated via micromolding. To accomplish this, a solution of PVP was poured into PDMS mold (4 × 4) under vacuum for 2 h (Moreira et al., 2020). Polyvinyl alcohol and chitosan were sequentially electrosprayed onto PVP microneedles surface via a traditional electrospinning equipment (Moreira et al., 2020). To formulate the microneedles loaded with nanoparticles or the drug, doxycycline was introduced into the chitosan solution before electrospraying and PVA was mixed with gold-core silica shell (AuMSS) nanorods (Moreira et al., 2020). Doxorubicin and AuMSS nanorods displayed good photothermal properties resulting in a temperature increase of 12°C when subjected to near-infrared irradiation (Moreira et al., 2020). In combination with pH, increased temperature can be used to modulate the release of doxorubicin (Moreira et al., 2020). Significantly, the combination of photothermal and chemotherapy led to increased cytotoxic action against cervical cancer cells (Moreira et al., 2020).

Dillon and colleagues also studied the feasibility of delivering sincalide and pentagastrin analogs from a dissolving microneedle system (Dillon et al., 2019). The peptide analogs were synthesized via fluorenylmethoxycarbonyl protecting group (Fmoc)-based solid phase methods (Dillon et al., 2019). In comparison with pentagastrin, the analog synthesized by these investigators was characterized by the lack of (1,1-dimethylethoxy) carbonyl at the C-terminus (Dillon et al., 2019). Compared to sincalide, the synthesized analog possessed similar amino acid sequence but there was no sulfonyl group linked to tyrosine (Dillon et al., 2019). The analogs were then incorporated into DMNs. The authors fabricated 18 × 18 MN array with each needle having a base width of 200 μm, length of 650 μm, and inter-needle spacing of 600 μm (Dillon et al., 2019). PVP, trehalose and glycerol (a plasticizer) were used to fabricate the MNs (Dillon et al., 2019). The peptide analogs were then added to the formulation. The polydimethylsiloxane microneedle molds and vacuum were used to prepare the MNs (Dillon et al., 2019). The authors carried out skin permeation testing to evaluate the ability of the MNs to transport peptide analogs through pig skin (Dillon et al., 2019). For the sincalide derivative, there was no significant difference between the release rate from the DMNs and the control formulation (Dillon et al., 2019). On the other hand, the quantity of pentagastrin analog transported increased significantly when DMNs were used in comparison to the control (Dillon et al., 2019).

Modern DMN fabrication techniques entail a reconstitution process for the encapsulated medications, and this can lead to drug degradation or constrained encapsulation for a feasible patch size (Kim, Choi et al., 2020; Kim, Yang et al., 2020). Kim and colleagues fabricated self-administrative powder-carrying MNs (PCMs), capable of placing insulin powder directly inside the skin without a reconstitution step (Kim, Choi et al., 2020; Kim, Yang et al., 2020). The authors used modified solvent-casting and molding techniques to prepare the MNs (Kim, Choi et al., 2020; Kim, Yang et al., 2020). Laser cutting was used to produce 5 × 5 MN arrays with a base diameter of 450 μm and a height pf 800 μm from stainless-steel. The polydimethylsiloxane (PDMS) pre-polymer base and curing agent, were mixed in a 10:1 ratio, and poured onto the master template (Kim, Choi et al., 2020; Kim, Yang et al., 2020). The resulting PDMS molds were stripped from the master template. Carboxymethyl cellulose (CMC) was used as the matrix material for the fabrication of the micro-shells. CMC was used as the backbone polymer because of its biodegradable, biocompatible, and viscous properties (Kim, Choi et al., 2020; Kim, Yang et al., 2020). To produce micro-shelled devices, CMC solution was dried at ambient temperature for 24 h. Micro-shells arrays were peeled from the mold and the empty micro-cavities filled with insulin powder (Kim, Choi et al., 2020; Kim, Yang et al., 2020). In comparison with DMNs possessing the same geometries, PCMs transported the required dose with no deleterious effect on insulin solubility and stability. Moreover, PCMs exhibited a prolonged release profile (Kim, Choi et al., 2020; Kim, Yang et al., 2020).

Autoimmunity linked to rheumatoid arthritis can be detected by using *anti*-citrullinated protein antibodies (ACPAs) (Kurowska et al., 2017). These antibodies cause autoimmune response characterized by the generation of T cells, dendritic cells and B cells in the lymph node. If this response takes place in the joint, inflammation occurs (Yao et al., 2019). The short-chain α-neurotoxin (NTX) in *Naja atra* venom is a competitive antagonist that binds to nicotinic acetylcholine receptors leading to neuromuscular blockade (Liu, Chou et al., 2020; Liu, Kong et al., 2020). More importantly, this neurotoxin is known to have anti-inflammatory effect and can potentially be utilized for the management of rheumatoid arthritis (Yao et al., 2019). Recently, Yao et al. fabricated neurotoxin-loaded DMNs (DMNs-NT) (Yao et al., 2019). The needles were prepared from polyvinylpyrrolidone (PVP) and chondroitin sulfate (CDNS) via a micro-molding technique. Carboxymethyl cellulose (CMC) was used to create the MN baseplate (Yao et al., 2019). The preparation of DMNs-NT entailed two-step centrifugation. The quadrangular pyramid-shaped MNs were mechanically robust with the upper region of the needle incorporating 15.4 ± 0.5 μg of drug per patch (Yao et al., 2019). Remarkably, the depth of skin penetration of the DMNs-NT in rats was greater than 70 μm and the cumulative penetration of neurotoxin (NT) from the DMNs was 95.8% in 4 h, whereas NT solution marginally crossed the skin (Yao et al., 2019). Furthermore, the MNs were fully dissolved in vivo after 10 min and the micro-conduits in the stratum corneum (SC) were sealed after 6 h (Yao et al., 2019). DMNs-NT reduced toe swelling in RA rats and decreased levels in serum levels of IL-1β and TNF-α resulting in decreased ankle joint damage (Yao et al., 2019).

Vora et al. investigated the effect of DMNs prepared from pullulan (PL) on the percutaneous transport of model compounds (methylene blue, insulin, fluorescein labeled bovine serum albumin, fluorescein sodium) (Vora et al., 2020). The authors used a micro-molding method to fabricate the DMNs. Insulin solution was added to PL solution and about 500 mg of the blend was poured into laser-engineered conical silicone micromolds (Vora et al., 2020). The filled molds were centrifuged at 2205 ×g for 15 min. Finally, the arrays were dried at room temperature for 48 h and stored at room temperature. The model protein/peptide (fluorescein labeled bovine serum albumin and insulin) and model dyes (methylene blue and fluorescein sodium) were loaded into PL DMNs (Vora et al., 2020). The authors used circular dichroism to prove the stability of the model compounds (fluorescein labeled bovine serum albumin and insulin) in the pullulan DMN arrays. The successful transport of the model compounds by pullulan DMNs across excised neonatal pig skin was also demonstrated (Vora et al., 2020).

Hypertrophic scar (HS) is linked to dermal fibro-proliferative disorders especially abnormal collagen deposition and excessive fibroblast proliferation (Song et al., 2020). HS can result in pain, disfigurement and pruritus (Ahuja & Chatterjee, 2014). The management of this disorder is challenging and there is still no universally accepted treatment regimen although intralesional triamcinolone injections are often used for less widespread hypertrophic scarring (Ahuja & Chatterjee, 2014). It has been observed that the use of this drug leads to reduced concentration of the proteinase inhibitors alpha2-antitrypsin and alpha2-macroglobulin in the scar (Ahuja & Chatterjee, 2014). This can result in less collagenase (matrix metalloproteinase 1) breakdown with significant implications regarding the control of abnormal and excessive collagen (Ahuja & Chatterjee, 2014). However, the success of intralesional triamcinolone therapy is also a function of the skill of the medical professionals administering the injections

(Lin et al., 2019). Despite using local anesthetics concurrently, traditional hypodermic injection is painful to patients with subsequent low drug compliance (Lin et al., 2019). Recently, Lin et al. used dissolving microneedles (DMNs) for the transdermal delivery of triamcinolone (Lin et al., 2019). Sodium hyaluronic acid (HA) was employed as the MN material and hydroxypropyl-β-cyclodextrin (HP-β-CD) was added to harden the MNs (Lin et al., 2019). These needles were then used to pierce hard and compact hypertrophic scar (Lin et al., 2019). The hardness of this combination was ascribed to the hydrogen bonding between HA and HP-β-CD which hindered the movement of molecular chains leading to increased material elastic modulus (Lin et al., 2019). Furthermore, HP-β-CD facilitated the loading of hydrophobic drugs into the DMN needle tips. Micromolding was the fabrication method used and the master mold was prepared from brass. It was composed of 125 MNs (300 μm in diameter, 1200 μm in height with a tip-to-tip distance of 900 μm and a tip diameter of 500 μm) (Lin et al., 2019). Dissolving microneedle array without the drug was prepared via a modified micromolding method that employed triple centrifugation (Lin et al., 2019). The PDMS mold was filled with a dispersion of HA/HP-β-CD with subsequent centrifugation. Following the elimination of excess HA/HP-β-CD dispersion, the female mold was again centrifuged. Finally, base solution, which was prepared by mixing PVP K90 and ethyl alcohol at a ratio of 1:2.7 (w/v), was poured onto the female mold and centrifuged to create the medication-free base portion of the DMN array (Lin et al., 2019). This same method was utilized for the preparation of triamcinolone acetonide-loaded DMNs (TAs) (TA-DMNs). To obtain TA-DMNs, different ratios of the TA-HP-β-CD-IC dispersion to HA were prepared to obtain dispersions which were then used for DMN fabrication (Lin et al., 2019).

Carvacrol (CAR) is a monoterpenoid phenol obtained from thyme and oregano essential oils (Marjanović et al., 2020). CAR has demonstrated excellent *anti*-biofilm and antimicrobial activities against several Gram-positive and Gram-negative bacteria (Marchese et al., 2018; Mir et al., 2020). Recently, Mir and et al. evaluated the influence of dissolving microneedles on the transdermal penetration of CAR (Mir et al., 2020). The authors first used poly (caprolactone) (PCL) to prepare a site-specific nanoparticle (NP) delivery system with a prolonged antimicrobial action. The CAR-loaded nanoparticles were used to prepare dissolving microneedles (Mir et al., 2020). The rationale was to overcome necrotic wound tissue and enhance painless drug delivery. To prepare MN arrays tip-loaded with NPs (CAR-PCL NPs-MNs), the authors mixed lyophilized CAR-PCL NPs with polyvinyl alcohol/polyvinylpyrrolidone (PVA/PVP) hydrogel (Mir et al., 2020). Next, MN molds were filled with CAR-PCL NPs/polymer gels and kept in a positive pressure chamber (Mir et al., 2020). To prepare the second group of microneedles, PVA/PVP hydrogels were poured into the 24 h-dried molds and kept under positive pressure of 3–4 bars for 15 min, with subsequent drying at 19°C for 48 h (Mir et al., 2020). Following dessication, DMNs were detached from the molds and evaluated for mechanical strength (Mir et al., 2020). Permeation experiments across neonatal pig skin showed that CAR-PCL NPs-MNs increased skin accumulation of CAR after 24 h (83.8% ± 5.15%) in comparison to the free CAR-MNs (7.3% ± 2.04%) (Mir et al., 2020). Antimicrobial action in an ex-vivo wound model was also demonstrated (Mir et al., 2020).

Chi et al. studied the effect of chitosan-based DMNs on wound healing (Chi et al., 2020). Vascular endothelial growth factor-loaded DMNs exhibited temperature-based controllable drug release profile (Chi et al., 2020). The authors used silicone mold for DMN fabrication. Chitosan (CS) pre-gel was prepared by dissolving CS powder in 2% acetic acid solution with subsequent ultrasonic homogenization (Chi et al., 2020). The conical microcavities of the mold were filled with the CS pre-gel the under vacuum (Chi et al., 2020). For the solidification of the CS gel, the authors immersed the entire mold into an aqueous solution of sodium hydroxide. To incorporate the drug, a combination of N-isopropylacrylamide (NIPAM) pre-gel solution mixed with VEGF (different concentrations) and 1% 2-hydroxy-2-methylpropiophenone (HMPP) was poured into the mold and subjected to vacuum for 30 min (Chi et al., 2020). After disposing the excess pre-gel, ultraviolet radiation was used to strengthen the NIPAM gel. The aqueous solution of CS was then placed into the mold and subjected to vacuum drying for MN base formation (Chi et al., 2020). Finally, the CSMNs were peeled from the mold (Chi et al., 2020). Large doses of VEGF were loaded into the microcavities of the CSMNA and released in a controlled manner into wound sites and this significantly facilitated healing (Chi et al., 2020). The authors ascribed the effect of CSMN to increased collagen deposition, inflammatory inhibition, granulation tissue formation and angiogenesis (Chi et al., 2020).

Uncontrolled spread of prostate stromal cells that results in increased size of the tissue is called benign prostatic hyperplasia (BPH) (Ciríaco et al., 2020). Dutasteride is a dual inhibitor of α-reductase isozymes (type I and type II) approved by the FDA for the management of prostatic hyperplasia

(Pinacho-Garcia et al., 2020). Recently, Giffen and coworkers evaluated dutasteride-loaded dissolving microarrays for once-weekly delivery in rats (Giffen et al., 2020). Dissolvable microneedles were fabricated using a nanomilled suspension and were applied to rats at doses up to 0.32 mg/kg. In the study, serum dutasteride was detected for approximately 14 days following a single administration (Giffen et al., 2020). In silico modeling of rodent data using a two-compartment intradermal model was carried out and it was predicted that, in humans, a once-weekly dose of 2 mg, delivered as DMNs, could transport cumulative and therapeutically significant levels of dutasteride in a fashion similar to that observed with oral administration (Giffen et al., 2020). In another set of experiments, Milewski _underand coworkers evaluated the effect of dissolving microneedles (MicroCor®) on the transdermal transport of Peptide A (Milewski et al., 2020). Peptide A is an investigational drug that is used as a daily subcutaneous injection. The authors prepared a 5600-needle dissolving microneedle system (~200 μm long microstructures, 2 cm^2 array) that was loaded with a therapeutic dose of Peptide A (Milewski et al., 2020). The dissolving microneedles were stable under room-temperature for 3 months (Milewski et al., 2020). In vivo experiments with Yucatan minipigs showed that the use of Peptide-A-loaded microneedles resulted in significant transdermal transport of the peptide leading to an absolute bioavailability of 74% (Milewski et al., 2020). The bioavailability value was comparable to that obtained with subcutaneous administration of the peptide. Moreover, only mild and temporary skin irritation was documented in the study (Milewski et al., 2020).

A new technique of fabricating etonogestrel (ENG)-loaded DMNs has been reported (He et al., 2020). The authors utilized N-methyl pyrrolidinone (NMP) as the solvent for making the needle-tips. ENG was encapsulated in the biodegradable polylactic acid-*co*-glycolic acid (PLGA) needle-tips via a tightly regulated casting-mold method thus averting the influence of high temperature on the stability of the medication (He et al., 2020). A polymeric combination made from polyvinyl alcohol (PVA) and polyvinylpyrrolidone K90 (PVP K90) (mass ratio 1:1) was used for the preparation of the backing layer of the MNs (He et al., 2020). The DMNs were prepared using 4 steps: (a) needle-tip solution loaded with ENG was used to fill a PDMS mold, and the mold was subsequently placed in a vacuum (b) the excess solution on the mold was removed and the filled mold moved to the oven for NMP removal and needle-tip solidification (c) About 50 μL of the PVA-PVP blend solution was placed into the mold which was then subjected to vacuum (d) The mold was dried for 6 h under a relative humidity that was below 10% (He et al., 2020). Following solidification, the needle-tips were robust enough to pierce the skin. The drug loading capacity was 153.0 ± 13.5 μg (He et al., 2020). Following the administration of the implantable MNs, it was possible to quantify the plasma ENG for 336 h and the $AUC_{0\rightarrow 48h}$ was responsible for only 37.8% of $AUC_{0\rightarrow \infty}$. Thus, the implantable MNs have the potential of being used as an extended-release drug delivery system (He et al., 2020).

Arbutin is a hydrophilic phenolic compound found predominantly in pears, bearberry and wheat (Aung et al., 2020). The compound has been utilized widely as a skin lightning cosmeceutical and it can decrease melanin synthesis via tyrosinase inhibition (Aung et al., 2020). Tyrosinase catalyzes melanin synthesis. Arbutin exists either as alpha-arbutin (4- hydroxyphenyl-α-D-glucopyranoside) or beta-arbutin (4- hydroxyphenyl-β-D-glucopyranoside) (Aung et al., 2020). Even though alpha-arbutin has shown a potent inhibitory action on human tyrosinase activity, it has a poor transdermal delivery profile which is ascribed to high hydrophilicity (Aung et al., 2020). Aung et al. fabricated dissolving microneedles (DMNs) and loaded them with alpha-arbutin with the aim of enhancing the trans-epidermal transport of the compound (Aung et al., 2020). DMNs were produced from hydroxypropyl methylcellulose (HPMC), polyvinylpyrrolidone K-90 (PVP K-90), Gantrez™ S-97, chitosan. The authors formulated DMNs from a combination of 40% w/w PVP K-90 and 8% w/w HPMC at the weight ratio of 1:1 (PVP/HPMC) (Aung et al., 2020). The deposited quantity of alpha-arbutin in the skin from HPMC/PVP DMNs and Gantrez™ S-97 DMNs was 312.23 μg and 107.76 μg respectively (Aung et al., 2020). On the other hand, the transdermal flux values of Gantrez™ S97 DMNs and HPMC/PVP DMNs across full-thickness neonatal porcine skin over 12 h were 66.21 $\mu g/cm^2/h$ and 74.24 $\mu g/cm^2/h$, respectively (Aung et al., 2020). HPMC/PVP Gantrez™ S-97 DMNs enhanced the percutaneous delivery of alpha-arbutin across the skin by about 4.7 and 2 times, respectively in comparison to gel formulations (Aung et al., 2020). HPMC/PVP DMNs containing alpha-arbutin transported more of the compound into mouse skin than a marketed formulation (Aung et al., 2020).

Rapamycin (sirolimus) is a lipophilic macrolide that acts as an immunosuppressant (Rezabakhsh et al., 2017). The drug acts via autophagy induction due to its ability to block mTOR (Rezabakhsh et al., 2017).

mTOR exists as a rapamycin sensitive mTOR Complex1 (mTORC1) which phosphorylates initiation factor 4E binding proteins (4E-BPs) and P70S6K1 (Rezabakhsh et al., 2017). The protein (mTOR) also functions as mTOR Complex2 (mTORC2), which phosphorylates Akt (protein kinase B) (Rezabakhsh et al., 2017). Recently, Mao et al. prepared dissolving microneedles from polyvinylpyrrolidone (PVP) and loaded the needles with rapamycin (Mao et al., 2020). The DMNs exhibited adequate mechanical strength, pierced the skin and transported the drug 200 μm into the tissue (Mao et al., 2020). The DMN stem completely dissolved and 80% of the drug was released within 10 min after injection into mouse skin (Mao et al., 2020). Rapamycin-loaded DMNs exhibited good anti-angiogenic effect which was demonstrated through human umbilical vein endothelial cell growth inhibition and the reduction of vascular endothelial growth factor secretion (Mao et al., 2020).

Diphtheria is a highly infectious disease caused by *Corynebacterium diphtheriae* (Adler et al., 2013). Diphtheria vaccine is usually administered intramuscularly in combination with tetanus and pertussis vaccines (Möller et al., 2019). Leone and coworkers investigated the influence of dissolving microneedles on the intradermal delivery of diphtheria toxoid (DT) (Leone, Du et al., 2020; Leone, Romeijin et al., 2020). The authors incorporated poly (I:C), gibbsite, DT and DT-adjuvant (full or fractional) into dissolving microneedles prepared from hyaluronic acid (Leone, Du et al., 2020; Leone, Romeijin et al., 2020). The authors incorporated 0.3% (w/v) DT (full dose) or 0.1% (w/v) DT (fractional dose) into the DMNs. Hyaluronic acid-diphtheria toxoid solution was poured into polydimethylsiloxane molds (Leone, Du et al., 2020; Leone, Romeijin et al., 2020). Following numerous centrifugation and vacuum cycles, the DMNs were dried overnight at 37°C (Leone, Du et al., 2020; Leone, Romeijin et al., 2020). Subsequently, a backplate without the antigen was fabricated using a mixture of the catalyst and vinylpolysiloxane base (1:1 ratio). A glue solution was then used to fill each cavity before curing. At last, the MNs were detached from the mold and kept in a desiccator at room temperature (Leone, Du et al., 2020; Leone, Romeijin et al., 2020). The DMNs pierced human skin, dissolved within 20 min and deposited the antigen-adjuvant dose, which lasted not less than 5 h in the skin (Leone, Du et al., 2020; Leone, Romeijin et al., 2020). Vaccination with DMNs resulted in greater immune response compared with hollow microneedles but there was no enhanced response following repeated intradermal delivery of fractional doses compared to a single-full dose administration (Leone, Du et al., 2020; Leone, Romeijin et al., 2020).

Elastic properties of the skin as well as DMN shape exert significant influence on the percutaneous transport efficiency of medications (Lahiji, 2019). To enhance delivery efficiency, Lahiji and coworkers fabricated a novel tissue interlocking DMN (TI-DMN) based on a hyaluronic acid backbone (Lahiji, 2019). TI-DMN was easy to make and significantly enhanced the percutaneous transport efficiency of encapsulated compounds in comparison with conventional DMNs (Lahiji, 2019). Increased tissue interlinking by TI-DMN was achieved via its wide body, sharp tip and narrow neck geometry (Lahiji, 2019). The MN preparation procedure consisted of two key steps: the formation of a primary layer shaped in the form of a wine glass and the creation of a sharp tip secondary layer (Lahiji, 2019). In the first step, the viscous hyaluronic acid (HA) polymer was poured on a thin solidified layer in a vertical position inside a centrifuge (Lahiji, 2019). Subsequently, a non-sticky Parafilm M coated plate was used to create a primary layer (Lahiji, 2019). Finally, the second HA polymer drop was poured on top of the primary layer with subsequent centrifugation resulting in the production of narrow-necked, wide-bodied and sharp-tip dissolving microneedles (Lahiji, 2019).

Hao et al. synthesized monomethoxy-poly (ethylene glycol)-polycaprolactone (MPEG-PCL) nanoparticles responsive to near-infrared light and loaded them with indocyanine green (ICG) and 5-fluorouracil (5-Fu) (Hao et al., 2020). The authors subsequently integrated the 5-Fu-ICG-MPEG-PCL with hyaluronic acid MNs to obtain 5-Fu-ICG-MPEG-PCL loaded hyaluronic acid MNs for the management of skin cancers (Hao et al., 2020). ICG is a photothermal agent while 5-Fu is recommended by the FDA for the management of melanoma (Hao et al., 2020). Hyaluronic acid, the microneedle carrier, possesses good skin penetration ability and is approved by FDA as a pharmaceutical adjuvant . The authors investigated the feasibility of using dissolvable microneedles to deliver 5-Fu-ICG-MPEG-PCL across the skin with concomitant control of the release profile of the incorporated drugs with near-infrared radiation (Hao et al., 2020). The aim of the research project was to enhance the cure rate of skin cancer, achieve a single-dose cure, and develop a new treatment option for this disease (Hao et al., 2020).

It has been shown that the capacity of DMNs to transport drugs across the skin can diminish due to incomplete insertion induced by the elastic characteristics and stiffness of the skin (Jung et al., 2020). To circumvent poor insertion of the DMNs, Jung et al. fabricated micropillar integrated DMNs (Jung et al., 2020). Even though micropillars were designed as applicators for the implantation of DMNs, they can also pierce the skin resulting in inflammation and skin damage (Jung et al., 2020). The authors fabricated an integrated DMN system with a safety rings capable of inserting DMNs into the skin with minimal risk of skin injury and high penetration efficiency (Jung et al., 2020). The separable micropillar integrated microneedles do not require uninterrupted fastening to the skin and can be peeled after administration, with the DMNs left in the skin (Jung et al., 2020).

In another study, Chen et al. fabricated core-shell structure microneedles for immune response enhancement via intralesional administration of indoleamine 2,3-dioxygenase (IDO) blockade and indocyanine green (a photosensitizer) (Chen, Han et al., 2020; Chen, Quan et al., 2020). The authors loaded indocyanine green into chitosan nanoparticles (ICG-NPs), with subsequent concentration on the tip shell of the MNs (Chen, Han et al., 2020; Chen, Quan et al., 2020). The MN core consisted of cross-linked poly (vinyl pyrrolidone) and poly (vinyl alcohol) gel loaded with 1-methyl-tryptophan (1-MT) (Chen, Han et al., 2020; Chen, Quan et al., 2020). The core-shell structure microneedles were used to induce tumor-associated antigens and damage-associated molecular patterns (Chen, Han et al., 2020; Chen, Quan et al., 2020). The MNs efficiently transformed near-infrared laser into heat for primary tumor ablation, facilitated immunostimulatory cytokine secretion as well as dendritic cell maturation via direct deposition of the ICG-NPs loaded tips into the melanoma site (Chen, Han et al., 2020; Chen, Quan et al., 2020). Animal experiments demonstrated that 80% of the tumor was eliminated with subsequent relapse-free survival in more than 4 months (Chen, Han et al., 2020; Chen, Quan et al., 2020). The combination therapy markedly curbed lung metastasis and regulated the growth of previously metastasized tumors (Chen, Han et al., 2020; Chen, Quan et al., 2020). The study demonstrated the feasibility of using microneedle-based photothermal therapy to sensitize tumor to IDO blockade and induce antitumor immunity (Chen, Han et al., 2020; Chen, Quan et al., 2020).

Electrospun fibers are versatile materials that can be optimized for drug delivery. These fibers possess intertwined openings, which create a tortuous path for particles (Creighton et al., 2020). Creighton and coworkers used electrospun fibers to prepare DMNs (Creighton et al., 2020). The authors created a two-layer container that had the shape of a MN array, placed electrospun fibers within the container patterns, and filled them with a polymer dispersion to fabricate integrated fiber MNs (Creighton et al., 2020). The authors used polycaprolactone fibers electrospun from an optimized solution onto a collector with conical patterns to form the integrated fiber microneedles (Creighton et al., 2020). Following electrospinning, an aqueous dispersion of poly (acrylic acid) was used with the collector to form a backing layer and a dissolvable system around the electrospun fibers (Creighton et al., 2020). The authors demonstrated that two-layer collectors with pattern dimensions down to 100 μm can be conveniently produced with commonly used laboratory apparatus (Creighton et al., 2020). The authors used finite element techniques in conjunction with experimental verification to demonstrate the feasibility of using fiber patterning with different materials and dimensions (Creighton et al., 2020).

Several researchers are formulating sustained-release depot systems for delivering antiretroviral medications. Rojekar and colleagues recently fabricated dissolving microneedles (Fig. 3.2) and loaded them with etravirine nanosuspension (ETR NS) (Rojekar et al., 2021). The dissolving microneedles (DMNs) loaded with ETR NS provided 12.84% ± 1.33%

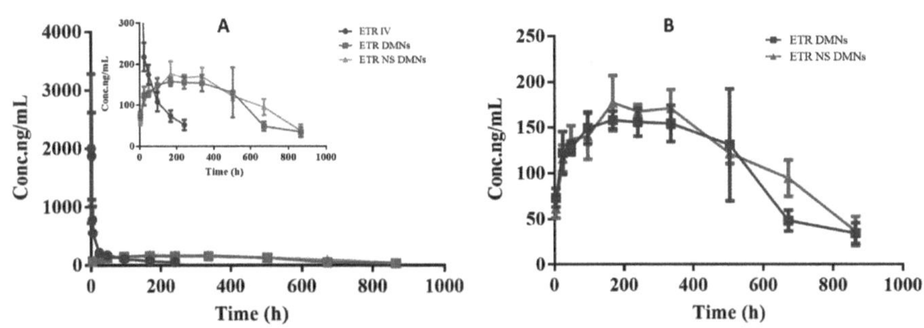

Figure 3.2 Plasma concentration—time profiles after a single 2.5-mg/kg dose of the ETR solution in female rats Etravirine (ETR) nanosuspension (NS) loaded dissolving microneedles (DMNs) and ETR Loaded DMNs (n = 5 & 4), showing the effect of dose and route of administration and (B) 36-days mean plasma concentration-time profiles of ETR after intradermal application of 4.8 mg dose of ETR NS loaded DMNs and ETR loaded DMNs patches in female rats (n = 5 & 4 per treatment group). *(From Rojekar, S., Vora, L. K., Tekko, I. A., Volpe-Zanutto, F., McCarthy, H. O., Vavia, P. R. &.Donnelly, R. F. (2021). Etravirine-loaded dissolving microneedle arrays for long-acting delivery.* European Journal of Pharmaceutics and Biopharmaceutics, 165, *41–51. https://doi.org/10.1016/j.ejpb.2021.04.024).*

ETR accumulation in ex-vivo neonatal pig skin after 6 h (Rojekar et al., 2021). In vivo rat pharmacokinetic data showed that the C_{max} displayed by DMNs loaded with ETR powder and ETR NS were 158 ± 10 ng/mL and 177 ± 30 ng/mL, respectively. DMN groups exhibited a higher T_{max}, t1/2, and mean residence time in comparison to ETR solutions administered intravenously (Rojekar et al., 2021).

Li et al. studied the effect of dissolving microneedle array (DMNA) with different needle-geometries on transcutaneous immunization with ovalbumin (Li et al., 2020). A two-stage molding procedure for loading MNs with ovalbumin (OVA) was utilized. MN dimensions and geometry correlated with the molds used for experiments (Li et al., 2020). The authors evaluated the dissolving microneedles with different geometries (Li et al., 2020). BCA assay was used to reveal the drug loading of the DMNs (27–100 μg). SDS-polyacrylamide gel electrophoresis was utilized to assess the stability of OVA in the DMNAs and the authors reported that OVA loaded into the DMNAs was stable during fabrication (Li et al., 2020). The immune responses generated in animal experiments by injections via hypodermic needles and DMNAs were compared via immunoglobulin G (IgG) determination (Li et al., 2020). DMNAs loaded with OVA generated more potent immune responses in comparison to injections via hypodermic needles (Li et al., 2020). The results suggested that: (a) DMNA structure has a significant effect on the insertion properties, mechanical strength and dissolution of the needles hence the more potent immune response; (b) cone-shaped dissolving microneedle arrays were shown to be effective for transcutaneous immunization (Li et al., 2020).

It has been reported that crosslinked hyaluronic acid is not appropriate for MN fabrication due to the low fluidity of crosslinked hyaluronic acid hydrogel (Kim, Choi et al., 2020; Kim, Yang et al., 2020). In an interesting report, MNs were prepared via crosslinked hyaluronic acid nanoparticles for sustained drug delivery. An important benefit with this class of compounds is that crosslinked hyaluronic acid (X-linked HA) is more processable (Kim, Choi et al., 2020; Kim, Yang et al., 2020). The authors carried out in vitro insertion studies into pig skin to study the puncture performance of the MNs fabricated from crosslinked hyaluronic acid nanoparticles (NPs) (Kim, Choi et al., 2020; Kim, Yang et al., 2020). After a pre-calculated attachment time, the residual height of the X-linked HA-NP-microneedle array patch (MAP) was quantified to assess the rate of dissolution (Kim, Choi et al., 2020; Kim, Yang et al., 2020). X-linked HA-NP-MAP and free hyaluronic acid microneedle array patch loaded with rhodamine

B isothiocyanate-dextran were applied to murine dorsal skin with subsequent assay of the relative fluorescent intensity over a certain period (Kim, Choi et al., 2020; Kim, Yang et al., 2020). The authors reported that the skin puncture capability of the X-linked HA-NPMAP was over 90% (Kim, Choi et al., 2020; Kim, Yang et al., 2020). The diameter of re-dispersed X-linked HA-NPs was identical to that of the preformed X-linked HANPs. The rate of dissolution was equivalent to that of the free HA-MAP (Kim, Choi et al., 2020; Kim, Yang et al., 2020). In animal studies, X-linked HA-NP-MAP was applied to the murine dorsal skin and the relative fluorescent intensity of X-linked HA-NP-MAP was more sustained than that of hyaluronic acid microneedle array patch (Kim, Choi et al., 2020; Kim, Yang et al., 2020).

In a recent study, Li and coworkers fabricated a microneedle patch with biodegradable and rapidly separable polylactic-*co*-glycolic acid and polylactic acid needles and this was used for the sustained release of a contraceptive hormone (levonorgestrel) (Li et al., 2019). The MNs efficiently pierced skin under compression and snapped off under shear within 5 s following the administration of the patch due to the bubble architecture between each microneedle and the patch backing (Li et al., 2019). The MN patch was well tolerated in rats with minor conspicuous evidence of use. Significantly, blood levels of the drug was kept above the therapeutic level for one month (Li et al., 2019). A more optimal control of female fertility can be achieved with progress in the development of rapidly separable microneedle patch for long-acting and self-administered contraceptive medications (Li et al., 2019).

References

Adler, N. R., Mahony, A., & Friedman, N. D. (2013). Diphtheria: Forgotten, but not gone. *Internal Medicine Journal, 43*(2), 206–210. https://doi.org/10.1111/imj.12049

Ahuja, R. B., & Chatterjee, P. (2014). Comparative efficacy of intralesional verapamil hydrochloride and triamcinolone acetonide in hypertrophic scars and keloids. *Burns, 40*(4), 583–588. https://doi.org/10.1016/j.burns.2013.09.029

Aung, N. N., Ngawhirunpat, T., Rojanarata, T., Patrojanasophon, P., Opanasopit, P., & Pamornpathomkul, B. (2020). HPMC/PVP dissolving microneedles: A promising delivery platform to promote trans-epidermal delivery of alpha-arbutin for skin lightning. *AAPS PharmSciTech, 21*(1). https://doi.org/10.1208/s12249-019-1599-1

Balmert, S. C., Carey, C. D., Falo, G. D., Sethi, S. K., Erdos, G., Korkmaz, E., & Falo, L. D. (2020). Dissolving undercut microneedle arrays for multicomponent cutaneous vaccination. *Journal of Controlled Release, 317*, 336–346. https://doi.org/10.1016/j.jconrel.2019.11.023

Chen, Z., Han, B., Liao, L., Hu, X., Hu, Q., Gao, Y., & Qiu, Y. (2020). Enhanced transdermal delivery of polydatin via a combination of inclusion complexes and dissolving microneedles for treatment of acute gout arthritis. *Journal of Drug Delivery Science and Technology, 55*, 101487. https://doi.org/10.1016/j.jddst.2019.101487

Chen, J., Huang, W., Huang, Z., Liu, S., Ye, Y., Li, Q., & Huang, M. (2018). Fabrication of tip-dissolving microneedles for transdermal drug delivery of meloxicam. *AAPS PharmSciTech, 19*(3), 1141–1151. https://doi.org/10.1208/s12249-017-0926-7

Chen, M., Quan, G., Wen, T., Yang, P., Qin, W., Mai, H., Sun, Y., Lu, C., Pan, X., & Wu, C. (2020). Cold to hot: Binary cooperative microneedle array-amplified photoimmunotherapy for eliciting antitumor immunity and the abscopal effect. *ACS Applied Materials and Interfaces, 12*(29), 32259–32269. https://doi.org/10.1021/acsami.0c05090

Chi, J., Zhang, X., Chen, C., Shao, C., Zhao, Y., & Wang, Y. (2020). Antibacterial and angiogenic chitosan microneedle array patch for promoting wound healing. *Bioactive Materials, 5*(2), 253–259. https://doi.org/10.1016/j.bioactmat.2020.02.004

Ciríaco, S. L., Carvalho, I. P. S., Alves Terceiro Neto, J., de Sousa Lima Neto, J., de Oliveira, D. H. B., Cunha, A. P. G. P., Cavalcante, Y. T. D., da Silva, D. T. C., da Silva, J. A., Mineiro, A. L. B. B., de Lima Chagas Moreno Fernandes, M. Z., & Carvalho, A. L. M. (2020). Development of microemulsion of tamsulosin and dutasteride for benign prostatic hyperplasia therapy. *Colloids and Surfaces B: Biointerfaces, 185*. https://doi.org/10.1016/j.colsurfb.2019.110573

Creighton, R. L., Phan, J., & Woodrow, K. A. (2020). In situ 3D-patterning of electrospun fibers using two-layer composite materials. *Scientific Reports, 10*(1). https://doi.org/10.1038/s41598-020-64846-z

Dillon, C., Hughes, H., O'Reilly, N. J., Allender, C. J., Barrow, D. A., & McLoughlin, P. (2019). Dissolving microneedle based transdermal delivery of therapeutic peptide analogues. *International Journal of Pharmaceutics, 565*, 9–19. https://doi.org/10.1016/j.ijpharm.2019.04.075

Donadei, A., Kraan, H., Ophorst, O., Flynn, O., O'Mahony, C., Soema, P. C., & Moore, A. C. (2019). Skin delivery of trivalent Sabin inactivated poliovirus vaccine using dissolvable microneedle patches induces neutralizing antibodies. *Journal of Controlled Release, 311–312*, 96–103. https://doi.org/10.1016/j.jconrel.2019.08.039

Dong, Y., Sun, Q., & Zhang, X. (2017). PD-1 and its ligands are important immune checkpoints in cancer. *Oncotarget, 8*(2), 2171–2186. https://doi.org/10.18632/oncotarget.13895

Giffen, P. S., Bhuiya, R., Brackenborough, K., Hobbs, M. J., Qian, L., & Burke, M. D. (2020). Controlled delivery of dutasteride using dissolvable microarrays: Initial formulation and in vivo evaluation. *Journal of Pharmaceutical Sciences, 109*(3), 1303–1311. https://doi.org/10.1016/j.xphs.2019.11.012

Hao, Y., Chen, Y. W., He, X. L., Yang, F., Han, R. X., Yang, C. L., Li, W., & Qian, Z. Y. (2020). Near-infrared responsive 5-fluorouracil and indocyanine green loaded MPEG-PCL nanoparticle integrated with dissolvable microneedle for skin cancer therapy. *Bioactive Materials, 5*(3), 542–552. https://doi.org/10.1016/j.bioactmat.2020.04.002

He, M., Yang, G., Zhao, X., Zhang, S., & Gao, Y. (2020). Intradermal implantable PLGA microneedles for etonogestrel sustained release. *Journal of Pharmaceutical Sciences, 109*(6), 1958–1966. https://doi.org/10.1016/j.xphs.2020.02.009

Huang, D., Chen, Y. S., & Rupenthal, I. D. (2018). Overcoming ocular drug delivery barriers through the use of physical forces. *Advanced Drug Delivery Reviews, 126*, 96–112. https://doi.org/10.1016/j.addr.2017.09.008

Ita, K. (2017). Dissolving microneedles for transdermal drug delivery: Advances and challenges. *Biomedicine and Pharmacotherapy, 93*, 1116–1127. https://doi.org/10.1016/j.biopha.2017.07.019

Ita, K., & Ashong, S. (2020). Percutaneous delivery of antihypertensive agents: Advances and challenges. *AAPS PharmSciTech, 21*(2). https://doi.org/10.1208/s12249-019-1583-9

Jin, X., Zhu, D. D., Chen, B. Z., Ashfaq, M., & Guo, X. D. (2018). Insulin delivery systems combined with microneedle technology. *Advanced Drug Delivery Reviews, 127*, 119–137. https://doi.org/10.1016/j.addr.2018.03.011

Jung, C. R., Lahiji, S. F., Jung, H., Kim, Y., & Kim, H. (2020). Rapidly separable micropillar integrated dissolving microneedles. *Pharmaceutics, 12*(6), 1–11. https://doi.org/10.3390/pharmaceutics12060581

Khandan, O., Famili, A., Kahook, M. Y., & Rao, M. P. (2012). Titanium-based, fenestrated, in-plane microneedles for passive ocular drug delivery. In *Conference proceedings : Annual international conference of the IEEE engineering in medicine and biology society. IEEE engineering in medicine and biology society. Conference, 2012* (pp. 6572–6575).

Kim, D. S., Choi, J. T., Kim, C. B., Shin, Y. R., Park, P. G., Kim, H., Lee, J. M., & Park, J. H. (2020). Microneedle array patch (MAP) consisting of crosslinked hyaluronic acid nanoparticles for processability and sustained release. *Pharmaceutical Research, 37*(3). https://doi.org/10.1007/s11095-020-2768-3

Kim, S., Yang, H., Eum, J., Ma, Y., Fakhraei Lahiji, S., & Jung, H. (2020). Implantable powder-carrying microneedles for transdermal delivery of high-dose insulin with enhanced activity. *Biomaterials, 232*, 119733. https://doi.org/10.1016/j.biomaterials.2019.119733

Kurowska, W., Kuca-Warnawin, E. H., Radzikowska, A., & Maslinski, W. (2017). The role of anti-citrullinated protein antibodies (ACPA) in the pathogenesis of rheumatoid arthritis. *Central European Journal of Immunology, 42*(4), 390–398. https://doi.org/10.5114/ceji.2017.72807

Lahiji, F. (2019). Tissue interlocking dissolving microneedles for accurate and efficient transdermal delivery of biomolecules. *Scientific Reports, 9*(1).

Lee, H., Song, C., Baik, S., Kim, D., Hyeon, T., & Kim, D.-H. (2018). Device-assisted transdermal drug delivery. *Advanced Drug Delivery Reviews*, 35–45. https://doi.org/10.1016/j.addr.2017.08.009

Leone, M., Mönkäre, J., Bouwstra, J. A., & Kersten, G. (2017). Dissolving microneedle patches for dermal vaccination. *Pharmaceutical Research, 34*(11), 2223–2240. https://doi.org/10.1007/s11095-017-2223-2

Leone, M., Romeijn, S., Du, G., Le Dévédec, S. E., Vrieling, H., O'Mahony, C., Bouwstra, J. A., & Kersten, G. (2020). Diphtheria toxoid dissolving microneedle vaccination: Adjuvant screening and effect of repeated-fractional dose administration. *International Journal of Pharmaceutics, 580*, 119182. https://doi.org/10.1016/j.ijpharm.2020.119182

Leone, M., Romeijn, S., Slütter, B., O'Mahony, C., Kersten, G., & Bouwstra, J. A. (2020). Hyaluronan molecular weight: Effects on dissolution time of dissolving microneedles in the skin and on immunogenicity of antigen. *European Journal of Pharmaceutical Sciences, 146*, 105269. https://doi.org/10.1016/j.ejps.2020.105269

Lewis, A. L., Chaft, J., Girotra, M., & Fischer, G. W. (2020). Immune checkpoint inhibitors: A narrative review of considerations for the anaesthesiologist. *British Journal of Anaesthesia, 124*(3), 251–260. https://doi.org/10.1016/j.bja.2019.11.034

Li, Y., Hu, X., Dong, Z., Chen, Y., Zhao, W., Wang, Y., Zhang, L., Chen, M., Wu, C., & Wang, Q. (2020). Dissolving microneedle arrays with optimized needle geometry for transcutaneous immunization. *European Journal of Pharmaceutical Sciences, 151*, 105361. https://doi.org/10.1016/j.ejps.2020.105361

Lin, S., Quan, G., Hou, A., Yang, P., Peng, T., Gu, Y., Qin, W., Liu, R., Ma, X., Pan, X., Liu, H., Wang, L., & Wu, C. (2019). Strategy for hypertrophic scar therapy: Improved delivery of triamcinolone acetonide using mechanically robust tip-concentrated dissolving microneedle array. *Journal of Controlled Release, 306*, 69–82. https://doi.org/10.1016/j.jconrel.2019.05.038

Li, W., Terry, R. N., Tang, J., Feng, M. R., Schwendeman, S. P., & Prausnitz, M. R. (2019). Rapidly separable microneedle patch for the sustained release of a contraceptive. *Nature Biomedical Engineering, 3*(3), 220–229. https://doi.org/10.1038/s41551-018-0337-4

Liu, C. C., Chou, Y. S., Chen, C. Y., Liu, K. L., Huang, G. J., Yuid, J. S., Wu, C. J., Liaw, G. W., Hsieh, C. H., & Chenid, C. K. (2020). Pathogenesis of local necrosis induced by Naja atra venom: Assessment of the neutralization ability of Taiwanese freeze-dried neurotoxic antivenom in animal models. *PLoS Neglected Tropical Diseases, 14*(2). https://doi.org/10.1371/journal.pntd.0008054

Liu, G.-S., Kong, Y., Wang, Y., Luo, Y., Fan, X., Xie, X., Yang, B.-R., & Wu, M. X. (2020). Microneedles for transdermal diagnostics: Recent advances and new horizons. *Biomaterials, 232*, 119740. https://doi.org/10.1016/j.biomaterials.2019.119740

Mao, J., Wang, H., Xie, Y., Fu, Y., Li, Y., Liu, P., Du, H., Zhu, J., Dong, L., Hussain, M., Li, Y., Zhang, L., Zhu, J., & Tao, J. (2020). Transdermal delivery of rapamycin with poor water-solubility by dissolving polymeric microneedles for anti-angiogenesis. *Journal of Materials Chemistry B, 8*(5), 928–934. https://doi.org/10.1039/c9tb00912d

Marchese, A., Arciola, C. R., Coppo, E., Barbieri, R., Barreca, D., Chebaibi, S., Sobarzo-Sánchez, E., Nabavi, S. F., Nabavi, S. M., & Daglia, M. (2018). The natural plant compound carvacrol as an antimicrobial and anti-biofilm agent: Mechanisms, synergies and bio-inspired anti-infective materials. *Biofouling, 34*(6), 630–656. https://doi.org/10.1080/08927014.2018.1480756

Marjanović, D. S., Zdravković, N., Milovanović, M., Trailović, J. N., Robertson, A. P., Todorović, Z., & Trailović, S. M. (2020). Carvacrol acts as a potent selective antagonist of different types of nicotinic acetylcholine receptors and enhances the effect of monepantel in the parasitic nematode *Ascaris suum. Veterinary Parasitology, 278*. https://doi.org/10.1016/j.vetpar.2020.109031

Milewski, M., Kapoor, Y., Ding, Z., Zhang, J., Ghartey-Tagoe, E., Manser, K., Nissley, B., Petrescu, I., Xu, L., Duffield, B., Chen, G., Yang, Q., Singh, P., & Smith, R. (2020). Stabilization and transdermal delivery of an investigational peptide using MicroCor® solid-state dissolving microstructure arrays. *Journal of Pharmaceutical Sciences, 109*(3), 1288–1296. https://doi.org/10.1016/j.xphs.2019.11.006

Mir, M., Permana, A. D., Ahmed, N., Khan, G. M., Rehman, A.u., & Donnelly, R. F. (2020). Enhancement in site-specific delivery of carvacrol for potential treatment of infected wounds using infection responsive nanoparticles loaded into dissolving microneedles: A proof of concept study. *European Journal of Pharmaceutics and Biopharmaceutics, 147*, 57–68. https://doi.org/10.1016/j.ejpb.2019.12.008

Möller, J., Kraner, M., Sonnewald, U., Sangal, V., Tittlbach, H., Winkler, J., Winkler, T. H., Melnikov, V., Lang, R., Sing, A., Mattos-Guaraldi, A. L., & Burkovski, A. (2019). Proteomics of diphtheria toxoid vaccines reveals multiple proteins that are immunogenic and may contribute to protection of humans against *Corynebacterium diphtheriae. Vaccine, 37*(23), 3061–3070. https://doi.org/10.1016/j.vaccine.2019.04.059

Moreira, A. F., Rodrigues, C. F., Jacinto, T. A., Miguel, S. P., Costa, E. C., & Correia, I. J. (2020). Poly (vinyl alcohol)/chitosan layer-by-layer microneedles for cancer chemo-photothermal therapy. *International Journal of Pharmaceutics, 576*, 118907. https://doi.org/10.1016/j.ijpharm.2019.118907

Pinacho-Garcia, L. M., Valdez, R. A., Navarrete, A., Cabeza, M., Segovia, J., & Romano, M. C. (2020). The effect of finasteride and dutasteride on the synthesis of neurosteroids by glioblastoma cells. *Steroids, 155*. https://doi.org/10.1016/j.steroids.2019.108556

Rezabakhsh, A., Ahmadi, M., Khaksar, M., Montaseri, A., Malekinejad, H., Rahbarghazi, R., & Garjani, A. (2017). Rapamycin inhibits oxidative/nitrosative stress and enhances angiogenesis in high glucose-treated human umbilical vein endothelial cells: Role of autophagy. *Biomedicine & Pharmacotherapy, 93*, 885–894. https://doi.org/10.1016/j.biopha.2017.07.044

Rojekar, S., Vora, L. K., Tekko, I. A., Volpe-Zanutto, F., McCarthy, H. O., Vavia, P. R., & Donnelly, R. F. (2021). Etravirine-loaded dissolving microneedle arrays for long-acting delivery. *European Journal of Pharmaceutics and Biopharmaceutics, 165*, 41–51. https://doi.org/10.1016/j.ejpb.2021.04.024

Rzhevskiy, A. S., Singh, T. R. R., Donnelly, R. F., & Anissimov, Y. G. (2018). Microneedles as the technique of drug delivery enhancement in diverse organs and tissues. *Journal of Controlled Release, 270*, 184–202. https://doi.org/10.1016/j.jconrel.2017.11.048

Sabri, A. H., Kim, Y., Marlow, M., Scurr, D. J., Segal, J., Banga, A. K., Kagan, L., & Lee, J. B. (2020). Intradermal and transdermal drug delivery using microneedles – fabrication, performance evaluation and application to lymphatic delivery. *Advanced Drug Delivery Reviews*, 195–215. https://doi.org/10.1016/j.addr.2019.10.004

Song, J., Li, X., & Li, J. (2020). Emerging evidence for the roles of peptide in hypertrophic scar. *Life Sciences, 241*. https://doi.org/10.1016/j.lfs.2019.117174

Sullivan, S. P., Koutsonanos, D. G., Del Pilar Martin, M., Lee, J. W., Zarnitsyn, V., Choi, S. O., Murthy, N., Compans, R. W., Skountzou, I., & Prausnitz, M. R. (2010). Dissolving polymer microneedle patches for influenza vaccination. *Nature Medicine, 16*(8), 915–920. https://doi.org/10.1038/nm.2182

Vora, L. K., Courtenay, A. J., Tekko, I. A., Larrañeta, E., & Donnelly, R. F. (2020). Pullulan-based dissolving microneedle arrays for enhanced transdermal delivery of small and large biomolecules. *International Journal of Biological Macromolecules, 146*, 290–298. https://doi.org/10.1016/j.ijbiomac.2019.12.184

Yang, B., Dong, Y., Shen, Y., Hou, A., Quan, G., Pan, X., & Wu, C. (2021). Bilayer dissolving microneedle array containing 5-fluorouracil and triamcinolone with biphasic release profile for hypertrophic scar therapy. *Bioactive Materials, 6*(8), 2400–2411. https://doi.org/10.1016/j.bioactmat.2021.01.014

Yang, P., Lu, C., Qin, W., Chen, M., Quan, G., Liu, H., Wang, L., Bai, X., Pan, X., & Wu, C. (2020). Construction of a core-shell microneedle system to achieve targeted co-delivery of checkpoint inhibitors for melanoma immunotherapy. *Acta Biomaterialia, 104*, 147–157. https://doi.org/10.1016/j.actbio.2019.12.037

Yao, G., Quan, G., Lin, S., Peng, T., Wang, Q., Ran, H., Chen, H., Zhang, Q., Wang, L., Pan, X., & Wu, C. (2017). Novel dissolving microneedles for enhanced transdermal delivery of levonorgestrel: In vitro and in vivo characterization. *International Journal of Pharmaceutics, 534*(1), 378–386. https://doi.org/10.1016/j.ijpharm.2017.10.035

Yao, W., Tao, C., Zou, J., Zheng, H., Zhu, J., Zhu, Z., Zhu, J., Liu, L., Li, F., & Song, X. (2019). Flexible two-layer dissolving and safing microneedle transdermal of neurotoxin: A biocomfortable attempt to treat rheumatoid arthritis. *International Journal of Pharmaceutics, 563*, 91–100. https://doi.org/10.1016/j.ijpharm.2019.03.033

Ye, Y., Yu, J., Wen, D., Kahkoska, A. R., & Gu, Z. (2018). Polymeric microneedles for transdermal protein delivery. *Advanced Drug Delivery Reviews, 127*, 106–118. https://doi.org/10.1016/j.addr.2018.01.015

CHAPTER 4

Anatomy and physiology of the eye

Introduction

The human eye (Fig. 4.1) is a complex organ (Willoughby et al., 2010) that is approximately 69—85 mm in circumference and 22—27 mm in anteroposterior diameter (Kels et al., 2015). It is a complicated sensory organ that weighs approximately 7.5 g which is less than 0.05% of the total weight of the human body (Suri et al., 2020). The total volume of the eye is approximately 6.5—7 mL (Nayak & Misra, 2018). The orbit of the eye has the shape of a pyramid with its posteromedial apex with its base anteriorly pointing in the direction of the optic canal (Jayaram & Calder, 2004). The eye has mean dimensions of 35 mm (height), 40 mm (width) and 25 mm (depth) (Jayaram & Calder, 2004). The eye globe (bulbus oculi) occupies the anterior and superolateral position within the orbit (Jayaram & Calder, 2004). The eyeball takes up about one-third (or less) of the orbital volume, while the remaining two-thirds consist of vasculature, muscles, fat and nerves (Malhotra et al., 2011). The eye consists of the anterior chamber made up of the cornea, iris, lens, ciliary body and the anterior sclera (Budak et al., 2018). The posterior chamber consists of the retina, choroid and the optic nerve (Budak et al., 2018). The anterior part of the eye include aqueous humor, ciliary body, tear, conjunctiva while the posterior portion also has retinal blood vessels, vitreous humor, Bruch's membrane, and the posterior sclera (Nayak & Misra, 2018). Tear, which is made up of three layers (mucin, aqueous and lipid), is the outermost precorneal thin film and it is about 3 μm thick (Nayak & Misra, 2018). The tear film protects, nourishes and lubricates the ocular surface (Swamynathan & Wells, 2020). Tear refers to a multilayered fluid comprising a superficial lipid layer obtained from the meibomian glands, central aqueous layer principally acquired from the lacrimal gland (with the remaining portion acquired from the conjunctival epithelium) and a basal glycocalyx made up of mucins (Swamynathan & Wells, 2020).

Microneedles
ISBN 978-0-323-91811-4
https://doi.org/10.1016/B978-0-323-91811-4.00005-5

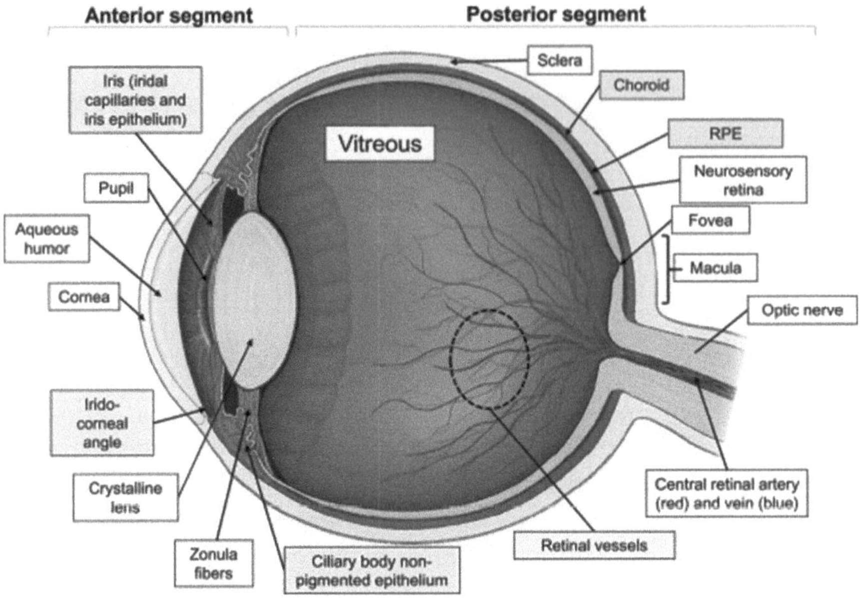

Fig. 4.1 Anatomy of the human eye Structures contributing to the blood-aqueous barrier, to the retinal pigment epithelium (RPE) forming the outer blood-retinal barrier and to the inner blood-retinal barrier are labeled in green [dark gray in print version], pink [dark gray in print version], and yellow [light gray in print version], respectively. *(From Matet, A., Behar-Cohen, F., Cassoux, N., Declèves, X., Cisternino, S., Sosnik, A., & Bendayan, R. (2020). Chapter 10 - Retinal and choroidal cancers: Blood-retinal barriers considerations in ocular chemotherapy. In:* Cancer sensitizing agents for chemotherapy *(Vol. 7, pp. 303–335). Academic Press. https://doi.org/10.1016/B978-0-12-816434-1.00010-3).*

Structure and function of the human eye

Aqueous humor (AH) refers to a clear colorless liquid made by ciliary body processes (Shi et al., 2020). It is produced from blood plasma but possesses more lactate and ascorbate and fewer proteins (Shi et al., 2020). It is made up of amino acids, carbohydrates, organic ions and water (Swamynathan & Wells, 2020). Usually, aqueous humor moves out of the anterior chamber through Schlemm's canal and the trabecular meshwork. It can also exit via a nontraditional pathway (uveoscleral outflow) (Shi et al., 2020). AH promotes immune quiescence of the anterior segment, clears waste products, nourishes ocular structures, maintains intraocular pressure and supports optical transparency (Shi et al., 2020). Vitreous humor (vitreous body) is a hydrophilic, viscoelastic, colorless, transparent and inert gel located between the retina and the crystalline lens (Pigaiani et al., 2020). It preserves anatomical expansion and tonus of the globe and also keeps the retina in place (Pigaiani et al., 2020). Vitreous humor volume is about 4–5 mL (Pigaiani et al., 2020). The uvea performs nutritive and vascular function while the sclera is a fibrous layer with the transparent cornea positioned anteriorly (Malhotra et al., 2011).

The sclera maintains the shape of the eye and creates a connective tissue coat that protects this organ from internal and external forces (Willoughby et al., 2010). The cornea protects deep-lying parts of the eye from mechanical injury and infections (Willoughby et al., 2010). It also refracts and transmits light to the retina and the lens (Willoughby et al., 2010). The uvea comprises the ciliary body, choroid and the iris while the retina is the sensory and neural layer of the eye (Malhotra et al., 2011). The anterior portion of the human eye is transparent, avascular and directly exposed to the environment (Cabrera et al., 2019). In contrast, the posterior segment of the eye is delicate, vascular, and not easily reachable. The retina, a multicellular layer that supports vision, is the location of many ophthalmic disorders (Cabrera et al., 2019). The conjunctival mucosa occupy the inner surface of the eyelids and the anterior portion of the bulbus oculi (Jayaram & Calder, 2004). Four recti muscles originate from a tendinous ring created from the periosteum at the orbital apex, which lies medial to the eyeball (Jayaram & Calder, 2004). This ring covers the medial region of the superior orbital fissure and the optic canal (Jayaram & Calder, 2004).

The cornea refers to the most anterior part of the eye and it is positioned in front of the iris and pupil (Willoughby et al., 2010). It is a transparent layer that encloses the anterior portion of the eye, and executes functions required for the maintenance of ocular health and clear vision (Zhang et al., 2017). These functions include the provision of focusing power via corneal refractive index and shape (Ray, 2015), and protecting other portions of the eye from dust, bacteria, and other damaging particles (Zhang et al., 2017). In adults, the diameter (horizontal) of the cornea is about 11.5–12.0 mm^2 and approximately 1.0 mm greater than the vertical diameter (DelMonte & Kim, 2011). To serve its principal function of letting light into the eye and providing two-thirds of the focusing power of this organ, the cornea must remain transparent (Al-Aqaba et al., 2019). The cornea has more nerves than any other tissue in the human body (Willoughby et al., 2010). Most corneal nerves are sensory, and they originate from the ophthalmic portion of the trigeminal nerve (Willoughby et al., 2010). The human cornea is made up of three cellular (stroma, epithelium, endothelium) and two interfacial (Descemet membrane, Bowman membrane) layers (DelMonte & Kim, 2011). The cornea is the main structural barrier of the eye that also protects the tissue against infections (DelMonte & Kim, 2011). In combination with tear film, this tissue provides a suitable anterior refractive surface for the eye. Its clarity derives from the structure and function of its constituents (DelMonte & Kim, 2011). The pre-pupillary cornea (optic zone), has a diameter of 4 mm and is located in the center of the cornea, anterior to the pupil. It is responsible for most of the cornea's refractive activity. The corneal epithelial surface forms the first barrier to the outside environment and it is a nonkeratinizing, stratified, and squamous layer that is remarkably uniform from limbus to limbus (DelMonte & Kim, 2011). The corneal epithelium encompasses one layer of basal cells, two to three layers of wing cells and two to three layers of superficial cells (Willoughby et al., 2010). The superficial epithelial cells has irregular surface because of ridge-like folds of the plasmalemma (microplicae) that connect with the tear film (Willoughby et al., 2010). The function of the tear film is to smoothen out microirregularities of the anterior surface of the epithelium (DelMonte & Kim, 2011). The air—tear film interface, in combination with the underlying cornea, is responsible for two-thirds of the overall refractive capacity of the eye (DelMonte & Kim, 2011). The corneal epithelium is highly stable and impermeable due to the availability of cell junctions (Willoughby et al., 2010). It is also strongly tethered to the basal lamina (Willoughby et al., 2010). Corneal basal lamina consists of nidogens, heparan sulfate proteoglycans, laminin, type IV collagen and other collagens (Saikia et al., 2018).

The corneal stroma consists of aligned cells guided by parallel collagen lamellae (Wang et al., 2017). The corneal stroma is responsible for most of the structural framework of the cornea and is made up of about 80%—85% of its thickness (DelMonte & Kim, 2011). The lamellar stroma which has a thickness of 500-mm grants anatomical integrity to the cornea (Willoughby et al., 2010). The corneal stroma has approximately 300 distinct lamellae with each of these consisting of parallel-oriented bundles of collagen fibers especially type I collagen (Lagali, 2020). The main stromal cell, the keratocyte, is located throughout the stroma with the cell body often positioned between stromal lamellae (Lagali, 2020). Keratocytes perform numerous functions such as collagen synthesis as well as collagen degradation. It also participates in wound healing by transforming into a (myo) fibroblast phenotype (Lagali, 2020). Stromal keratocytes secrete proteoglycans and collagen which are important for hydration and corneal transparency

(Willoughby et al., 2010). The Bowman's layer, which is an acellular zone of 10–15 mm beneath the basal lamina separates the epithelium from the stroma (Willoughby et al., 2010). The stroma is distinct from other collagenous structures mainly because it is transparent, and this derives from the precise organization of the extracellular matrix and the stromal fibers (DelMonte & Kim, 2011). Collagen fibers are organized as fibrils with subsequent packing of the fibrils into lamellae (DelMonte & Kim, 2011). The stroma contains 200 to 250 different lamellae and each layer is organized at right angles relative to fibers in adjoining lamellae (DelMonte & Kim, 2011).

The most anterior portion of the stroma consists of the Bowman's layer, which is a 10 μm layer of randomly oriented, compacted collagen fibrils that connect with the underlying corneal stromal lamellae (Lagali, 2020). Bowman's layer does not regenerate but facilitates a more rapid and complete anterior stromal and anterior corneal nerve regeneration as well as healing following damage to the eye (Lagali, 2020). The Bowman layer (BL) is not considered a genuine membrane but rather it is recognized as the acellular condensate of the most anterior segment of the stroma. This smooth layer aids in the maintenance of corneal shape (DelMonte & Kim, 2011). In BL, proteoglycans and collagen fibrils (mostly types I and III) are randomly organized (Sharma et al., 2018). These fibrils have a diameter of 20–30 nm. The BL has a mean radius of 7.34 mm and becomes thinner with age (Sharma et al., 2018). Apart from keratocytes and other cell populations, the corneal stroma is made up of proteoglycans, which are formed by keratocytes and comprise keratan sulfate and dermatan (Lagali, 2020). These constituents interface with collagen fibrils to maintain corneal transparency and fibril assembly (Lagali, 2020). The proteoglycan population of the stroma can regenerate since they are synthesized by keratocytes (Lagali, 2020). Most of the stromal extracellular matrix is made up of collagen fibrils organized in 200–250 parallel lamellae that run from limbus to limbus (Willoughby et al., 2010). Fibrils impart mechanical strength to the cornea (Willoughby et al., 2010).

Corneal nerves are present throughout the corneal stroma, and the density of these nerves is greater in the anterior stroma within the subbasal and subepithelial nerve plexi (Lagali, 2020). The trigeminal ganglion has the shape of a canoe with a mean number of 27,400 pseudo-unipolar neurons (neurons from which a single axon emerges and divides into two with one extending to the peripheral end organ and the other going to the central nervous system) (Al-Aqaba et al., 2019). Nerve fibers from the trigeminal ganglion neurons extend supra-choroidally and branch out, creating a plexus at the corneoscleral limbus (Labetoulle et al., 2019). The corneal sensory fibers, which constitute most of the corneal nerve fibers, originate from cell bodies of the ophthalmic segment of the trigeminal nerve (Al-Aqaba et al., 2019). A small proportion of neurons (approximately 1.5% of gangial neurons) located in the neural or medial crest section of the trigeminal ganglion serve the cornea (Al-Aqaba et al., 2019). About 40 thick nerve stacks cross into the cornea, and these are equally distributed around the limbus (Labetoulle et al., 2019). Nerve trunks, (without a myelin sheath for transparency) enter the corneal stroma radially from the limbal plexus and then move in the anterior direction, terminating as nerve endings that function as mechanical, chemical, and temperature "nociceptors" delivering sensory information (temperature, touch and pain) from the ocular surface to the brain (Labetoulle et al., 2019). Radially-oriented and large nerve bundles innervate the cornea, moving forward in the anterior direction and creating branches that provide nerves for the anterior and mid-stroma (Ali et al., 2016). About 33—90 fascicles, containing 900—1500 axons, enter the human cornea, while 20—40 fascicles enter the cornea of other mammals (Ali et al., 2016). The most posterior layer of the cornea is called the endothelium and it regulates deturgescence (relative dehydration) and aids the maintenance of corneal transparency via the provision of transport and barrier functions (Ali et al., 2016). The endothelium creates a hexagonal monolayer, which originates from neural crest cells that are in turn derived from the optic cup (Feizi, 2018).

The endothelium is constituted by polygonal cells that interface with aqueous humor and also wrap the posterior surface of Descemet's membrane (DM) (Zavala et al., 2013). Corneal endothelial cell (CEC) monolayer provides a barrier function via adherens junctions and tight junctions (Ali et al., 2016). This monolayer is relevant for corneal clarity. CECs are typically arrested in the cell cycle G1 phase and under normal physiological conditions neither regenerate nor proliferate (Feizi, 2018). As a consequence, loss of corneal endothelial cells can lead to both residual cell migration and compensatory enlargement (Feizi, 2018). The endothelium does not proliferate in contrast to the epithelium. The later can renew itself (Feizi, 2018). Cell injury forces residual endothelial cells to enlarge and migrate resulting in the maintenance of corneal transparency (Feizi, 2018). Descemet's membrane (DM), which is located between the posterior stroma and the corneal endothelium is the basement membrane of the posterior cornea (Saikia et al., 2018). It is part of the endothelial-Descemet's membrane complex which is relevant for corneal physiology (Saikia et al., 2018). The endothelial-DM complex permits important nutrients to cross into the stroma but hinders the transport of transforming growth factor beta from the aqueous humor into the stroma (Saikia et al., 2018). DM thickness is about 10 μm in adults and 3 μm in children. It comprises two layers: a posterior non-banded layer which thickens with age and is constantly produced as well as an anterior banded layer made up of proteoglycans and collagen lamellae (Saikia et al., 2018). Four key ingredients make up the DM lattice: (a) nodes of the lattice, (b) rod-like structures linking the nodes, (c) amorphous materials (d) fine filaments distributed in the interstices (Saikia et al., 2018).

The human sclera is the white outer coat that encloses the eye globe (Saikia et al., 2018). It is a tough, opaque, and fibrous tissue that attaches the eye to extraocular muscles, contributes to the maintenance of the shape of this organ, offers protection against internal and external forces, and shields the inner structures of the eye (Consejo & Iskander, 2016; Consejo et al., 2019). The sclera has three key anatomical layers: (a) the superficial vascularized episclera (b) the avascular scleral stroma and (c) the lamina fusca (Atta et al., 2020). The sclera is a very resilient and structurally complex connective tissue that performs multiple functions critical to vision (Boote et al., 2020). It forms approximately 85% of the human eyeball outer tunic. This tissue is a rigid layer that helps in maintaining ocular tension and also shields other anatomic structures of the eye (Atta et al., 2020). It is a dense, connective tissue that defines the shape and size of the eye (Harper & Summers, 2015). It partakes in vision by distorting the lens-iris-diaphragm as well as the retina (Atta et al., 2020). The sclera can withstand the force created by intraocular pressure, supports the retina, protects the eye from external injury and furnishes this organ with a pathway for aqueous outflow (Harper & Summers, 2015). The sclera possesses specialized domains for the ingress and egress of relevant blood vessels and nerves, the location of the cornea, and the attachment of extraocular muscles (Harper & Summers, 2015). Regardless of diverse structural regions, the sclera has to control the shape of the eye during important events that require globe deformation (accommodation, eye movements, and fluctuations in ocular tension) (Harper & Summers, 2015).

The human sclera encompasses about 50% collagen by weight, made up of mostly type I collagen, which is about 95% of the available collagen (Harper & Summers, 2015). Collagen fibrils within the sclera are organized into fairly interconnected and irregularly arranged lamellae (Harper & Summers, 2015). The scleral lamellar organization is somewhat analogous to the organization of collagen in the cornea but scleral collagen fibrils are very diverse in diameter, the lamellae differ in thickness, and lamella orientation is irregular relative to adjoining lamella (Harper & Summers, 2015). Apart from type I collagen, the sclera tissue also has collagen types XIII and XII, VIII, VI, V, IV and III (Harper & Summers, 2015). Proteoglycans control collagen fibril assembly as well as arrangement and abound throughout the entire scleral extracellular matrix (Harper & Summers, 2015). Proteoglycans are made up of a core protein tethered to glycosaminoglycan side chains (repeating disaccharide units such as dermatan sulfate, keratan sulfate chondroitin sulfate and heparan sulfate) (Harper & Summers, 2015). Sulfate residues on the glycosaminoglycan side chains impart substantial negative charge to proteoglycans and this is relevant for interactions with other extracellular proteins (e.g. collagen) and water (Harper & Summers, 2015). The sclera interacts with extraocular muscles and is similar to tendons from cellular and anatomical standpoints (Atta et al., 2020).

The conjunctiva is a translucent and thin tissue that lies adjacent to the sclera with pore radius that is about 3.0—5.5 nm (Suri et al., 2020). The conjunctival epithelium is relatively more permeable to macromolecules and has a larger surface area (twenty times more) compared to the cornea (Suri et al., 2020). The large surface area is attributed to wider intercellular spaces (Suri et al., 2020). The conjunctiva is a stratified, non-keratinizing epithelium bestrewn with goblet cells (GCs) that envelops the outer eyeglobe and the inner eyelids (Swamynathan & Wells, 2020). This tissue stretches to the limbus which is adjacent to the cornea (Swamynathan & Wells, 2020). The conjunctival epithelium is devoid of a well-arranged basement layer and consists of 2—3 cell layers. There are few individual GC cells in the lid wiper and bulbar segments but there is an array of these cells in the nasal portion of the conjunctiva (Swamynathan & Wells, 2020). GCs within the conjunctival bulbar region represent the key source of the mucin-containing stratum of the pre-corneal tear film (Doughty, 2020). The limbus refers to a thin band of tissue that surrounds the cornea (Osei-Bempong et al., 2013). It acts as a depot for limbal stem cells which is responsible for the maintenance and regeneration of the corneal epithelium. The limbus also protects the conjunctiva and its arteries, veins and capillaries (Osei-Bempong et al., 2013). This tissue demarcates the transparent cornea from the opaque sclera (which is in turn covered by the translucent conjunctiva) (Buskirk, 1989). It is an interesting part of the eye due to the harboring in this region of the aqueous outflow apparatus and the supply of blood to the peripheral cornea (Alabi et al., 2018). At the outer surface, the limbus separates the conjunctiva from the corneal epithelium while at the inner surface, demarcates the corneal endothelium from the trabecular meshwork (Williams et al., 2018). The limbus is the site of diverse but physiologically relevant activities related to vision. It accommodates the key pathways of aqueous humor outflow, required, among other functions, for the sustenance of intraocular pressure and also provides nourishment for the peripheral cornea (Consejo & Iskander, 2016).

The uvea, which is the middle vascular coat of the eye (Nickla & Wallman, 2010) is made up of the iris, the ciliary body and the choroid (Goel et al., 2013). The iris is a pigmented and circular diaphragm that lies in a coronal plane anteriorly to the lens (Bloom et al., 2020). Iris color is ascribed to the fluctuating quantities of pheomelanin (red/yellow melanin) and eumelanin (brown/black melanin) synthesized by melanocytes (Kolb, 1995). Variation in the iris color of human eyes derive from the disparate quantities of melanin granules within a constant number of melanocytes in the superficial stroma (Imesch et al., 1997). The lens, which refracts incoming light is located between the posterior and anterior chambers of the eye. This transparent tissue situated behind the iris, is held by zonule fibers (ligaments) tethered to the anterior region of the ciliary body (Kolb, 1995). Like the lens in a camera, the basic function of the eye lens is to transmit and direct light onto the retina (Hejtmancik & Shiels, 2015). Even though the cornea provides approximately 80% of total refraction, the lens adjusts focused light onto the retina (Hejtmancik & Shiels, 2015). To carry out this function, the human lens contains one of the highest protein levels of any tissue (Hejtmancik & Shiels, 2015). Proteins constitute almost 60% of the total mass of the lens (Hejtmancik & Shiels, 2015). The lens is enclosed by a capsule derived from collagen, on which the anterior-facing basal poles of the epithelial cells rest (Hejtmancik & Shiels, 2015). The basal poles of the fiber cells facing the rear portion of the eye also rest on this collagenous capsule (Hejtmancik & Shiels, 2015). The relaxation or contraction of zonule fibers, which is attributed to the activities of the ciliary muscles, alters lens shape (Kolb, 1995). This process is termed accommodation and permits us to form retinal image (Kolb, 1995). The lens is connected peripherally to the ciliary body, while its unbound central area forms a hole called the pupil (Bloom et al., 2020). The anterior chamber occupies the area between the cornea and the iris (Ray, 2015). It is filled with a plasma-like fluid (aqueous humor) that determines the shape of this region of the eye (Al-Aqaba et al., 2019; Ray, 2015).

The ciliary body covers the inner portion of the anterior chamber of the eye coating (Pašta, 2013). It is made up of the ciliary muscle and processes (Pašta, 2013). On the surface of the ciliary body is the black ciliary epithelium, from which aqueous humor is generated (Pašta, 2013). The ciliary body has the shape of a ring along the inner wall of the eyeball and it extends anteriorly from the iris to the ora serrata posteriorly (Goel et al., 2013). The ciliary body is responsible for two important functions: eye accommodation via the regulation of lens shape and the secretion of aqueous humor into the anterior chamber by the epithelium cells (Tserevelakis et al., 2017). This tissue is predominantly made up of smooth muscles that are arranged in longitudinal radial and circular modes (Goel et al., 2013). It comprises the ciliary processes and ciliary muscle. Ciliary processes are directed inwards as radial ridges and are about 70 in number (Goel et al., 2013).

The ciliary body produces aqueous humor, which maintains intraocular pressure and feeds the avascular tissues in the anterior segment with nutrients (Suri et al., 2020). Ciliary body, which is formed from smooth muscles, also facilitates waste drainage from the cornea and the lens (Suri et al., 2020). The ciliary body is relaxed when we see distant objects and narrowed when we see near objects (Xie et al., 2018). The ciliary processes include a central core of stroma which is enveloped by a two-layered epithelium. The inner epithelial layer is bathed by aqueous humor (Goel et al., 2013). This non-pigmented epithelial layer produces aqueous humor via active transport (ultrafiltration and diffusion) (Goel et al., 2013). The aqueous humor soaks the avascular portions of the eye (trabecular meshwork, anterior vitreous, posterior surface of the cornea and lens) before egressing into the episcleral veins via the canal of Schlemm (Goel et al., 2013). The ciliary processes produce neuropeptides, antioxidant enzymes, glycoproteins of the vitreous body and aqueous humor (Napier & Kidson, 2005). Aqueous humor is synthesized by the pars plicata (Agrawal & Martin, 2008) while zonular fibers link the lens to the ciliary body (Napier & Kidson, 2005). In combination with the ciliary muscle, zonula fibers assist in aqueous humor outflow and lens accommodation via the trabecular meshwork (TM) (Napier & Kidson, 2005). The TM is an architecturally complex, vascular tissue spanning the entire length of the Schlemm's canal and connecting Schwalbe's line to the scleral spur/ciliary muscle (Stamer & Clark, 2017). TM is made up of three areas: (a) the uveal meshwork, (b) the corneoscleral meshwork and (c) the juxtacanalicular meshwork (Stamer & Clark, 2017). Ciliary processes form an elegant array of radial folds around the eye. Each fold is made up of an inner capillary core, enclosed by a loose stroma and surrounded by a two-layered epithelium (the inner and outer ciliary epithelia) (Napier & Kidson, 2005).

The choroid refers to a pigmented vascular tissue that spans anteriorly from the ora serrata and posteriorly to the optic nerve head (Sezer et al., 2016). It is made up of resident immunocompetent cells, fibroblasts, melanocytes, blood vessels, as well as supporting elastic and collagenous connective tissue (Nickla & Wallman, 2010). The choroid is about 0.15 mm thick anteriorly and 0.22 mm posteriorly (Sezer et al., 2016). This tissue constitutes the posterior region of the uveal tract, which is anteriorly contiguous with the iris and the ciliary body (Sezer et al., 2016). The choroid is made up of choroidal vessels, the choriocapillaris and the Bruch's membrane (BM) (Sezer et al., 2016). The innermost layer of the choroid is formed by the complex pentalaminar BM while the outermost layer is the suprachoroid (Sezer et al., 2016). Between the choroid and the sclera lies the suprachoroidal space (Nickla & Wallman, 2010). The 5-laminar BM forms a single functional unit with the choriocapillaris and the retinal pigment epithelium (RPE) (Booij et al., 2010). The 5 layers of BM are: the (a) the basement membrane of the choriocapillaris (b) outer collagenous zone (c) elastin layer, (d) inner collagenous zone and (e) basement membrane of the RPE (Fang et al., 2017). The BM mediates the exchange of oxygen, nutrients, several biomolecules and waste products (Booij et al., 2010). It separates the RPE from the choriocapillaris (Fields et al., 2020).

The choroid provides blood to the outer retina and is mainly made up of blood vessels (Nickla & Wallman, 2010). It is also responsible for changes in choroidal thickness, thermoregulation, and the production of growth factors (Nickla & Wallman, 2010). The choroid facilitates the drainage of aqueous humor from the anterior chamber through the uveoscleral pathway (Nickla & Wallman, 2010). The choriocapillaris forms a thin sheet proximate to Bruch's membrane (Nickla & Wallman, 2010) and constitutes a highly interconnected network of capillaries (Nickla & Wallman, 2010). The outer layer of Bruch's membrane is created from the fibrous basement membrane of the capillary endothelial cells (Nickla & Wallman, 2010). The choriocapillaris is approximately 7 μm in the periphery but about 10 μm thick at the fovea, where there is a high density of capillaries (Nickla & Wallman, 2010). The capillaries originate from arterioles in the Sattler's layer. Each arteriole gives rise to lobular or hexagonal-shaped region of a layer of capillaries imbuing the choriocapillaris with a patch-like structure (Nickla & Wallman, 2010). The choroid predominantly comprises veins, arteries, and arterioles. It declines in diameter from the outer region to the inner domain as the vessels form branches (Ramirez et al., 2012). The arterial vessels refer to some recurring branches of the main arterial circle of the iris and branches of the posterior ciliary arteries (Ramirez et al., 2012). The vessels decrease in size while branching toward the capillary bed and there are three strata of decreasing size: Haller's layer, Sattler's layer, and the choriocapillaris (Ramirez et al., 2012).

The retina refers to a sheet of neural tissue found at the posterior portion of the eye (Miltner & La Torre, 2019). It transforms light into an array of action potentials that is then sent to the brain for visual processing (Miltner & La Torre, 2019). This tissue covers the vitreous cavity and the interior surface of the eye (Willoughby et al., 2010). The neural retina comprise six key classes of neurons: ganglion cells, amacrine cells, horizontal cells, bipolar cells and photoreceptors, which trap and process light signals (Willoughby et al., 2010). The tissue also includes the Müllerian glia that function as the organizational scaffold of the neural retina. Six main neural groups in the retina are grouped into three layers: the retinal ganglion cell layer, the inner nuclear layer, and the outer nuclear layer (Miltner & La Torre, 2019). The retina has a distinctive central structure with a highly ordered synaptic arrangement for high acuity vision as well as a remarkable neuronal spatial distribution (Hoshino et al., 2017). The fovea centralis, which is the central portion of the retina, occupies 1.5 mm and includes an avascular pit made up of long- and medium-wavelength cone photoreceptors with a few short-wavelength cones, but there are no rods (Hoshino et al., 2017). Each fovea cone is linked to only one ganglion and one bipolar cell thus imparting a greater resolution to the cone system (Sung & Chuang, 2010). In other areas of the retina, each ganglion cell and bipolar cell serve several photoreceptors (Sung & Chuang, 2010). The fovea (or macula) occupies the center of the retina and is responsible for high spatial acuity vision which is required for face recognition and reading (Cowan et al., 2020).

Rods and cones constitute the two classes of photoreceptors (Sung & Chuang, 2010). Cones are the receptors responsible for vision in bright light as well as color vision (photopic visual system) (May-Simera et al., 2017) and they derive their name from the fact that their light-sensing outer segments are conical in shape (Sung & Chuang, 2010). Three classes of cones (M−, L-, and S), which are variable in their spectral sensitivity can be identified in the human retina (May-Simera et al., 2017). The detection of light and its transformation into an electric signal (phototransduction) occurs in the outer segments of cones and rods (Vinberg et al., 2018). The outer segment of the cone includes an orderly bundle of membranous laminella that create a highly intricate surface membrane (Sung & Chuang, 2010). The outer segment also adjoins plasma membrane (Sung & Chuang, 2010). Rods are extremely sensitive and can detect single photons of light facilitating the detection of photostimuli in dimly lit conditions (scotopic visual system) (May-Simera et al., 2017; Vinberg et al., 2018). Rods contain a single rod opsin and are ultrasensitive to blue-green light (500 nm) (Hendrickson et al., 2008). In contrast to other cells in the retina, there is only one kind of rod photoreceptor, and it is the major cell type in the retina (Sung & Chuang, 2010). There are about 5 million bipolar cells, 130 million photoreceptors, and approximately one million ganglion cells in the human retina (Sung & Chuang, 2010). There are more rods (approximately 20 times higher) than cones and these rods are distributed throughout the retina except in the fovea area (Sung & Chuang, 2010).

Bipolar cells (BCs) are neurons that link the inner retina to the outer retina (Euler et al., 2014). These specialized cells provide an "additional" stratum of processing that is normally absent in other sensory organs. Distinct (usually more than ten) classes of BCs convert photoreceptor signal via polarity (ON vs., OFF), chromatic preference, and kinetics (sustained vs. transient responses) (Euler et al., 2014). Each of the approximately 13 specific classes of bipolar cells systematically transforms photoreceptor input in a peculiar manner, thus creating specific channels that encode features like chromatic composition, temporal profile, contrast and polarity (Euler et al., 2014). BCs initially respond at their dendrites via different mechanisms including diverse contact morphologies, receptor classes, and secondary messenger systems, as well as inputs from horizontal cells (Euler et al., 2014). Extra capacity for signal modulation is present in the axonal terminal system, in which lateral inputs from amacrine cells as well as local ionic currents are responsible for determining a bipolar cell's eventual output (Euler et al., 2014). Single bipolar cells may contribute differential input to distinct postsynaptic circuits. Bipolar cell output signals are the fundamental "bricks" from which the microcircuits of the inner retina obtain a sketch of the visual environment (Euler et al., 2014).

Horizontal cells (HCs) are a class of retinal interneurons that control information flow from the BCs to the ganglion cells (RGCs) in the inner plexiform layer (IPL) and from the photoreceptors to the bipolar cells in the outer plexiform layer of the retina respectively (Chaya et al., 2017). HCs, which are electrically linked to one another via gap junctions, obtain glutamatergic input from photoreceptors (PRs) and generate feedforward as well as feedback signals to BCs and PRs respectively (Chaya et al., 2017). Three distinct feedback mechanisms for inhibition have been suggested: hemichannel-assisted ephaptic feedback, GABA-assisted feedback, and pH-assisted feedback (Chaya et al., 2017). HCs fine-tune photoreceptor output and contribute in different ways toward early visual processing (Chapot et al., 2017). HCs contribute to color opponency, contrast improvement, and the formation of center—surround receptive fields in BCs and cone photoreceptors (Chapot et al., 2017). Amacrine cells (ACs) refer to non-spiking, axonless and inhibitory interneurons in the retina (Goodson et al., 2018; Poleg-Polsky et al., 2018). They produce ganglion cell output neurons in the inner plexiform layer (IPL) and synapses with glutamatergic bipolar interneurons (Goodson et al., 2018). The somas of amacrine cells are mostly situated in the inner nuclear layer, but a small proportion is restricted to the ganglion cell layer (Goodson et al., 2018). ACs constitute approximately 8% of retinal cells, but there are more than 30 distinct subclasses within this cell group (Goodson et al., 2018). Amacrines are typically classified into three main groups based on the expression of GlyT1 (glycinergic, ~43%), GAD65/67 (GABAergic, ~43%), or neither of these markers (nGnG, ~15%) (Goodson et al., 2018). Synaptic inputs and outputs enter on and leave from the same dendritic branches in amacrine cells, allowing the processing of visual information through several parallel channels within the same cell, often without the participation of the soma (Poleg-Polsky et al., 2018).

Retinal ganglion cells (RGCs) refer to neurons that indirectly receive visual information from retinal photoreceptors (Aparicio et al., 2017). Retinal ganglion cell axons take off from the retina and move to the retinorecipient areas of the brain (Clements & Wright, 2018). RGC axons form synapses in the image-forming regions of the brain (lateral geniculate nucleus and the superior colliculus) as well as the non-image forming zones (the medial terminal nucleus and the suprachiasmatic nucleus) (Clements & Wright, 2018). RCGs constitute the only retinal output, and they are larger than the majority of the other neurons (Miltner & La Torre, 2019). Furthermore, they have elaborate dendritic arbors that obtain inputs from the preceding retinal circuitry, and convey these signals to the retinorecipient regions of the brain (Miltner & La Torre, 2019). RGCs direct their axons toward the center of the retina, where they assemble as the optic nerve gathers all the axons at the optic nerve head (Miltner & La Torre, 2019). Subsequently, axonal fibers transmit information to the superior colliculi, suprachiasmatic nuclei, lateral geniculate and other pretectal nuclei of the thalamus (Miltner & La Torre, 2019).

Müller cells (MCs) are radial glial cells that provide metabolic and homeostatic support of retinal neurons (Arrigo et al., 2020; Li et al., 2019; Reichenbach & Bringmann, 2013). By enabling water, ion and bicarbonate movement across cells, MCs regulate the chemical content of the extracellular space fluid (Arrigo et al., 2020; Li et al., 2019; Reichenbach & Bringmann, 2013). Müller cells supply anti-oxidative and trophic support to neurons and photoreceptors. They also control the permeability of the blood-retinal barrier (Reichenbach & Bringmann, 2013). Müller cells take part in the retinoid cycle, prevent glutamate toxicity, recycle neurotransmitters, redistribute ions via spatial buffering, and regulate nutrient supplies via several mechanisms (Coughlin et al., 2017). The optic disc is the point of entry for arteries providing blood for the inner layers of the transparent region of the retina and the point of exit for the ganglion cell nerve fibers and central veins (Quinn et al., 2019). The arteries form the so-called "arcade" that includes the macula with perifovea, parafovea, fovea and foveola (Quinn et al., 2019). The macula lutea (macula), so named because of its yellow color, occupies the most sensitive area of the retina, offering the highest sharp-sightedness (Nguyen et al., 2019). There are no vessels in this region. There are also carotenoids (zeaxanthin and lutein) that constitute the macular pigments and are responsible for the yellow color (Nguyen et al., 2019). These macular pigments are known to possess blue-light filtering properties (Nguyen et al., 2019). Fovea refers to the small depression in the center of the macula (GeethaRamani & Balasubramanian, 2018) with a high density of receptors. It is responsible for the fine spatial details of vision (Poletti et al., 2013). It is specialized for sharp-sighted and mediates much of our conscious visual experience (McGregor, 2019). Increase in the number of midget ganglion cells of the retina can enhance foveal sharp-sightedness (McGregor, 2019).

There are several routes of drug delivery to the eye (Fig. 4.2). In the next chapter, we will examine the use of microneedles for ocular drug delivery.

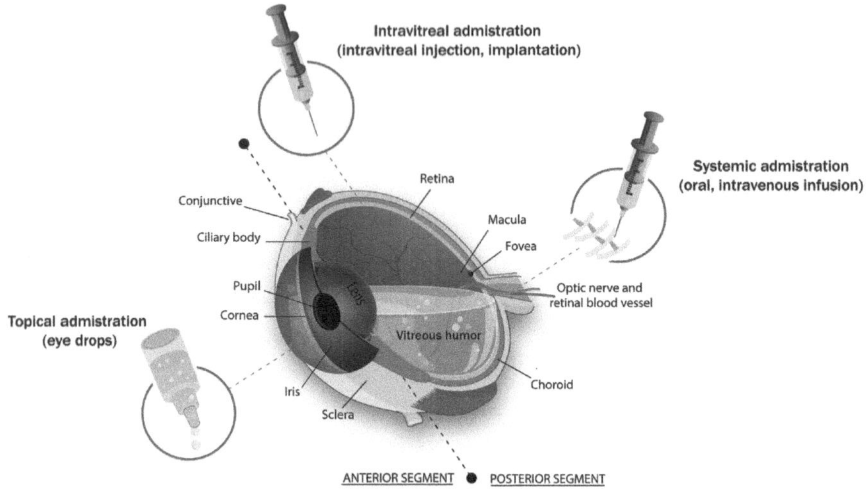

Fig. 4.2 Anatomical components of the eye and frequently used routes of drug administration. Different routes for ocular drug delivery. *(From Ilochonwu, B. C., Urtti, A., Hennink, W. E., & Vermonden, T. (2020). Intravitreal hydrogels for sustained release of therapeutic proteins.* Journal of Controlled Release, 326, *419–441. https://doi.org/10.1016/j.jconrel.2020.07.031).*

References

Agrawal, P., & Martin, K. R. (2008). Ciliary body position variability in glaucoma patients assessed by scleral transillumination. *Eye, 22*(12), 1499–1503. https://doi.org/10.1038/eye.2008.79

Al-Aqaba, M. A., Dhillon, V. K., Mohammed, I., Said, D. G., & Dua, H. S. (2019). Corneal nerves in health and disease. *Progress in Retinal and Eye Research, 73*. https://doi.org/10.1016/j.preteyeres.2019.05.003

Alabi, E., Hutchings, N., Bizheva, K., & Simpson, T. (2018). Relación entre las mediciones del diámetro y la profundidad de vasos sanguíneos en el limbo obtenidas mediante tomografía de coherencia óptica de ultra-alta resolución. *Journal of Optometry, 11*(1), 57–65. https://doi.org/10.1016/j.optom.2017.02.003

Ali, M., Raghunathan, V. K., Li, J. Y., Murphy, C. J., & Thomasy, S. M. (2016). Biomechanical relationships between the corneal endothelium and Descemet's membrane. *Experimental Eye Research, 152*, 57–70. https://doi.org/10.1016/j.exer.2016.09.004

Aparicio, J. G., Hopp, H., Choi, A., Mandayam Comar, J., Liao, V. C., Harutyunyan, N., & Lee, T. C. (2017). Temporal expression of CD184(CXCR4) and CD171(L1CAM) identifies distinct early developmental stages of human retinal ganglion cells in embryonic stem cell derived retina. *Experimental Eye Research, 154*, 177–189. https://doi.org/10.1016/j.exer.2016.11.013

Arrigo, A., Perra, C., Aragona, E., Giusto, D., Doglioni, C., Pierro, L., Giordano Resti, A., Bandello, F., & Battaglia Parodi, M. (2020). Extrafoveal Müller cells detection in vivo in the human retina: A pilot study based on optical coherence tomography. *Experimental Eye Research, 199*, 108183. https://doi.org/10.1016/j.exer.2020.108183

Atta, G., Tempfer, H., Kaser-Eichberger, A., Guo, Y., Schroedl, F., Traweger, A., & Heindl, L. M. (2020). The lymphangiogenic and hemangiogenic privilege of the human sclera. *Annals of Anatomy, 230*. https://doi.org/10.1016/j.aanat.2020.151485

Bloom, J., Czyz, M. M., & Anatomy, C. N. (2020). *Head and neck, eye iris sphincter muscle*. Treasure Island (FL): StatPearls Publishing [Internet] https://www.ncbi.nlm.nih.gov/books/NBK532252/2020.

Booij, J. C., Baas, D. C., Beisekeeva, J., Gorgels, T. G. M. F., & Bergen, A. A. B. (2010). The dynamic nature of Bruch's membrane. *Progress in Retinal and Eye Research, 29*(1), 1−18. https://doi.org/10.1016/j.preteyeres.2009.08.003

Boote, C., Sigal, I. A., Grytz, R., Hua, Y., Nguyen, T. D., & Girard, M. J. A. (2020). Scleral structure and biomechanics. *Progress in Retinal and Eye Research, 74*, 100773. https://doi.org/10.1016/j.preteyeres.2019.100773

Budak, G., Dash, S., Srivastava, R., Lachke, S. A., & Janga, S. C. (2018). Express: A database of transcriptome profiles encompassing known and novel transcripts across multiple development stages in eye tissues. *Experimental Eye Research, 168*, 57−68. https://doi.org/10.1016/j.exer.2018.01.009

Van Buskirk, E. M. (1989). The anatomy of the limbus. *Eye, 3*(2), 101−108.

Cabrera, F. J., Wang, D. C., Reddy, K., Acharya, G., & Shin, C. S. (2019). Challenges and opportunities for drug delivery to the posterior of the eye. *Drug Discovery Today, 24*(8), 1679−1684. https://doi.org/10.1016/j.drudis.2019.05.035

Chapot, C. A., Euler, T., & Schubert, T. (2017). How do horizontal cells 'talk' to cone photoreceptors? Different levels of complexity at the cone−horizontal cell synapse. In *Journal of physiology* (Vol. 595, pp. 5495−5506). Blackwell Publishing Ltd. https://doi.org/10.1113/JP274177. Issue 16.

Chaya, T., Matsumoto, A., Sugita, Y., Watanabe, S., Kuwahara, R., Tachibana, M., & Furukawa, T. (2017). Versatile functional roles of horizontal cells in the retinal circuit. *Scientific Reports*, 7(1). https://doi.org/10.1038/s41598-017-05543-2

Clements, R., & Wright, K. M. (2018). Retinal ganglion cell axon sorting at the optic chiasm requires dystroglycan. *Developmental Biology, 442*(2), 210−219. https://doi.org/10.1016/j.ydbio.2018.08.010

Consejo, A., & Iskander, D. R. (2016). Corneo-scleral limbus demarcation from 3D height data. *Contact Lens and Anterior Eye, 39*(6), 450−457. https://doi.org/10.1016/j.clae.2016.05.001

Consejo, A., Llorens-Quintana, C., Bartuzel, M. M., Iskander, D. R., & Rozema, J. J. (2019). Rotation asymmetry of the human sclera. *Acta Ophthalmologica, 97*(2), e266−e270.

Coughlin, B. A., Feenstra, D. J., & Mohr, S. (2017). Müller cells and diabetic retinopathy. *Vision Research, 139*, 93−100. https://doi.org/10.1016/j.visres.2017.03.013

Cowan, C. S., Renner, M., De Gennaro, M., Gross-Scherf, B., Goldblum, D., Hou, Y., Munz, M., Rodrigues, T. M., Krol, J., Szikra, T., Cuttat, R., Waldt, A., Papasaikas, P., Diggelmann, R., Patino-Alvarez, C. P., Galliker, P., Spirig, S. E., Pavlinic, D., Gerber-Hollbach, N., … Roska, B. (2020). Cell types of the human retina and its organoids at single-cell resolution. *Cell, 182*(6), 1623−1640. https://doi.org/10.1016/j.cell.2020.08.013. e34.

DelMonte, D. W., & Kim, T. (2011). Anatomy and physiology of the cornea. *Journal of Cataract and Refractive Surgery, 37*(3), 588−598. https://doi.org/10.1016/j.jcrs.2010.12.037

Doughty, M. J. (2020). On the in vivo assessment of goblet cells of the human bulbar conjunctiva by confocal microscopy − a review. *Contact Lens and Anterior Eye, 43*(4), 315−321. https://doi.org/10.1016/j.clae.2020.01.004

Euler, T., Haverkamp, S., Schubert, T., & Baden, T. (2014). Retinal bipolar cells: Elementary building blocks of vision. *Nature Reviews Neuroscience, 15*(8), 507–519. https://doi.org/10.1038/nrn3783

Fang, Y., Jonas, J. B., Yokoi, T., Cao, K., Shinohara, K., & Ohno-Matsui, K. (2017). Macular Bruch's membrane defect and dome-shaped macula in high myopia. *PloS One, 12*(6). https://doi.org/10.1371/journal.pone.0178998

Feizi, S. (2018). Corneal endothelial cell dysfunction: Etiologies and management. *Therapeutic Advances in Ophthalmology*. https://doi.org/10.1177/2515841418815802, 251584141881580.

Fields, M. A., Del Priore, L. V., Adelman, R. A., & Rizzolo, L. J. (2020). Interactions of the choroid, Bruch's membrane, retinal pigment epithelium, and neurosensory retina collaborate to form the outer blood-retinal-barrier. *Progress in Retinal and Eye Research, 76*, 100803. https://doi.org/10.1016/j.preteyeres.2019.100803

GeethaRamani, R., & Balasubramanian, L. (2018). Macula segmentation and fovea localization employing image processing and heuristic based clustering for automated retinal screening. *Computer Methods and Programs in Biomedicine, 160*, 153–163. https://doi.org/10.1016/j.cmpb.2018.03.020

Goel, R., Murthy, K. R., Srikanth, S. M., Pinto, S. M., Bhattacharjee, M., Kelkar, D. S., Madugundu, A. K., Dey, G., Mohan, S. S., Krishna, V., Prasad, T. S. K., Chakravarti, S., Harsha, H. C., & Pandey, A. (2013). Characterizing the normal proteome of human ciliary body. *Clinical Proteomics, 10*(1). https://doi.org/10.1186/1559-0275-10-9

Goodson, N. B., Nahreini, J., Randazzo, G., Uruena, A., Johnson, J. E., & Brzezinski, J. A. (2018). Prdm13 is required for Ebf3+ amacrine cell formation in the retina. *Developmental Biology, 434*(1), 149–163. https://doi.org/10.1016/j.ydbio.2017.12.003

Harper, A. R., & Summers, J. A. (2015). The dynamic sclera: Extracellular matrix remodeling in normal ocular growth and myopia development. *Experimental Eye Research, 133*, 100–111. https://doi.org/10.1016/j.exer.2014.07.015

Hejtmancik, J. F., & Shiels, A. (2015). Overview of the lens. In *Progress in molecular biology and translational science* (Vol. 134, pp. 119–127). Elsevier B.V. https://doi.org/10.1016/bs.pmbts.2015.04.006

Hendrickson, A., Bumsted-O'Brien, K., Natoli, R., Ramamurthy, V., Possin, D., & Provis, J. (2008). Rod photoreceptor differentiation in fetal and infant human retina. *Experimental Eye Research, 87*(5), 415–426. https://doi.org/10.1016/j.exer.2008.07.016

Hoshino, A., Ratnapriya, R., Brooks, M. J., Chaitankar, V., Wilken, M. S., Zhang, C., Starostik, M. R., Gieser, L., La Torre, A., Nishio, M., Bates, O., Walton, A., Bermingham-McDonogh, O., Glass, I. A., Wong, R. O. L., Swaroop, A., & Reh, T. A. (2017). Molecular anatomy of the developing human retina. *Developmental Cell, 43*(6), 763–779.e4. https://doi.org/10.1016/j.devcel.2017.10.029

Imesch, P. D., Wallow, I. H. L., & Albert, D. M. (1997). The color of the human eye: A review of morphologic correlates and of some conditions that affect iridial pigmentation. In *Survey of ophthalmology* (Vol. 41, pp. S117–S123). Elsevier USA. https://doi.org/10.1016/s0039-6257(97)80018-5. Issue 2.

Jayaram, H., & Calder, I. (2004). Eye and orbit. *Anaesthesia and Intensive Care Medicine, 5*(9), 300–301. https://doi.org/10.1383/anes.5.9.300.49895

Kels, B. D., Grzybowski, A., & Grant-Kels, J. M. (2015). Human ocular anatomy. *Clinics in Dermatology, 33*(2), 140–146. https://doi.org/10.1016/j.clindermatol.2014.10.006

Kolb, H. (1995). In H. Kolb, E. Fernandez, & R. Nelson (Eds.), *Gross anatomy of the eye. Webvision: The organization of the retina and visual system [internet]*. Salt Lake City (UT): University of Utah Health Sciences Center. Available from: https://www.ncbi.nlm.nih.gov/books/NBK11534/, 2007.

Labetoulle, M., Baudouin, C., Calonge, M., Merayo-Lloves, J., Boboridis, K. G., Akova, Y. A., Aragona, P., Geerling, G., Messmer, E. M., & Benítez-del-Castillo, J. (2019). Role of corneal nerves in ocular surface homeostasis and disease. *Acta Ophthalmologica, 97*(2), 137–145. https://doi.org/10.1111/aos.13844

Lagali, N. (2020). Corneal stromal regeneration: Current Status and future therapeutic potential. *Current Eye Research, 45*(3), 278–290. https://doi.org/10.1080/02713683.2019.1663874

Li, X., Liu, J., Hoh, J., & Liu, J. (2019). Müller cells in pathological retinal angiogenesis. *Translational Research, 207*, 96–106. https://doi.org/10.1016/j.trsl.2018.12.006

Malhotra, A., Minja, F. J., Crum, A., & Burrowes, D. (2011). Ocular anatomy and cross-sectional imaging of the eye. *Seminars in Ultrasound, CT and MRI, 32*(1), 2–13. https://doi.org/10.1053/j.sult.2010.10.009

May-Simera, H., Nagel-Wolfrum, K., & Wolfrum, U. (2017). Cilia - the sensory antennae in the eye. *Progress in Retinal and Eye Research, 60*, 144–180. https://doi.org/10.1016/j.preteyeres.2017.05.001

McGregor, J. E. (2019). Restoring vision at the fovea. *Current Opinion in Behavioral Sciences, 30*, 210–216. https://doi.org/10.1016/j.cobeha.2019.10.003

Miltner, A. M., & La Torre, A. (2019). Retinal ganglion cell replacement: Current status and challenges ahead. *Developmental Dynamics, 248*(1), 118–128. https://doi.org/10.1002/dvdy.24672

Napier, H. R. L., & Kidson, S. H. (2005). Proliferation and cell shape changes during ciliary body morphogenesis in the mouse. *Developmental Dynamics, 233*(1), 213–223. https://doi.org/10.1002/dvdy.20302

Nayak, K., & Misra, M. (2018). A review on recent drug delivery systems for posterior segment of eye. *Biomedicine and Pharmacotherapy, 107*, 1564–1582. https://doi.org/10.1016/j.biopha.2018.08.138

Nguyen, K. H., Patel, B., & Tadi, P. (2019). In *StatPearls [internet]*. Treasure Island (FL): StatPearls Publishing and A.f. Anatomy, Head and Neck, Eye Retina https://www.ncbi.nlm.nih.gov/books/NBK542332/.

Nickla, D. L., & Wallman, J. (2010). The multifunctional choroid. *Progress in Retinal and Eye Research, 29*(2), 144–168. https://doi.org/10.1016/j.preteyeres.2009.12.002

Osei-Bempong, C., Figueiredo, F. C., & Lako, M. (2013). The limbal epithelium of the eye - a review of limbal stem cell biology, disease and treatment. *BioEssays, 35*(3), 211–219. https://doi.org/10.1002/bies.201200086

Pašta, J. (2013). Laser therapy in ophthalmology. In *Lasers for medical applications: Diagnostics, therapy and surgery* (pp. 395–458). Elsevier Ltd. https://doi.org/10.1533/9780857097545.4.395

Pigaiani, N., Bertaso, A., De Palo, E. F., Bortolotti, F., & Tagliaro, F. (2020). Vitreous humor endogenous compounds analysis for post-mortem forensic investigation. *Forensic Science International, 310*, 110235. https://doi.org/10.1016/j.forsciint.2020.110235

Poleg-Polsky, A., Ding, H., & Diamond, J. S. (2018). Functional compartmentalization within starburst amacrine cell dendrites in the retina. *Cell Reports, 22*(11), 2898–2908. https://doi.org/10.1016/j.celrep.2018.02.064

Poletti, M., Listorti, C., & Rucci, M. (2013). Microscopic eye movements compensate for nonhomogeneous vision within the fovea. *Current Biology, 23*(17), 1691–1695. https://doi.org/10.1016/j.cub.2013.07.007

Quinn, N., Csincsik, L., Flynn, E., Curcio, C. A., Kiss, S., Sadda, S. V. R., Hogg, R., Peto, T., & Lengyel, I. (2019). The clinical relevance of visualising the peripheral retina. *Progress in Retinal and Eye Research, 68*, 83–109. https://doi.org/10.1016/j.preteyeres.2018.10.001

Ramirez, J. M., Salazar, J. J., de, R., Rojas, B., Gallego, B. I., Ramrez, A. I., & Trivio, A. (2012). *Choroidal vessel wall: Hypercholesterolaemia-induced dysfunction and potential role of statins*. https://doi.org/10.5772/47794

Ray, N. J. (2015). Biophysical chemistry of the ageing eye lens. *Biophysical Reviews,* 7(4), 353–368. https://doi.org/10.1007/s12551-015-0176-4

Reichenbach, A., & Bringmann, A. (2013). New functions of Müller cells. *Glia, 61*(5), 651–678. https://doi.org/10.1002/glia.22477

Saikia, P., Medeiros, C. S., Thangavadivel, S., & Wilson, S. E. (2018). Basement membranes in the cornea and other organs that commonly develop fibrosis. *Cell and Tissue Research, 374*(3), 439–453. https://doi.org/10.1007/s00441-018-2934-7

Sezer, T., Altınısık, M., Koytak, İ. A., & Özdemir, M. H. (2016). The choroid and optical coherence tomography. *Turk Oftalmoloiji Dergisi, 46*(1), 30–37. https://doi.org/10.4274/tjo.10693

Sharma, B., Dubey, A., Prakash, G., & Vajpayee, R. B. (2018). Bowman's layer transplantation: Evidence to date. *Clinical Ophthalmology, 12,* 433–437. https://doi.org/10.2147/OPTH.S141127

Shi, M., Zhang, L., Ye, E. A., Wang, A., Li, G., & Chen, L. (2020). Aqueous humor induces lymphatic regression on the ocular surface. *Ocular Surface, 18*(3), 505–510. https://doi.org/10.1016/j.jtos.2020.03.002

Stamer, W. D., & Clark, A. F. (2017). The many faces of the trabecular meshwork cell. *Experimental Eye Research, 158,* 112–123. https://doi.org/10.1016/j.exer.2016.07.009

Sung, C. H., & Chuang, J. Z. (2010). The cell biology of vision. *Journal of Cell Biology, 190*(6), 953–963. https://doi.org/10.1083/jcb.201006020

Suri, R., Beg, S., & Kohli, K. (2020). Target strategies for drug delivery bypassing ocular barriers. *Journal of Drug Delivery Science and Technology, 55,* 101389. https://doi.org/10.1016/j.jddst.2019.101389

Swamynathan, S. K., & Wells, A. (2020). Conjunctival goblet cells: Ocular surface functions, disorders that affect them, and the potential for their regeneration. *Ocular Surface, 18*(1), 19–26. https://doi.org/10.1016/j.jtos.2019.11.005

Tserevelakis, G. J., Avtzi, S., Tsilimbaris, M. K., & Zacharakis, G. (2017). Delineating the anatomy of the ciliary body using hybrid optical and photoacoustic imaging. *Journal of Biomedical Optics, 22*(6). https://doi.org/10.1117/1.JBO.22.6.060501

Vinberg, F., Chen, J., & Kefalov, V. J. (2018). Regulation of calcium homeostasis in the outer segments of rod and cone photoreceptors. *Progress in Retinal and Eye Research, 67,* 87–101. https://doi.org/10.1016/j.preteyeres.2018.06.001

Wang, S., Ghezzi, C. E., Gomes, R., Pollard, R. E., Funderburgh, J. L., & Kaplan, D. L. (2017). In vitro 3D corneal tissue model with epithelium, stroma, and innervation. *Biomaterials, 112,* 1–9. https://doi.org/10.1016/j.biomaterials.2016.09.030

Williams, R., Lace, R., Kennedy, S., Doherty, K., & Levis, H. (2018). Biomaterials for regenerative medicine approaches for the anterior segment of the eye. *Advanced Healthcare Materials,* 7(10), 1701328. https://doi.org/10.1002/adhm.201701328

Willoughby, C. E., Ponzin, D., Ferrari, S., Lobo, A., Landau, K., & Omidi, Y. (2010). Anatomy and physiology of the human eye: Effects of mucopolysaccharidoses disease on structure and function - a review. *Clinical and Experimental Ophthalmology, 38*(1), 2–11. https://doi.org/10.1111/j.1442-9071.2010.02363.x

Xie, Y., Wang, M., Cong, Y., Cheng, M., Wang, S., & Wang, G. (2018). The pilocarpine-induced ciliary body contraction affects the elastic modulus and collagen of cornea and sclera in early development. *Biomedicine and Pharmacotherapy, 108,* 1816–1824. https://doi.org/10.1016/j.biopha.2018.10.032

Zavala, J., López Jaime, G. R., Rodríguez Barrientos, C. A., & Valdez-Garcia, J. (2013). Corneal endothelium: Developmental strategies for regeneration. *Eye, 27*(5), 579–588. https://doi.org/10.1038/eye.2013.15

Zhang, L., Anderson, M. C., & Liu, C.-Y. (2017). The role of corneal stroma: A potential nutritional source for the cornea. *Journal of Nature and Science, 3*(8).

CHAPTER 5

Microneedle-assisted drug delivery to the eye

Introduction

According to the World Health Organization, more than 285 million patients suffer from eye diseases globally, while 39 million people are completely blind (Suri et al., 2020). This number will rise in the next 10–20 years (Suri et al., 2020). There are over 3.3 million Americans aged 40 years and above who are either have low vision (having best-corrected visual acuity less than 6/12 (<20/40) in the better-seeing eye or are legally blind (having best-corrected visual acuity of 6/60 or worse (=20/200) in the better-seeing eye) not counting those who are classified as being blind) ("Https://Www.Cdc.Gov/Visionhealth/Basics/Ced/Index.Html," 2021). Several disorders impacting the posterior segment of the eye can result in irreversible visual damage (Agban et al., 2019). Diseases that affect the posterior segment of the human eye and account for most cases of vision loss globally include age-related macular degeneration (AMD), retinitis pigmentosa (RP), and diabetic retinopathy (DR) (Bhatnagar et al., 2018). Sustained wearing of contact lens, diabetes mellitus, and upswing in aging population are linked to the rising prevalence of eye diseases (Than et al., 2018). From the standpoint of drug delivery, the eyeball can be divided into two portions: (a) the anterior segment which is made up of the aqueous humor, trabecular meshwork, cornea, lens, and the iris (b) the posterior segment that include the macula, optic nerve, vitreous humor, sclera, retina and the choroid (Lee et al., 2020). Anterior uveitis, conjunctivitis, and cataract are categorized as anterior segment diseases include while diseases of the posterior segment include age-related macular degeneration, retinitis pigmentosa, and diabetic retinopathy (Gupta & Yadav, 2019). Glaucoma is known to target several tissues located in both the posterior (e.g., optic nerve head) and anterior (e.g., trabecular meshwork) segments of the human eye (Izzotti et al., 2015).

Microneedles
ISBN 978-0-323-91811-4
https://doi.org/10.1016/B978-0-323-91811-4.00011-0

Ocular drug delivery is complicated because of the existence of continuous eye drainage and several tissue barriers (Park et al., 2016). The sclera and cornea (dense tissue layers) restrict drug transport into the interior regions of the human eye (Park et al., 2016; Thakur et al., 2021). Eye drops are commonly used for the management of pathological conditions that affect the anterior segment of the eye, including glaucoma and corneal diseases (Park et al., 2018). Nevertheless, topical administration of eye drops can lead to low bioavailability (1%—7%) and poor drug transport across the anterior chamber (Chiang et al., 2018). Furthermore, it is important to administer eye drops topically every hour for several days with the objective of managing infectious diseases (an example is keratitis) in the cornea (Lee et al., 2018). Poor compliance is frequently observed with such drug administration schedules which can in turn cause perpetual scar formation in the cornea (Lee et al., 2018). Furthermore, this regimen is not convenient and topical eye drops are less useful for the therapy of diseases affecting the posterior segment of the human eye (Chiang et al., 2018).

The periocular route (peribulbar, retrobulbar, posterior juxtascleral, subtenon and subconjunctival) can be utilized for the delivery of medications to the posterior region of the human eye (Agban et al., 2019; Nayak and Misra, 2018). This approach entails drug administration in the form of a depot into spaces and tissues that enclose the eye, from where they can slowly be transported across the sclera into posterior segment tissues (Agban et al., 2019). As an alternative approach, a more invasive procedure including the intravitreal drug administration can be used for patients with retinal diseases (Park et al., 2018). Approximately 6 million injections were administered intravitreally in the United States in 2016 (Hartman & Kompella, 2018). Interestingly, the number of intravitreal injections is expected to rise with the introduction of new therapeutic agents (Hartman & Kompella, 2018). Conventional drug administration into the eye has several limitations: repeated drug administration is required and this can result in patient non-compliance (Gupta & Yadav, 2019). Another drawback of the traditional approach is the low efficiency and poor selectivity of pharmacotherapies for many vision-threatening diseases (Kim et al., 2015). This can result in undesirable side effects thus limiting treatment efficacy (Kim et al., 2015) The fabrication of novel ocular drug delivery systems is challenging because the eye is a delicate organ (Gupta & Yadav, 2019).

At the moment, various approaches such as intravitreal as well as intrascleral injections, topical eye drops and oral tablets are utilized for the management of eye disorders (Gupta & Yadav, 2019). Even though contact lenses and biodegradable ocular inserts are being designed for corneal drug delivery, inefficiency and invasiveness of these approaches need to be addressed (Lee et al., 2018). Injections are efficient since they are capable of delivering the appropriate amount of medication to the relevant region of the eye (Gupta & Yadav, 2019). Traditional hypodermic needles are utilized for intraocular injections (Gupta & Yadav, 2019). There are different routes of administering these intraocular injections: intracorneal, intravitreal, periocular and subconjunctival (Gupta & Yadav, 2019). Even though intravitreally-administered drugs have improved the clinical outcomes of retinal disorders like AMD and DR, it is troubling that intravitreal injections can cause side effects (Park et al., 2018). Intravitreal injections are invasive and the repeated administration can result in non-compliance and reduced bioavailability (Gupta & Yadav, 2019). For instance, intravitreal injection can lead to retinal detachment, endophthalmitis, and ocular bleeding (Park et al., 2018). Furthermore, most ocular medications do not act in the vitreous humor or on the corneal surface (Kim et al., 2015).

Adverse reactions have been reported in 8%–53% of glaucoma patients that received locally administered anti-glaucoma agents (Kim et al., 2015). In addition, undesirable contact with eye lens aided cataract formation in 7–14% of patients receiving steroids intravitreally (Kim et al., 2015). Glaucoma and AMD are the two major risk factors for sightlessness in industrialized countries (Kim et al., 2015). Most drugs against AMD act on the choroid and/or retina while medications that are used for the management of glaucoma act on the trabecular meshwork or the ciliary body (Kim et al., 2015). Successful management of eye diseases depends not only on the development of new therapeutic agents but also on effective methods of administration to achieve desired concentrations at the target tissue (Hartman & Kompella, 2018). Over the last 15 years, researchers have been aggressively exploring new means for delivering drugs, cells, implants, and other fluids to treat various ocular diseases and prevent blindness (Hartman & Kompella, 2018). Although the initial thought of an injection into the eye sends chills running up and down the spine for most people, especially those with normal vision, most successful and promising clinical trials and experimental studies have focused on testing their drug product using three different eye injection techniques, intravitreal (IVT), subretinal (SR), and suprachoroidal (SC) (Hartman & Kompella, 2018).

Microneedle-mediated ocular drug delivery

Fig. 5.1 shows a diagram of a primate eye showing different modes of adeno-associated viral (AAV) delivery, including intravitreal, subretinal (transretinal), and suprachoroidal or subretinal (Yiu et al., 2020). Yiu et al. used transscleral microneedles to inject AAV serotype 8 (AAV8) into the

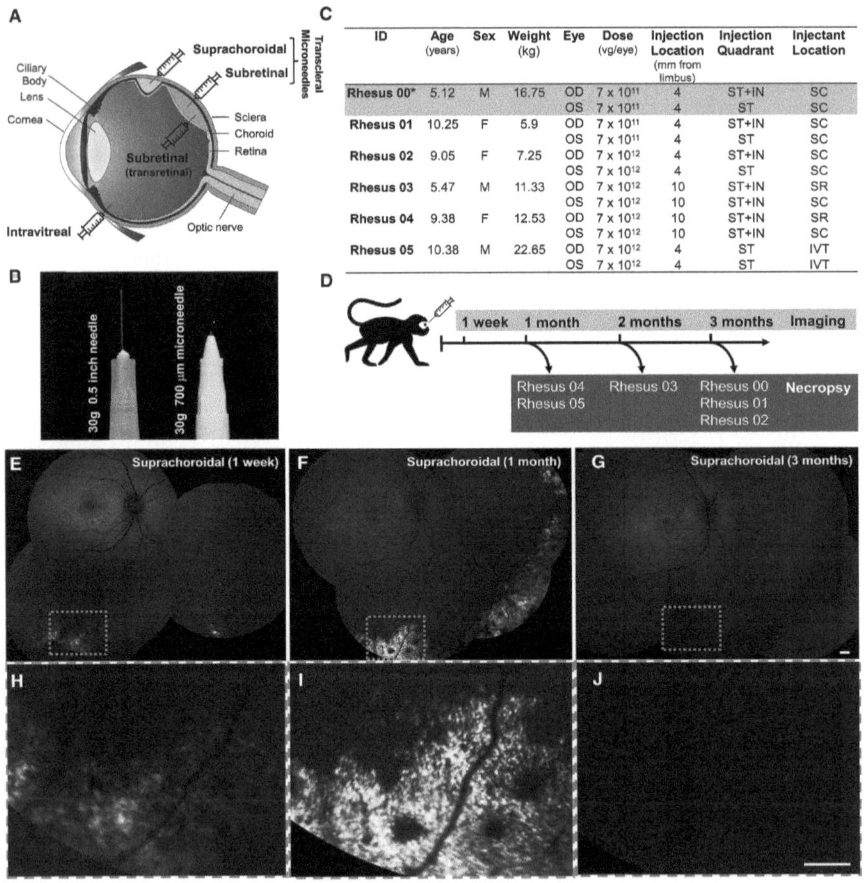

ID	Age (years)	Sex	Weight (kg)	Eye	Dose (vg/eye)	Injection Location (mm from limbus)	Injection Quadrant	Injectant Location
Rhesus 00*	5.12	M	16.75	OD	7×10^{11}	4	ST+IN	SC
				OS	7×10^{11}	4	ST	SC
Rhesus 01	10.25	F	5.9	OD	7×10^{11}	4	ST+IN	SC
				OS	7×10^{11}	4	ST	SC
Rhesus 02	9.05	F	7.25	OD	7×10^{12}	4	ST+IN	SC
				OS	7×10^{12}	4	ST	SC
Rhesus 03	5.47	M	11.33	OD	7×10^{12}	10	ST+IN	SR
				OS	7×10^{12}	10	ST+IN	SC
Rhesus 04	9.38	F	12.53	OD	7×10^{12}	10	ST+IN	SR
				OS	7×10^{12}	10	ST+IN	SC
Rhesus 05	10.38	M	22.65	OD	7×10^{12}	4	ST	IVT
				OS	7×10^{12}	4	ST	IVT

Fig. 5.1 Experimental design and in vivo GFP transgene expression in rhesus macaque eyes after suprachoroidal injection of AAV8 (A) Diagram of a primate eye showing different modes of AAV delivery, including intravitreal, subretinal (transretinal), and suprachoroidal or subretinal injections using transscleral microneedles. (B) Photograph comparing a standard 30-gauge needle with the transscleral microneedle used in this study. (C) Table outlining study animals and eyes, demographics, AAV8 dose, location and quadrant of injections, and the anatomic location of the injectant. *Rhesus 00 was found to have pre-existing AAV-neutralizing antibodies and was subsequently excluded. (D) Schematic of time points for ocular imaging and necropsy of study animals. (E–G) Scanning laser ophthalmoscopy (SLO) montages and (H–J) magnified

subretinal or suprachoroidal space in rhesus macaques (Yiu et al., 2020). The authors showed that microneedle-assisted transscleral subretinal injection of AAV8 resulted in minimal intraocular inflammation as well as focal gene expression with transduction of photoreceptors and retinal pigment epithelium (Yiu et al., 2020). This method of transscleral viral delivery can be carried out without vitreoretinal surgery, resulting in diffuse or focal transgene expression (Yiu et al., 2020).

Several investigators have used microneedles for the delivery of medications to different regions of the eye. Bhatnagar et al. prepared dissolving microneedles and used them for the rapid delivery of besifloxacin across the cornea (Bhatnagar et al., 2018). The MNs were produced from polyvinylpyrrolidone and polyvinyl alcohol via the micromolding method (Bhatnagar et al., 2018). Besifloxacin, was incorporated into a 6×6 microneedle array containing 103.4 ± 8.5 μg of the drug (Bhatnagar et al., 2018). The MNs pierced the cornea up to a depth of 200 μm and fully dissolved in the tissue within 5 min. The administration of microneedles markedly enhanced besifloxacin transport and deposition into the cornea in comparison with the free drug solution (Bhatnagar et al., 2018). Furthermore, besifloxacin-loaded microneedles exhibited higher antibacterial activity in *Staphylococcus aureus*-infected cornea compared to free drug solution (Bhatnagar et al., 2018).

Microneedles (MNs) can be used for minimally invasive drug delivery to different ocular tissues (Park et al., 2018). Biodegradable MNs are beneficial compared to other types of MNs including solid, coated and hollow particularly when sustained drug administration is required (Lee et al., 2018). Among different approaches for delivering drugs using MNs (Lee et al., 2018). The "poke with patch" (solid MNs) approach is not optimal for the cornea because the corneal surface is surrounded by the tear fluid that is subject to continual turnover (Lee et al., 2018). Biodegradable microneedles can pierce the epithelium of the cornea and enter the corneal

views of the *yellow-dashed boxes* [light gray dashed boxes in print version] in (E–G) showing GFP expression at 1 week (E and H), 1 month (F and I), and 3 months (G and J) after AAV injections. Scale bars, 1 mm. *F*, female; *IN*, inferonasal; *IVT*, intravitreal; *M*, male; *OD*, right eye; *OS*, left eye; *SC*, suprachoroidal; *SR*, subretinal; *ST*, superotemporal; *vg*, viral genomes. *(From Yiu, G., Chung, S. H., Mollhoff, I. N., Nguyen, U. T., Thomasy, S. M., Yoo, J., Taraborelli, D., & Noronha, G. (2020). Suprachoroidal and subretinal injections of AAV using transscleral microneedles for retinal gene delivery in nonhuman primates.* Molecular Therapy - Methods and Clinical Development, 16, *179–191. https://doi.org/10.1016/j.omtm.2020.01.002).*

stroma by overcoming barriers formed by the corneal epithelium and the washout by the tear film (Lee et al., 2018). Moreover, localized delivery can be obtained via MNs (Lee et al., 2018). Microneedles can also be utilized for suprachoroidal drug transport (Lee et al., 2018). This can be achieved by using transscleral microneedles that possess adequate length to cross the sclera without piercing the retina (Hartman & Kompella, 2018). When compared to traditional hypodermic needles, microneedles offer reliable and less invasive drug administration into the SCS (Hartman & Kompella, 2018). A hollow microneedle comparable in thickness to the sclera and conjunctiva can be chosen so that the microneedle does not pierce the choroid and retina thus avoiding unintended intravitreal injection (Chiang et al., 2018). Precise positioning of the medication in the SCS can potentially curtail damage to the underlying retinal strata (Hartman & Kompella, 2018). The microneedle is usually placed perpendicular to the hard stop at the hub at the microneedles' base and the scleral surface (Chiang et al., 2018). This is done for error-free control of microneedle insertion depth (Chiang et al., 2018). The microneedle is frequently inserted for about 1 min to minimize vitreal reflux (Chiang et al., 2018). Suprachoroidal and intra-scleral administration of MNs depend upon either partial penetration or full insertion of a MN through scleral tissue (Park et al., 2016). The sclera, an acellular tissue of collagen fibers ingrained in a matrix of glycosaminoglycan, occupies a significant proportion of the ocular surface (Park et al., 2016).

Minimally invasive permeation of the scleral tissue by MNs is important for therapeutically useful ocular drug delivery (Park et al., 2016). It is also critical to evaluate the energy and force needed to puncture the sclera (Park et al., 2016). This is useful for the design of MNs with the required mechanical strength (Park et al., 2016). Fracture toughness and puncture resistance of the sclera was recently measured via a double MN insertion method (Park et al., 2016). Fracture toughness refers to the ability to prevent crack propagation of a material (Zhang et al., 2021). J-integral is frequently utilized as a fracture toughness index (Zhang et al., 2021). It refers to a path-independent integral around any closed contour that encompasses the crack tip and this integral is used for the quantitative description of the average strength of the stress and strain field under elastoplastic or elastic conditions (Zhang et al., 2021):

$$J = \int_{\Gamma} W dy - \vec{T} \frac{\partial \vec{u}}{\partial x} \partial s \tag{5.1}$$

where W is the total strain energy density, Γ is any counterclockwise path around the crack tip, $T\rightarrow$

Fracture toughness can also be computed from Eq. (5.2), (Zhang et al., 2021):

$$J_{IC} = \int \frac{(F_1 - F_2)du}{\int dA} \tag{5.2}$$

where J_{IC} is fracture toughness.

Puncture resistance denotes the maximum force required for a puncture probe to penetrate through a specimen clamped between two plates in a sample holder (Thang Nguyen et al., 2005). It refers to the amount of energy needed to create a new fracture area (Park et al., 2016). Puncture resistance is normally expressed either as the energy, i.e. the area under the load–deformation curve up to the point of puncture or the load at puncture (Lange et al., 2002). Puncture resistance factor K is the product of the force at break F and the work to break W (Lange et al., 2002):

$$K = F \times W \tag{5.3}$$

To mimic the conditions of microneedle insertion into the sclera, force displacement curves derived from microneedle -insertion tests were utilized for the computation of fracture toughness and puncture resistance of the tissue (Park et al., 2016). The reliance of mechanical properties on the pre-strain of the sclera, insertion speeds, and microneedle sizes were then evaluated for a better understanding of the influence of the insertion conditions (Park et al., 2016). The utilization of microneedle was beneficial because of greater sensitivity to sub-tissue structure during the measurement as well as the minimal invasiveness of the approach (Park et al., 2016).

The vitreous humor functions as a natural reservoir, slowly releasing a medication for over one month, depending on the employed formulation (Chiang et al., 2018). Nevertheless, the therapeutic agent can also diffuse isotropically through the vitreous humor toward non-target areas of the eye (e.g., the ciliary body and the lens), leading to adverse events (Chiang et al., 2018). In addition, drug penetration from the vitreous to the retinal pigment epithelium and the choroid can also be hampered by an intact blood-retinal barrier (Chiang et al., 2018). Researchers can deliver medications to ophthalmic tissues that are situated above the ciliary body anteriorly, the choroid and close to the retina posteriorly by injecting medications into the suprachoroidal space (SCS) (Kim et al., 2015). The SCS, which is found between the choroid and the sclera, can extend to hold a drug formulation or a fluid (Kim et al., 2015). The SCS occupies the region between the scleral wall of the eye and the choroidal vasculature (Yiu et al., 2020). In the human eye, this space is not conspicuous except in cases when suprachoroidal fluid accumulation is detected via ultrasonography (Yiu et al., 2020). The SCS maintains intraocular pressure (IOP) through the uveoscleral outflow (Chiang et al., 2018). The position of the SCS next to the sites of drug action for diseases such as glaucoma (ciliary body), diabetic retinopathy, wet AMD and uveitis (retina and/or choroid) can facilitate enhanced medication levels in these tissues (Kim et al., 2015). Suprachoroidal drug delivery requires placement of drug product directly beneath the sclera and above the choroid in the virtual SCS (Hartman & Kompella, 2018). With the aim of not piercing the choroidal vasculature or underlying structures in any significant manner, needles used for this purpose are ideally of a length comparable to scleral thickness at less than 1 mm (Hartman & Kompella, 2018). Given this necessity, needle lengths have decreased to a micron size, that is, below 1 mm (Hartman & Kompella, 2018). Due to the feasibility of fabricating microneedles as well as nanoneedles and the success of intravitreal injections, researchers are increasingly using the suprachoroidal route for drug delivery (Kompella et al., 2020).

Kim et al. used microneedles for targeted and minimally invasive delivery of microparticles and nanoparticles into the suprachoroidal space (Kim et al., 2015). The researchers inserted single hollow MNs into the sclera of rabbits and administered non-biodegradable fluorescently tagged microparticles and nanoparticles that were suspended and delivered into the ocular SCS (Kim et al., 2015). The MNs were prepared from 33-gauge needle cannulas which were shortened to about 750 μm. Laser beam was used to shape the bevel of the needles (Kim et al., 2015). The MNs were electropolished, cleaned with deionized water and attached to glass syringes (Kim et al., 2015). Saline formulations of the particles were distributed over 29%−42% of the suprachoroidal space (SCS) (Kim et al., 2015) Hyaluronic acid was added by the authors to increase particle spread in the SCS (Kim et al., 2015). The authors were able to achieve targeted particle delivery within the suprachoroidal space by employing polymeric drug delivery systems that localized the particles next to the ciliary body (strong non-Newtonian dispersions) or spread them over the entire choroidal surface (moderate non-Newtonian dispersions) (Kim et al., 2015).

In another interesting set of experiments, Park and colleagues investigated the feasibility of using precisely-controlled microneedle pen (MNP) for the delivery of medications into the sclera at different depths

(Park et al., 2018). The authors also examined how distinct depths of insertion can influence drug delivery into the sclera and the posterior region of the eye (Park et al., 2018). Microneedle pen for drug transport into the sclera was produced by using an MN together with an applicator made up of a slot, spring, trigger, stopper and an end tip (Park et al., 2018). The authors chose the spring load in the MNP for impact insertion into the scleral tissue. To evaluate insertion depths, three different spring types (with spring constants of 360.2 ± 10.3, 73.5 ± 5.1, and 1561.4 ± 10.4 N/m) were utilized (Park et al., 2018). The authors linked the end tip (700 μm diameter) to a trigger in a manner that enabled them to slide back and forth together (Park et al., 2018). The spring was positioned in the slot of the MNP body and the MN was shot by unloading the trigger. The speed and force of MN insertion into the sclera were controlled via the spring constant (Park et al., 2018). By controlling insertion speeds, the authors used the microneedle pen to achieve depth-controlled scleral delivery, and the utility of this approach was confirmed in ex vivo experiments (Park et al., 2018). It was revealed that the MNP system could be used for the scleral drug delivery to beagle eyes (Park et al., 2018).

Lee et al. fabricated biodegradable and detachable MN that can be utilized for sustained drug administration to the interior region of the cornea (Lee et al., 2018). The MN represents a hybrid MN comprising a biodegradable tip loaded with medication and a supporting base (Lee et al., 2018). The researchers administered the hybrid MN to the cornea via impact insertion, leaving only the biodegradable tip loaded with the drug within the cornea so that the medication can be released within a certain time (Lee et al., 2018). The detachability of the MN tips was optimized and their dimension tightly controlled via concentration-controlled molding (Lee et al., 2018). The supporting base and detachable tip were put together via pressure-assisted transfer molding to create a hybrid MN. The authors carefully optimized the injection dwell time, the MN tip dimension, and the insertion depth to obtain an effective intracorneal injection of the drug-loaded tip (Lee et al., 2018). The authors used the detachable hybrid MN in an *Acanthamoeba keratitis* model in which the drug-tip was effectively delivered to the inside of the mouse cornea in vivo (Lee et al., 2018). The therapeutic efficacy of the detachable biodegradable MN tips was confirmed in the MN treated cases over 7 days (Lee et al., 2018).

It is known that *Fusarium* spp., *Aspergillus spp., and Candida* spp. can cause keratitis fungal infection of the cornea (Mahmoudi et al., 2018). Voriconazole, ketoconazole, miconazole, itraconazole, fluconazole, econazole, and amphotericin B are the commonly used anti-fungal agents for the management of this disease (Roy et al., 2019). These anti-fungal drugs possess poor aqueous solubility and display low partitioning within the corneal mucosal layer (Roy et al., 2019). Significantly, eye drops get diluted and are rapidly cleared from the corneal surface via lacrimal drainage and secretions. This leads to poor corneal availability of these medications (Roy et al., 2019). Recently, Roy et al. studied the influence of microneedles on the availability of Amphotericin B (AmB) within the anterior chamber and the corneal stroma (Roy et al., 2019). The authors developed a AmB-loaded microneedle ocular patch (MOP) (Roy et al., 2019). The MOP design modeled the curvature of commonly marketed contact lens. The authors fabricated the MOP via a layered casting approach by utilizing a dissolvable polymer composite (Roy et al., 2019). An inverse polydimethylsiloxane mold was made from a stainless-steel template. The MN array consisting of 5×5 pyramidal needles measuring 255 ± 2.23 μm in width and 552 ± 1.64 μm in length were positioned at the center of stainless-steel cavities. Dissolving MOP was fabricated from a combination of PVP and PVA. The authors dissolved PVP (15% w/v) and PVA (15% w/v) in de-ionized water at 60°C (Roy et al., 2019). The mixture was then placed into the PDMS mold and air dried for 24 h. The MOP was subsequently kept in a desiccator until it was needed for more experiments (Roy et al., 2019). The AmB—loaded MOP had a compression strength of 35.1 ± 6.7 N and needed an insertional force of 1.07 ± 0.17 N into excised human cornea (Roy et al., 2019). There was remarkable increase in the corneal retention of AmB following the use of MOP in comparison with free AmB or liposomal AmB (Roy et al., 2019). Significantly, the use of AmB loaded MOP led to a reduction in fungal load (*Candida albicans*) within the cornea as observed in both in-vivo rabbit infection model and ex-vivo human cornea model (Roy et al., 2019).

Than and coworkers developed an eye patch featuring detachable microneedles (Than et al., 2018). The patch with double-layered MNs was formulated via a simple micromolding method (Than et al., 2018). To prepare these eye patches, centrifugation was used to cast crosslinked methacrylated hyaluronic acid (MeHA) aqueous solution into plasma treated polydimethylsiloxane micromolds (Than et al., 2018). Following air-drying at room temperature, un-modified HA solution was applied and centrifuged to fill the holes (Than et al., 2018). The researchers then dried the molds for an additional 12 h and added HA solution to create a supporting substrate followed by overnight drying (Than et al., 2018). The

MN patches were then gently detached from the molds, and subjected to low-intensity ultraviolet radiation for 3 min (Than et al., 2018). The MNs consist of an outer MeHA layer and a HA inner core. The MNs can pierce the wet cornea surface because the dissolving HA is covered by MeHA (Than et al., 2018). It was revealed that the microneedles can function as ingrained micro-depots for controlled drug transport (Than et al., 2018). Significantly, the biphasic drug release profile engendered by the micro-reservoirs can enhance therapeutic efficacy. The authors employed corneal neovascularization disease model to demonstrate that the MN-assisted transport of an anti-angiogenic monoclonal antibody (DC101) led to approximately 90% decrease of the neovascular area (Than et al., 2018). Moreover, fast release of diclofenac followed by a prolonged release of DC101 resulted in a synergistic effect (Than et al., 2018).

Macular edema, AMD, and uveitis are the key risk factors for blindness in developed countries (Jung et al., 2019). Medications can be administered into the suprachoroidal space for enhanced targeting of tissues in the posterior ocular segment (Jung et al., 2019). The delivery of microparticle increased from 20% to 45% and posterior drug targeting improved when 1 μm latex microparticles loaded with collagenase were administered into the SCS of rabbit eyes ex vivo using microneedles (Jung et al., 2019). The authors optimized collagenase concentration for decreased tissue damage and enhanced transport into the posterior region. The effect of collagenase injection within SCS went up and stabilized 4 h following administration (Jung et al., 2019). Concurrent administration of microparticles and collagenase resulted in a higher effect and enhanced delivery in the SCS in comparison to sequential injection (Jung et al., 2019). The injection of collagenase into the SCS of rabbit eyes in vivo led to transiently lowered intraocular pressure, with no apparent adverse events on the choroid, sclera, and retina (Jung et al., 2019).

The use of MNs for drug delivery to the sclera or cornea is complicated due to the absence of supporting pressure that is needed for dependable and predictable MN insertion (Song et al., 2015). Another obstacle is the round ocular surface (Song et al., 2015). The administration of MNs to a specific site on a small ocular tissue (less than a few millimeters in size) is a key impediment when small animals such as rats or mice are used for in vivo microneedle-mediated studies (Song et al., 2015). Traditional sheet-shaped MNs, roller, patch or band are not useful with minute and curved eye tissues (Song et al., 2015). It is therefore important to design microneedles and applicators appropriate for ocular tissues of small animals (Song et al., 2015).

Song et al. developed a "transfer-molding" technology that repetitively produces single or an array of MNs placed on a small rod (Song et al., 2015). This rod is subsequently introduced into a miniature target area of the ocular tissue (Song et al., 2015). A spring-loaded microneedle applicator system, MN pen (MNP), was also custom-manufactured to improve MN insertion reliability (Song et al., 2015). The authors applied the microneedle pens to the cornea. The MNPs were used for the successful delivery of rhodamine dye to the stromal layer of the cornea via a minute entry area of only approximately 1000 μm^2 (Song et al., 2015). In comparison with 30G needle tips, MNPs facilitated minimally invasive and localized rhodamine transport devoid of perforation (Song et al., 2015). The authors conducted MNP-assisted delivery of sunitinib malate to the cornea in a suture-induced angiogenesis model (Song et al., 2015). In comparison with transport with a 30G needle coated with sunitinib malate, only drug administration via MNPs efficiently blocked corneal neovascularization in vivo (Song et al., 2015). Sunitinib malate is typically used as an antagonist of in vitro angiogenesis (Gu et al., 2020).

Retinal vein cannulation is a complex surgical procedure which requires the drug injection into the retinal veins for the purpose of managing occlusions (Gonenc et al., 2017). Successful vein cannulation is predicated upon the identification of the time of venous puncture, micro-vessel cannulation, and the maintenance of cannulation during drug delivery (Gonenc et al., 2017). Sensing of tool-tissue interaction forces as well as advances in medical robotics can be used to overcome some of these problems (Gonenc et al., 2017). Furthermore, researchers can utilize these techniques to reduce surgeon effort, decrease the number of complications, avert tissue trauma and ultimately facilitate successful cannulation of the retinal vein (Gonenc et al., 2017). A system that couples a force-sensing microneedle to a handheld micromanipulator ("Micron") was recently developed (Gonenc et al., 2017). The authors fabricated a motorized force-sensing microneedle (MN) with a tip outer diameter of 70 μm (Gonenc et al., 2017) an incorporated the MN array into a lightweight modular unit equipped with relevant sensors (an encoder and fiber Bragg gratings) and a linear motor capable of being actuated independently (Gonenc et al., 2017). Diverse tool handles (robotic or manual) can be integrated with no interference with their operation. The typical tool has a bent (45° relative to the shaft of the tool) and sharp beveled (15°) tip (Gonenc et al., 2017). The researchers used this system to assess two different techniques of precisely quantifying the moment of venous puncture (Gonenc et al., 2017). The instance of venous puncture is a function of the tracked position of the needle tip as well as measuring tool-tissue interaction forces (Gonenc et al., 2017). The novel control technique actively compensates inadvertent motions of the operator apart from the additional tremor canceling function. Furthermore, the cannulation device is kept securely inside the vein after cannulation (Gonenc et al., 2017). Puncture detection algorithm and active positive holding component can result in prolonged cannulation which is most conspicuous in the more minute veins (Gonenc et al., 2017).

Thakur and colleagues fabricated MNs from polyvinylpyrrolidone of distinct molecular weights (Park et al., 2016; Thakur et al., 2021). The MNs were loaded with fluorescein sodium, fluorescein isothiocyanate–dextran (MW of 70 kDa) and fluorescein isothiocyanate–dextran (150 kDa) (Thakur et al., 2016). The MNs were prepared and their in vitro permeation, dissolution time, penetration depth (using confocal imaging and OCT), insertion and fracture forces (in the cornea and sclera), evaluated (Thakur et al., 2016). Laser-engineered silicon micromolds comprising 9 (3 × 3) conical MN, with a base width of 300 μm, an interspacing of 50 μm and an average height of 800 μm, were employed for the preparation of the microneedle arrays (Thakur et al., 2016). PVP (with a drug) and hydrogel formulation of PVP (without drug) was placed in the silicon molds and centrifuged at 5000 × g for 5 min to create the MNs shafts. The authors left the MNs to settle and produce a highly viscous gel (Thakur et al., 2016). Subsequently, PVP K90 gel was added and centrifuged again at 5000 × g for 5 min to create the MN baseplate. The MN arrays were subsequently detached from the molds and evaluated visually for formulation homogeneity and mechanical strength (Thakur et al., 2016). The average drug mass of the MNs (in the MN shafts) was from 0.96 to 9.91 μg. High MW PVP was used to fabricate MNs capable of withstanding higher forces with slight decrease in the height of the needle (Thakur et al., 2016). PVP MNs exhibited accelerated dissolution from 10 to 180 s, which was a function of PVP's molecular weight (Thakur et al., 2016). When MNs were used, there was significant increase in macromolecule transport across the corneal and scleral tissues in vitro compared to topically applied aqueous solutions (Thakur et al., 2016). Confocal microscopy proved that the macromolecules formed tissue depots which resulted in prolonged penetration. Nevertheless, the utilization of MNs did not substantially increase the transport of small molecules but MN administration can facilitate drug accumulation within the corneal and scleral tissues (Thakur et al., 2016).

The targeting of the vascular endothelial growth factor with medications is effective in only 30–60% of cases and several intraocular injections are typically needed (Galvin et al., 2016). Sometimes, these injections result in iatrogenic infections (Galvin et al., 2016). Recently, hyaluronan MNs were used to deliver the anti-angiogenic drug quininib in vivo (Galvin et al., 2016). The fabrication steps of HA microneedles loaded with quininib entailed desolvation from quininib-HA solution followed by cross-linking with 4-arm-PEG-amine followed by lyophilization (Galvin et al., 2016). Hollow needle-shaped particle ultrastructure were shown by scanning electron microscopy while electrophoretic light scattering was used to quantify the zeta potential (Galvin et al., 2016). The authors used HPLC to assay the pharmacokinetic profile and the loading efficiency of the drug released from the MNs in vitro. Drug loading into these MNs was 90% (Galvin et al., 2016). It was observed that 20% of the drug was released over 4 months (Galvin et al., 2016). In contrast, 60% of quininib was released in vitro in the presence hyaluronidase over the same period (Galvin et al., 2016). It was shown with zebrafish hyaloid vasculature assays that the drug released from these MNs markedly blocked ocular developmental angiogenesis in comparison to a control formulation (Galvin et al., 2016). Prolonged improvement of retinal vascular permeability was shown via a custom-made cysteinyl leukotriene induced rodent model (Galvin et al., 2016). The use of quininib-HA microneedles led to a prolonged release of the drug. In addition, the MNs were safe in animal experiments and quininib released from these MNs blocked retinal vascular permeability and angiogenesis in vivo (Galvin et al., 2016).

Glaucoma is an optic neuropathy defined by retinal ganglion cell damage and visual field defects (Mabuchi et al., 2020). The progressive depletion of retinal ganglion cells results in optic disc cupping and the thinning of the neuroretinal rim, a cardinal feature of glaucoma characterized by the loss of contrast sensitivity and visual field degeneration (Fenwick et al., 2020). Glaucoma's primary modifiable risk factor is intraocular pressure (IOP) and the progression of the disease can reduce considerably if IOP is decreased by 20%–50% of its baseline value (Fenwick et al., 2020). Glaucoma is subdivided into two categories: primary open-angle glaucoma (POAG) and primary angle-closure glaucoma (PACG) based on the anterior chamber angle structure (Fenwick et al., 2020). This chamber occupies the space between the iris and the peripheral cornea. It includes the Schlemm's canal through which aqueous humor exits the eye (Fenwick et al., 2020). As inferred from the name, anterior chamber angle is wide open in POAG, and IOP increase, if detected, is a painless and gradual process (Fenwick et al., 2020). On the other hand, the anterior chamber angle in PACG is either completely closed or constricted resulting in either a swift increase in IOP with blurred vision, nausea, eye pain and headaches, termed acute primary angle closure or asymptomatic chronic PACG (Fenwick et al., 2020).

Primary open-angle glaucoma (POAG) is the most common type of glaucoma, and it is categorized into normal tension glaucoma (NTG), in which the intraocular pressure (IOP) is within the statistically normal range (IOP≤21 mmHg) or high tension glaucoma (HTG), in which an increased IOP is the key characteristic (Mabuchi et al., 2020). Kim and coworkers used a hollow microneedle to deliver antiglaucoma drugs to the supraciliary space and showed that the use of MNs can lead to significant dose sparing of the medication in comparison with eye drops (Kim, Edelhauser et al., 2014; Kim, Groddniklaus et al., 2014). A single, hollow 33-gauge microneedle, 700−800 μm long, was injected into the sclera and used for antiglaucoma drug infusion into the supraciliary space of rabbits (Kim, Edelhauser et al., 2014; Kim, Groddniklaus et al., 2014). Sulprostone and brimonidine were delivered at various doses via topical and supraciliary administration (Kim, Edelhauser et al., 2014; Kim, Groddniklaus et al., 2014). The medications were delivered unilaterally, and rebound tonometry was used to determine the IOP of the two eyes for 9 h after injection and to evaluate pharmacological effects (Kim, Edelhauser et al., 2014; Kim, Groddniklaus et al., 2014). Supraciliary transport of both brimonidine and sulprostone decreased IOP by approximately 3 mm Hg bilaterally compared with topical administration (Kim, Edelhauser et al., 2014; Kim, Groddniklaus et al., 2014). There was a 100-fold dose sparing when both drugs were delivered via supraciliary microneedle-assisted injection. In addition, the kinetics of IOP increase immediately following intravitreal and supraciliary injection of placebos were comparable (Kim, Edelhauser et al., 2014; Kim, Groddniklaus et al., 2014).

The cornea is a transparent and an avascular tissue but systemic diseases and local factors can alter corneal transparency leading to corneal neovascularization (CNV) (Vivanco-Rojas et al., 2020). It is important to note that CVS can cause the visual acuity reduction, tissue scarring, edema, lipid deposition, and, in many cases, blindness (Vivanco-Rojas et al., 2020). In the last 20 years, the main treatment of neovascular disease has been endothelial cell growth inhibition (Wang et al., 2020). Although several *anti*-VEGF inhibitors have been utilized for the management of neovascular ophthalmopathy, the effect is still poor because of the short duration of action, poor response and adverse side effects (Wang et al., 2020). Recently, Kim and colleagues investigated the influence of coated microneedles on the intrastromal transport of bevacizumab (Kim, Edelhauser, et al., 2014). The authors coated stainless steel microneedles (400 μg long) with the drug. A silk suture was then positioned in the cornea about 1 mm from the limbus to bring about corneal neovascularization (Kim, Grossniklaus, et al., 2014). New Zealand white rabbits were grouped based on the mode of ocular drug administration: untreated, subconjunctival bevacizumab injection, topical eye drops and microneedle-assisted delivery (Kim, Grossniklaus, et al., 2014). Pharmacotherapy was launched 4 days following suture placement and the neovascularization area was quantified every day via digital photography for 18 days. The researchers reported that when the eyes were treated once with 4.4 μg bevacizumab following the application of microneedles, there was a 44% decrease in neovascularization by day 18 in comparison with untreated eyes (Kim, Grossniklaus, et al., 2014). Rabbit eyes treated once with 2500 μg of the drug via subconjunctival injection had results that were similar to those treated with microneedles (Kim, Grossniklaus, et al., 2014). In contrast, eyes that were treated once with the drug administered via the subconjunctival route did not display any significant effect in comparison with untreated eyes (Kim, Grossniklaus, et al., 2014). The authors observed decreased neovascularization area (6% decrease by day 18) compared with untreated eyes. Bevacizumab administered via eye drops three times daily for 2 weeks was markedly less effective compared to the same medication administered following a single microneedle application (Kim, Grossniklaus, et al., 2014).

AMD, DR, and uveitis are the main disorders of the posterior eye segment (Patel et al., 2011). The delivery of medications to the posterior segment of the eye is difficult due to the organ's natural barriers and poor access to target tissues (Patel et al., 2011). The main target tissues are the choroid and retina although other tissues can also be targeted (Patel et al., 2011). Patel and coworkers inserted a single, hollow microneedle into the sclera, and infused microparticle and nanoparticle suspensions into the suprachoroidal space (SCS) to enhance drug delivery (Patel et al., 2011). Ex vivo studies were carried out on pigs, whole rabbit, and human eyes. The authors used fluorescence microscopy, microcomputed tomography and brightfield microscopy to track particle transport (Patel et al., 2011). Microneedles delivered microparticle and nanoparticle suspensions as well as sulforhodamine B into the SCS of porcine, rabbit, and human eyes

(Patel et al., 2011). The authors infused volumes of up to 35 μL into the SCS. Microneedle pressure, length and particle size had significant effect on particle or sulforhodamine B transport into the suprachoroidal space (Patel et al., 2011). Applied pressures of 250−300 kPa and needle lengths of 800−1000 μm resulted in the most reliable delivery (Patel et al., 2011).

Tight junctions of the corneal epithelium, blinking of the eyelid, mucus and lacrimation block genetic vectors from entering the eye (Matthaei et al., 2012). Genes can be delivered to subepithelial layers via corneal intrastromal injection. However, this approach is challenging due to the miniaturized dimensions of the cornea in a mouse model (Matthaei et al., 2012). Furthermore, it is difficult to measure the area of the cornea covered by a single injection. Matthaei et al. investigated the effect of different microneedles and injection volumes on the intrastromal delivery of India Ink into the mouse cornea (Matthaei et al., 2012). The authors injected the dye intrastromally into murine eye using tri-surface 25° beveled 35 G needles, 30° beveled 33 G needles, or 25° beveled glass needles (Matthaei et al., 2012). 35 G needles were attached to a 10 μL syringe, 33 G needles were attached to a 2.5 μL syringe and glass needles were hand-pulled from 1 mm diameter glass capillaries (beveled to 25°) with an inner tip diameter of about 50 μm (Matthaei et al., 2012). A dual ferrule adaptor was used to attach glass needles to a 2.5 μL syringe (Matthaei et al., 2012). Computer-assisted measurement of the affected corneal area was carried out with corneoscleral buttons. The authors evaluated the intrastromal position of the administered dye histologically (Matthaei et al., 2012). The mean corneal area covered following intrastromal injections of 1 and 2 μL of the dye was $5.0 \pm 1.4\ mm^2$ (mean ± SD) and $7.7 \pm 1.4\ mm^2$ respectively (Matthaei et al., 2012). The mean percentage of total corneal area covered was from 65% to 81% for 2 μL injections and from 39% to 53% for 1 μL injections (Matthaei et al., 2012). Perforation rates were 44% for tri-surface beveled 35 G needles and 8% for 30° beveled 33 G needles (Matthaei et al., 2012). When the authors injected the dye with 33 G needles, the distribution area was the highest. Although intrastromal damage to needle tips was documented in 25% of glass needles, no perforation was detected with this class of needles (Matthaei et al., 2012).

Microneedles can be applied to the ocular surface (Fig. 5.2). In a recent publication, the application of different types of microneedles to the ocular surface was shown. The use of traditional (solid, hollow and coated) as well as emerging (dissolving, hydrogel forming and backward-facing barbed) microneedles for drug delivery to the eye was documented (Meng et al., 2020).

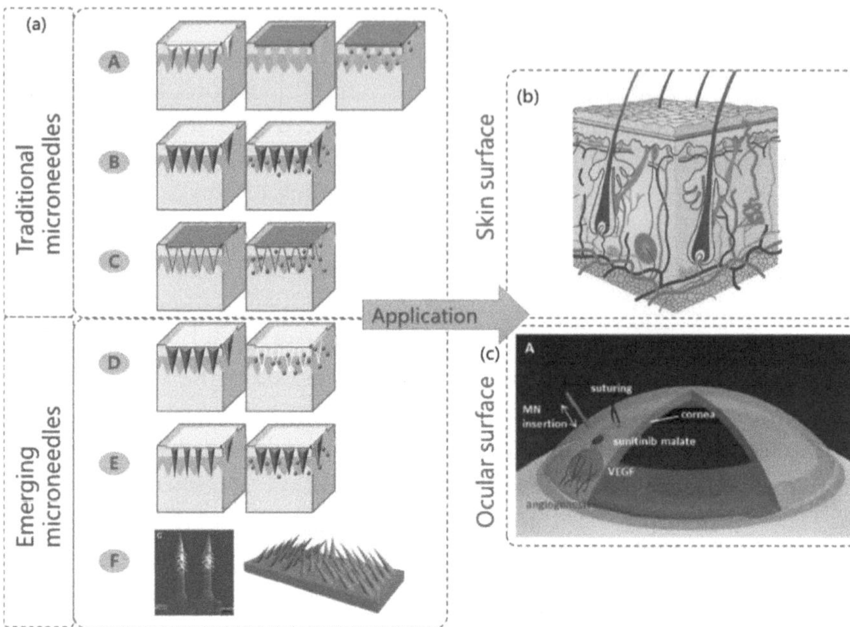

Fig. 5.2 Illustration of different types of microneedles and their applications in skin and ocular surface (A) Two types of microneedles including traditional microneedles (A – Solid MNs; B – Coated MNs; C – Hollow MNs) and emerging microneedles (D – Dissolving MNs; E – Hydrogel forming MNs; F – Backward-facing barbed MNs). Microneedles were applied in (B) skin and (C) ocular surface. *(From Meng, X., Zhang, Z., & Li, L. (2020). Micro/nano needles for advanced drug delivery.* SI: Biomaterials, *30(5), 589–596. https://doi.org/10.1016/j.pnsc.2020.09.016).*

The authors highlighted specialized microneedles such as those with backward-facing curved barbs (Meng et al., 2020).

While eye drops can be used for drug delivery into the anterior segment, this approach cannot lead to the delivery of therapeutically relevant drug quantities to the posterior segment (Jiang et al., 2007). Long diffusion distances from the ocular surface to the posterior segment, counter-directional intraocular flow of aqueous humor, low corneal permeability, conjunctival blood flow and precorneal tears can result in rapid drug elimination (Jiang et al., 2007). Jiang and colleagues studied the influence of hollow microneedles on intrascleral drug delivery (Jiang et al., 2007). The authors introduced individual hollow MNs into human cadaver sclera with subsequent infusion of fluorescently tagged nanoparticles or microparticles as well as sulforhodamine solution (Jiang et al., 2007). The authors measured and histologically imaged the infused fluid volume as a function of the presence of spreading enzymes (hyaluronidase and collagenase), needle retraction depth, infusion pressure and scleral thickness (Jiang et al., 2007). Partial retraction of the MN over a distance of 200−300 μm facilitated infusion of 10−35 μL of fluid into the sclera but fluid infusion was very slow following MN insertion into the tissue without retraction (Jiang et al., 2007). Infusion pressure and scleral thickness did not have substantial influence on fluid delivery (Jiang et al., 2007). Microparticles were delivered into the sclera only in the presence of collagenase and hyaluronidase but nanoparticle suspensions were directly delivered, which pointed to the influence of collagen fibers and scleral glycosaminoglycans as rate limiting barriers (Jiang et al., 2007).

Traditional melanoma therapies are usually accompanied by toxicity and side effects because of the recurrent and aggressive properties of melanoma cells (Kim et al., 2015). Polymeric microneedles can be used for less-invasive topical chemotherapies but the sustained release of the loaded medication is difficult (Kim et al., 2020). Furthermore, microneedle size is relatively large for the exceptionally sensitive, curvilinear and small cornea in the case of ocular melanoma (Kim et al., 2020). Kim et al. developed a miniaturized, bioresorbable and porous-silicon (p-Si) microneedles with covalently tethered drug payload at comparable doses to those of traditional polymeric MNs (Kim et al., 2020). The p-Si needles included a water-soluble film as a transient flexible holder that can be tightly interfaced with the irregular surface of living tissues, and will completely dissolve in saline solution within 60 s (Kim et al., 2020). The porous-silicon needles remain entrenched inside ocular tissues and then gradually degrade leading to prolonged drug release. The authors showed in a murine melanoma model that the microneedles can be used for unobtrusive topical drug administration with low side effects (Kim et al., 2020).

References

Agban, Y., Thakur, S. S., Mugisho, O. O., & Rupenthal, I. D. (2019). Depot formulations to sustain periocular drug delivery to the posterior eye segment. *Drug Discovery Today, 24*(8), 1458−1469. https://doi.org/10.1016/j.drudis.2019.03.023

Bhatnagar, S., Saju, A., Cheerla, K. D., Gade, S. K., Garg, P., & Venuganti, V. V. K. (2018). Corneal delivery of besifloxacin using rapidly dissolving polymeric microneedles. *Drug Delivery and Translational Research, 8*(3), 473−483. https://doi.org/10.1007/s13346-017-0470-8

CDC. (2021). *CDC vision health.* https://www.cdc.gov/visionhealth/basics/ced/index.html. (Accessed 9 June 2021).

Chiang, B., Jung, J. H., & Prausnitz, M. R. (2018). The suprachoroidal space as a route of administration to the posterior segment of the eye. *Advanced Drug Delivery Reviews, 126*, 58−66. https://doi.org/10.1016/j.addr.2018.03.001

Fenwick, E. K., Man, R. E., Aung, T., Ramulu, P., & Lamoureux, E. L. (2020). Beyond intraocular pressure: Optimizing patient-reported outcomes in glaucoma. *Progress in Retinal and Eye Research*, 100801. https://doi.org/10.1016/j.preteyeres.2019.100801

Galvin, O., Srivastava, A., Carroll, O., Kulkarni, R., Dykes, S., Vickers, S., Dickinson, K., Reynolds, A. L., Kilty, C., Redmond, G., Jones, R., Cheetham, S., Pandit, A., & Kennedy, B. N. (2016). A sustained release formulation of novel quininib-hyaluronan microneedles inhibits angiogenesis and retinal vascular permeability in vivo. *Journal of Controlled Release, 233*, 198–207. https://doi.org/10.1016/j.jconrel.2016.04.004

Gonenc, B., Chae, J., Gehlbach, P., Taylor, R. H., & Iordachita, I. (2017). Towards robot-assisted retinal vein cannulation: A motorized force-sensing microneedle integrated with a handheld micromanipulator †. *Sensors, 17*(10), 2195. https://doi.org/10.3390/s17102195

Gupta, P., & Yadav, K. S. (2019). Applications of microneedles in delivering drugs for various ocular diseases. *Life Sciences, 237*. https://doi.org/10.1016/j.lfs.2019.116907

Gu, J., Zhang, Y., Han, Z., Gao, L., Cui, J., Sun, Y., Niu, Y., You, B., Huang, C.-P., Chang, C., Wang, X., & Yeh, S. (2020). Targeting the ERβ/Angiopoietin-2/Tie-2 signaling-mediated angiogenesis with the FDA-approved anti-estrogen Faslodex to increase the Sunitinib sensitivity in RCC. *Cell Death and Disease, 11*(5), 367. https://doi.org/10.1038/s41419-020-2486-0

Hartman, R. R., & Kompella, U. B. (2018). Intravitreal, subretinal, and suprachoroidal injections: Evolution of microneedles for drug delivery. *Journal of Ocular Pharmacology and Therapeutics, 34*(1–2), 141–153. https://doi.org/10.1089/jop.2017.0121

Izzotti, A., Ceccaroli, C., Longobardi, M. G., Micale, R. T., Pulliero, A., La Maestra, S., & Sacca, S. C. (2015). Molecular damage in glaucoma: From anterior to posterior eye segment. The MicroRNA role. *MicroRNA, 4*(1), 3–17. https://doi.org/10.2174/2211536604666150707124640

Jiang, J., Gill, H. S., Ghate, D., McCarey, B. E., Patel, S. R., Edelhauser, H. F., & Prausnitz, M. R. (2007). Coated microneedles for drug delivery to the eye. *Investigative Ophthalmology and Visual Science, 48*(9), 4038–4043. https://doi.org/10.1167/iovs.07-0066

Jung, J. H., Park, S., Chae, J. J., & Prausnitz, M. R. (2019). Collagenase injection into the suprachoroidal space of the eye to expand drug delivery coverage and increase posterior drug targeting. *Experimental Eye Research, 189*. https://doi.org/10.1016/j.exer.2019.107824

Kim, Y. C., Edelhauser, H. F., & Prausnitz, M. R. (2014). Targeted delivery of anti-glaucoma drugs to the supraciliary space using microneedles. *Investigative Ophthalmology and Visual Science, 55*(11), 7387–7397. https://doi.org/10.1167/iovs.14-14651

Kim, Y. C., Grossniklaus, H. E., Edelhauser, H. F., & Prausnitz, M. R. (2014b). Intrastromal delivery of bevacizumab using microneedles to treat corneal neovascularization. *Investigative Ophthalmology and Visual Science, 55*(11), 7376–7386. https://doi.org/10.1167/iovs.14-15257

Kim, H., Lee, H. S., Jeon, Y., Park, W., Zhang, Y., Kim, B., Jang, H., Xu, B., Yeo, Y., Kim, D. R., & Lee, C. H. (2020). Bioresorbable, miniaturized porous silicon needles on a flexible water-soluble backing for unobtrusive, sustained delivery of chemotherapy. *ACS Nano, 14*(6), 7227–7236. https://doi.org/10.1021/acsnano.0c02343

Kim, Y. C., Oh, K. H., Edelhauser, H. F., & Prausnitz, M. R. (2015). Formulation to target delivery to the ciliary body and choroid via the suprachoroidal space of the eye using microneedles. *European Journal of Pharmaceutics and Biopharmaceutics, 95*, 398–406. https://doi.org/10.1016/j.ejpb.2015.05.020

Kompella, U. B., Hartman, R. R., & Patil, M. A. (2020). Extraocular, periocular, and intraocular routes for sustained drug delivery for glaucoma. In *Progress in retinal and eye Research*. https://doi.org/10.1016/j.preteyeres.2020.100901

Lange, J., Mokdad, H., & Wyser, Y. (2002). Understanding puncture resistance and perforation behavior of packaging laminates. *Journal of Plastic Film and Sheeting, 18*(4), 231–244. https://doi.org/10.1177/8756087902033865

Lee, K. J., Goudie, M. J., Tebon, P., Sun, W., Luo, Z., Lee, J., Zhang, S., Fetah, K., Kim, H. J., Xue, Y., Darabi, M. A., Ahadian, S., Sarikhani, E., Ryu, W. H., Gu, Z., Weiss, P. S., Dokmeci, M. R., Ashammakhi, N., & Khademhosseini, A. (2020). Non-transdermal microneedles for advanced drug delivery. *Advanced Drug Delivery Reviews, 165–166*, 41–59. https://doi.org/10.1016/j.addr.2019.11.010

Lee, K. J., Song, H. B., Cho, W., Kim, J. H., Kim, J. H., & Ryu, W. H. (2018). Intracorneal injection of a detachable hybrid microneedle for sustained drug delivery. *Acta Biomaterialia, 80*, 48–57. https://doi.org/10.1016/j.actbio.2018.09.039

Mabuchi, F., Mabuchi, N., Sakurada, Y., Yoneyama, S., Kashiwagi, K., Iijima, H., Yamagata, Z., Takamoto, M., Aihara, M., Iwata, T., Hashimoto, K., Sato, K., Shiga, Y., Nishiguchi, K. M., Nakazawa, T., Akiyama, M., Kawase, K., Ozaki, M., & Araie, M. (2020). Genetic variants associated with the onset and progression of primary open-angle glaucoma. *American Journal of Ophthalmology, 215*, 135–140. https://doi.org/10.1016/j.ajo.2020.03.014

Mahmoudi, S., Masoomi, A., Ahmadikia, K., Tabatabaei, S. A., Soleimani, M., Rezaie, S., Ghahvechian, H., & Banafsheafshan, A. (2018). Fungal keratitis: An overview of clinical and laboratory aspects. *Mycoses, 61*(12), 916–930. https://doi.org/10.1111/myc.12822

Matthaei, M., Meng, H., Bhutto, I., Xu, Q., Boelke, E., Hanes, J., & Jun, A. S. (2012). Systematic assessment of microneedle injection into the mouse cornea. *European Journal of Medical Research, 17*, 19. https://doi.org/10.1186/2047-783X-17-19

Meng, X., Zhang, Z., & Li, L. (2020). Micro/nano needles for advanced drug delivery. *SI: Biomaterials, 30*(5), 589–596. https://doi.org/10.1016/j.pnsc.2020.09.016

Nayak, K., & Misra, M. (2018). A review on recent drug delivery systems for posterior segment of eye. *Biomedicine and Pharmacotherapy, 107*, 1564–1582. https://doi.org/10.1016/j.biopha.2018.08.138

Park, S. H., Jo, D. H., Cho, C. S., Lee, K. J., Kim, J. H., Ryu, S., Joo, C., Kim, J. H., & Ryu, W. H. (2018). Depthwise-controlled scleral insertion of microneedles for drug delivery to the back of the eye. *European Journal of Pharmaceutics and Biopharmaceutics, 133*, 31–41. https://doi.org/10.1016/j.ejpb.2018.09.021

Park, S. H., Lee, K. J., Lee, J. Y., Yoon, J. H., Jo, D. H., Kim, J. H., Kang, K., & Ryu, W. H. (2016). Microneedle-based minimally-invasive measurement of puncture resistance and fracture toughness of sclera. *Acta Biomaterialia, 44*, 286–294. https://doi.org/10.1016/j.actbio.2016.08.011

Patel, S. R., Lin, A. S. P., Edelhauser, H. F., & Prausnitz, M. R. (2011). Suprachoroidal drug delivery to the back of the eye using hollow microneedles. *Pharmaceutical Research, 28*(1), 166–176. https://doi.org/10.1007/s11095-010-0271-y

Roy, G., Galigama, R. D., Thorat, V. S., Mallela, L. S., Roy, S., Garg, P., & Venuganti, V. V. K. (2019). Amphotericin B containing microneedle ocular patch for effective treatment of fungal keratitis. *International Journal of Pharmaceutics, 572*, 118808. https://doi.org/10.1016/j.ijpharm.2019.118808

Song, H. B., Lee, K. J., Seo, I. H., Lee, J. Y., Lee, S.-M., Kim, J. H., Kim, J. H., & Ryu, W. (2015). Impact insertion of transfer-molded microneedle for localized and minimally invasive ocular drug delivery. *Journal of Controlled Release: Official Journal of the Controlled Release Society, 209*, 272–279. https://doi.org/10.1016/j.jconrel.2015.04.041

Suri, R., Beg, S., & Kohli, K. (2020). Target strategies for drug delivery bypassing ocular barriers. *Journal of Drug Delivery Science and Technology, 55*, 101389. https://doi.org/10.1016/j.jddst.2019.101389

Thakur, R. R. S., Adwan, S., Tekko, I., Soliman, K., & Donnelly, R. F. (2021). Laser irradiation of ocular tissues to enhance drug delivery. *International Journal of Pharmaceutics, 596*, 120282. https://doi.org/10.1016/j.ijpharm.2021.120282

Thakur, R. R. S., Tekko, I. A., Al-Shammari, F., Ali, A. A., McCarthy, H., & Donnelly, R. F. (2016). Rapidly dissolving polymeric microneedles for minimally invasive intraocular drug delivery. *Drug Delivery and Translational Research, 6*(6), 800–815. https://doi.org/10.1007/s13346-016-0332-9

Thang Nguyen, C., Vu-Khanh, T., & Lara, J. (2005). A study on the puncture resistance of rubber materials used in protective clothing. *Journal of ASTM International, 2*(4), 12147. https://doi.org/10.1520/jai12147

Than, A., Liu, C., Chang, H., Duong, P. K., Cheung, C. M. G., Xu, C., Wang, X., & Chen, P. (2018). Self-implantable double-layered micro-drug-reservoirs for efficient and controlled ocular drug delivery. *Nature Communications, 9*(1). https://doi.org/10.1038/s41467-018-06981-w

Vivanco-Rojas, O., García-Bermúdez, M. Y., Iturriaga-Goyon, E., Rebollo, W., Buentello-Volante, B., Magaña-Guerrero, F. S., Bates, P., Pérez-Torres, A., & Garfias, Y. (2020). Corneal neovascularization is inhibited with nucleolin-binding aptamer, AS1411. *Experimental Eye Research, 193*, 107977. https://doi.org/10.1016/j.exer.2020.107977

Wang, X., Tang, L., Zhang, Z., Li, W., & Chen, Y. (2020). Keratocytes promote corneal neovascularization through VEGFr3 induced by PPARα-inhibition. *Experimental Eye Research, 193*, 107982. https://doi.org/10.1016/j.exer.2020.107982

Yiu, G., Chung, S. H., Mollhoff, I. N., Nguyen, U. T., Thomasy, S. M., Yoo, J., Taraborelli, D., & Noronha, G. (2020). Suprachoroidal and subretinal injections of AAV using transscleral microneedles for retinal gene delivery in nonhuman primates. *Molecular Therapy - Methods and Clinical Development, 16*, 179–191. https://doi.org/10.1016/j.omtm.2020.01.002

Zhang, Y., Shi, J., & Zheng, J. (2021). A method of fracture toughness JIC measurement based on digital image correlation and acoustic emission technique. *Materials and Design, 197*, 109258. https://doi.org/10.1016/j.matdes.2020.109258

CHAPTER 6

Microneedles for diagnostic testing

Introduction

The concentration of small molecules, nucleic acids, inflammatory mediators, proteins and lipids in the skin can be used for human health assessment (Paliwal, 2013; Paliwal, Hwang, Tsai, & Mitragotri, 2013). Nevertheless, skin chemistry, is seldom used in clinical practice for diagnosis (Paliwal, 2013; Paliwal, Hwang, Tsai, & Mitragotri, 2013). This is mainly because of the deficiency of simple and standardized techniques for routine assay of skin biomolecules (Paliwal, 2013; Paliwal, Hwang, Tsai, & Mitragotri, 2013). The most frequently used fluid for skin chemistry is the interstitial fluid (ISF) that is present in the dermis (Dervisevic et al., 2020). Recently, several investigators have utilized transdermal assay of the ISF for the biomarker and metabolite quantitation (e.g. cytokine, cortisol, alcohol, lactic acid, glucose) (Dervisevic et al., 2020; Zhang et al., 2019). The sampling of ISF for real-time assay requires techniques that are sensitive, rapid, painless, minimally invasive and easy to use by patients (Dervisevic et al., 2020). Scientists have designed numerous methods for transdermal monitoring. Microneedles (MNs) can be used to overcome some challenges with current techniques (Dervisevic et al., 2020). Microneedles are short enough to penetrate only the superficial skin layers, avoiding nerve endings and major blood vessels, while still accessing analyte-rich ISF or capillary blood (Ng, Dixon, Lau, & Moghimi, 2020). Broadly speaking, current microneedle diagnostic systems operate in one of the following ways: (a) they extract biofluids or target analytes from the skin for in vitro testing or (b) they perform in situ biosensing of target analytes in the skin (Ng, Dixon, Lau, & Moghimi, 2020). ISF is an attractive target because it contains many of the same constituents as blood plasma and suction blister fluid at clinically relevant concentrations. Electrochemical biosensors on microneedle platforms can be used for biomarker assay of the ISF (Madden et al., 2020).

Microneedles
ISBN 978-0-323-91811-4
https://doi.org/10.1016/B978-0-323-91811-4.00006-7

The incorporation of microneedles into a biosensor can be used for continuous monitoring or bio-chemical detection of analytes in the dermal ISF (Madden et al., 2020). This approach is minimally invasive. Even though research is predominantly focused on microneedles for the painless and minimally invasive drug delivery, promising results have been documented following the use of MNs for diagnosis (Babity et al., 2018). Needle-based collection of blood for laboratory analyses is popular because of low cost and high efficiency (Liu et al., 2020). Nevertheless, challenges do remain: (a)Several people (3.5%−10% of the global population) experience needle phobia that prevents them from seeking emergency or even routine medical care; (b) the reuse of inappropriately sterilized or unsterilized needles is widespread in developing countries and can cause the transmission of blood-borne infections (Liu et al., 2020); (c) the utilization of syringes and hypodermic by untrained personnel for self-administration is challenging with regard to waste management and safety (Liu et al., 2020). Even for an easy-to-use blood glucose meter, recurrent finger pricks often lead to discomfort and puncture-related pain (Liu et al., 2020).

Fig. 6.1 shows microneedle diagnostic platforms based on biofluid extraction using hollow (A) or solid (B) microneedles, specific target analyte capture (C) and electrochemical sensing (D) (Dixon et al., 2021). Microneedles can detect biomarkers in/from the skin in a minimally invasive fashion thus facilitating (near-) real-time diagnosis (Dixon et al., 2021). These devices can be used for sensing, biofluid extraction, and analyte-capturing (Dixon et al., 2021).

Point-of-care tests which are carried out at or near a patient, complements traditional laboratory diagnosis by supplying cost-effective, portable and user-friendly tools (Liu et al., 2020). Point-of-care test devices based on MNs are capable of quantifying different analytes in a minimally invasive manner (Liu et al., 2020). Researchers can improve electrochemical sensing by designing standard electrodes in the form of MNs (Dardano et al., 2019). MNs can be effortlessly functionalized, miniaturized, and rationally designed for specific biosensing objectives. It is noteworthy that MNs provide a painless access to body fluids (Dardano et al., 2019). They can pierce the skin and detect target chemicals within the ISF (Dardano et al., 2019). Metabolites, pharmaceuticals, and electrolytes represent targets which have been successfully quantified with electrochemical MN devices in human serum as well as in artificial ISF samples (Madden et al., 2020). Although several devices have been tested in preliminary clinical trials, integrated MN sensors are yet to advance to a

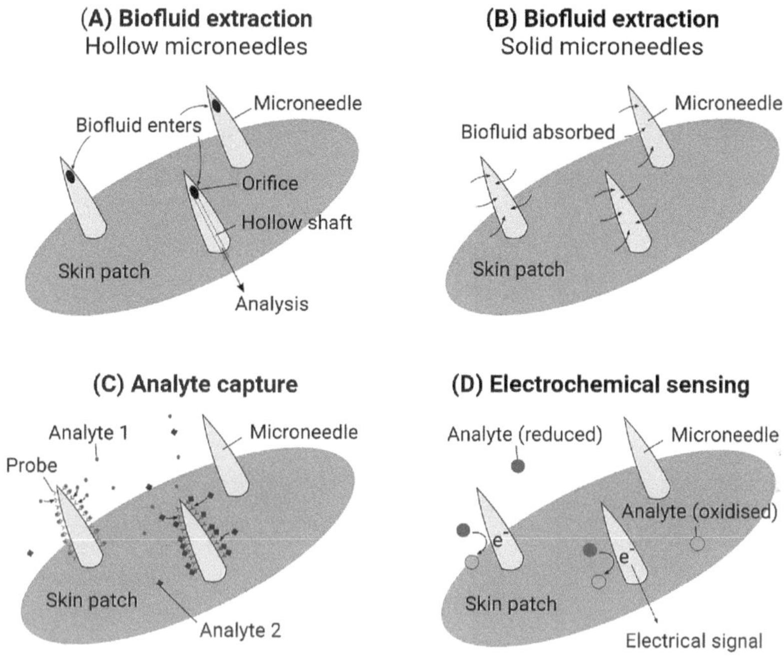

Figure 6.1 Microneedle diagnostic platforms The diagnosticsystems are based on biofluid extraction using hollow (A) or solid (B) microneedles, specific target analyte capture (C) or electrochemical sensing (D). *(From Dixon, R. V., Skaria, E., Lau, W. M., Manning, P., Birch-Machin, M. A., Moghimi, S. M., & Ng, K. W. (2021). Microneedle-based devices for point-of-care infectious disease diagnostics.* Acta Pharmaceutica Sinica B. *https://doi.org/10.1016/j.apsb.2021.02.010).*

mature stage of development (Madden et al., 2020). Moreover, only few studies have combined MN sensors with a communication module, miniaturized potentiostat and a power source (Madden et al., 2020).

Schematic images depicting various designs of microneedle arrays interfacing microfluidic on-chip analysis chambers are shown in Fig. 6.2.

Hollow microneedles have a hollow shaft and a surface orifice (Ng, Dixon, Lau, & Moghimi, 2020). Biofluid uptake through the orifice into the hollow microneedle shaft occurs by capillary action or can be induced through a microfluidic system, usually by applying negative pressure (Ng, Dixon, Lau, & Moghimi, 2020). Solid microneedles, on the other hand, do not have a surface orifice or a hollow shaft. The application of solid microneedles results in the migration of ISF to the skin's surface (Ng, Dixon, Lau, & Moghimi, 2020). ISF can then be absorbed into a sheet of paper for subsequent biomarker analysis (Ng, Dixon, Lau, & Moghimi, 2020).

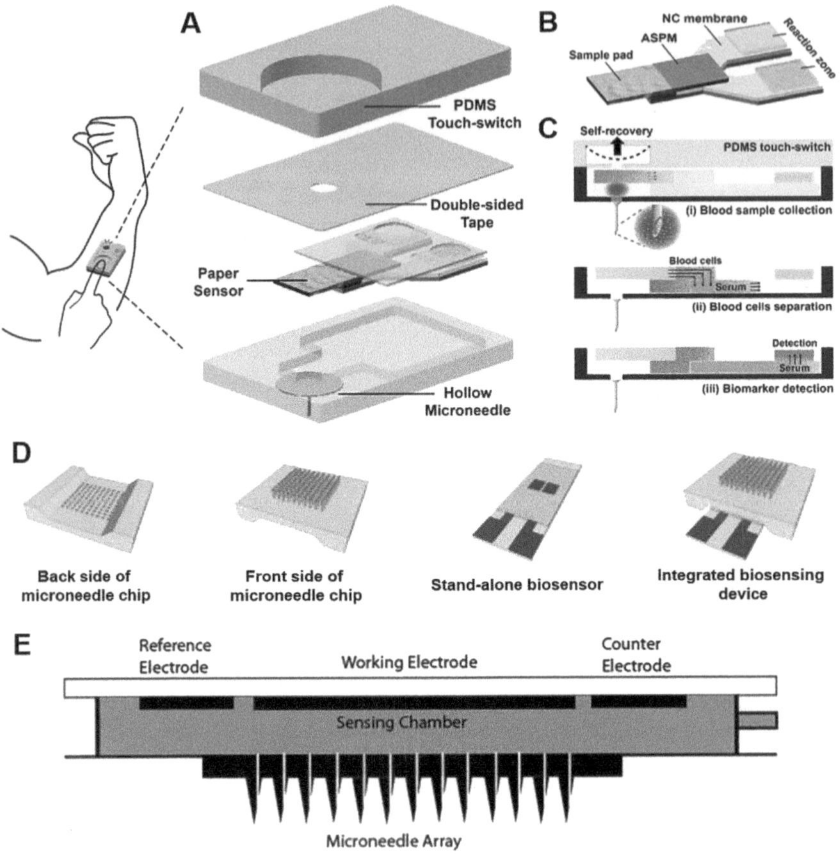

Figure 6.2 Schematic images depicting various designs of microneedle arrays interfacing microfluidic on-chip analysis chambers. (A)–(C) Schematic of a one-touch-activated blood multidiagnostic tool (OBMT). (A) OBMT complete device design and structure. (B) OBMT paper-based multiplex sensor. (C) Operating principle of the OBMT device. (D) Schematic of silicon dioxide transdermal biosensor showing the front side and backside of the microneedle chip, the "stand-alone" sensor section and the full integrated device. (E) Schematic of a continuous glucose monitoring hollow silicon microneedle-based device with glucose sensing chamber. *(From Dixon, R. V., Skaria, E., Lau, W. M., Manning, P., Birch-Machin, M. A., Moghimi, S. M., & Ng, K. W. (2021). Microneedle-based devices for point-of-care infectious disease diagnostics.* Acta Pharmaceutica Sinica B. *https://doi.org/10.1016/j.apsb.2021.02.010).*

Hydrogel-forming microneedles swell upon insertion into the dermal interstitium and absorb ISF (Ng, Dixon, Lau, & Moghimi, 2020). Downstream processing of the hydrogel microneedle array by centrifugation liberates the analytes of interest, which can then be quantified using conventional techniques such as high-performance liquid chromatography (Ng, Dixon, Lau, & Moghimi, 2020). Porous microneedle arrays contain interconnected pores in the body of the microneedles, which allow them to absorb skin biofluids by capillary action (Ng, Dixon, Lau, & Moghimi, 2020). The biofluids can then be recovered from the microneedles by centrifugation (Ng, Dixon, Lau, & Moghimi, 2020).

Microneedle-based diagnostics

It is challenging to reach the human perilymph because of the minute fluid volume (about 150 μL per ear) (Early et al., 2019). This fluid is essentially covered by hard cochlear bone. The highly sensitive architecture of the vestibular system and the inner ear that are bathed in perilymph can be destroyed if more than a few microliters of the fluid is collected at a time (Early et al., 2019). Whenever the perilymph is used for diagnosis, the benefit of doing so must outweigh the risk of inducing hearing impairment (Early et al., 2019). Early et al. utilized MNs for the collection of perilymph sample from the inner ear (Early et al., 2019). Wire electrical discharge machining was used for microneedle fabrication (Early et al., 2019). The distance from needle tip to the shelf was 670 μm, for the maintenance of a minimum 500 μm safety buffer between the osseus spiral lamina and the needle tip (Early et al., 2019). In this device, microliter aspiration control and controlled depth of perforation was utilized for safe biopsy of inner ear fluids (Early et al., 2019). Of eleven MN devices investigated using human temporal bones, nine were used for the successful aspiration of diagnostically relevant perilymph samples while seven samples showed alignment between visual, tactile and electrical tracking of round window membrane perforation (Early et al., 2019). Sample purity of the perilymph (5 μL) was 69% (Early et al., 2019). The successful adoption of microneedles for biopsy of inner earliquid can become a routine area of clinical otologic care (Early et al., 2019).

One key research area of medical engineering technologies is continuous glucose monitoring (Takeuchi & Kim, 2018). MNs can be used: (a) as a biological fluid collector and (b) as a sensing probe (Takeuchi & Kim, 2018). As biosensors, microneedles are typically coated or integrated with conductive materials. When these devices are employed as fluid collectors, researchers use certain geometrical designs, such as a porous and hollow MNs that work via negative pressure or capillary action (Takeuchi & Kim, 2018). These functionalized MNs are used for the extraction of blood or interstitial fluids for ex vivo analysis (Takeuchi & Kim, 2018). A long-term accurate assay by the sensing probe is important for microneedle-based continuous glucose monitoring. Microneedles as biological fluid collectors can also be integrated with ex vivo diagnostic systems for a more efficient sampling of ISF (Takeuchi & Kim, 2018).

Double-blind placebo-controlled food challenge is the gold-standard diagnostic test for food allergy, but it is challenging when several foods are considered as potential allergens (Sun et al., 2013). Furthermore, this test is difficult to carry out in young children, and can be dangerous because the readout may be an anaphylactic or allergic reaction (Sun et al., 2013). Skin prick testing can be used to predict double-blind placebo-controlled food challenge since it is based on allergen-mediated mast cell reactions (Sun et al., 2013). Approximately 30 allergens can be tested in a daily due to the drawbacks of currently used techniques. Sun et al. fabricated PVP microneedles that can concurrently deliver several peptides or proteins of diagnostic interest to the skin (Sun et al., 2013). This method is beneficial because it is inexpensive, user-friendly and can be used to test thousands of food allergen epitopes in parallel (Sun et al., 2013). The authors demonstrated effective delivery of proteins to the human intradermal space (Sun et al., 2013). Allergen proteins were incorporated into the MNs with the possibility of testing allergenic proteins or their constituent peptides in a one assay (Sun et al., 2013). PVP—protein mixture was poured onto PDMS molds to create biodegradable microneedles loaded with rhodamine B or fluorescein (Sun et al., 2013). The investigators inserted the microneedles into human foreskin used with wide-field microscopy to monitor the MNs over time (Sun et al., 2013). There was marked MN dissolution after 2.5 min, and after 5 min the MNs dissolved completely. The authors used fluorescence microscopy to image the skin following application of MNs loaded with rhodamine B-labeled casein (Sun et al., 2013). Fluorescence was detected at about 187 μm beneath the epidermis (Sun et al., 2013).

Recently, silicon micromolding method was used to fabricate carbon loaded polystyrene MNs and the ability of the these devices to function as transdermal biosensors was subsequently assessed (Hegarty et al., 2019). Anodization was used to increase the density of quinone groups for analysis. The authors employed pH dependent redox transitions as the platform for a sensor that does not need reagents (Hegarty et al., 2019). Nernstian properties determine the quinone oxidation process (Hegarty et al., 2019). The effectiveness of the MN electrode patch for analysis was examined by using tomato skin to mimic human skin (Hegarty et al., 2019). Despite matrix complexity, the MN sensor response was similar to conventional/commercial pH probes (Hegarty et al., 2019).

In another interesting study, MN-based biosensor was utilized for continuous lactate monitoring in the dermal ISF (Bollella et al., 2019). The normal blood levels of lactate in healthy people are from 0.5 to 2 mM at rest and these can increase to 15 mM during exercise and 25 mM under intensive physical activity (Bollella et al., 2019). Lactate concentration can also be elevated in conditions, such as liver disease, diabetes, pulmonary embolism, endotoxic shock and cardiac diseases (Bollella et al., 2019). Furthermore, lactate acidifies the cancer cell microenvironment and can be used as a marker in cancer diagnostics (Bollella et al., 2019). Electron transfer of lactate oxidase can be accomplished by using gold microneedles functionalized with nanocarbons (Bollella et al., 2019). Bollella et al. prepared a MN device from polycarbonate (Bollella et al., 2019). The polycarbonate structure ($1.0 \times 1.0 \times 0.2$ cm) was made up of 64 MNs arranged as 4×4 arrays that were perpendicular to the baseplate (Bollella et al., 2019). The surface of the MN electrode was altered via electrodeposition, electropolymerization and lactate oxidase immobilization (Bollella et al., 2019). The microneedle-based lactate oxidase biosensor was used to detect lactate (Bollella et al., 2019). The efficiency of lactate assay was evaluated in human serum and artificial interstitial fluid that both contained lactate. The findings from the study showed that the MN-based lactate sensor can be utilized for monitoring in real time and this is important in clinical care as well as sports medicine (Bollella et al., 2019). The same principle underpins the use of microneedles for glucose sensing.

Regulation of blood glucose level can be improved through the use of continuous glucose sensing (Chen et al., 2015). Chen and coworkers demonstrated the utility of microporous polyvinylidene fluoride (PVDF) membranes placed in between several layers of nanomaterials for continuous glucose sensing in vivo (Chen et al., 2015). The authors used layer-by-layer deposition to coat needle electrodes with platinum nanoparticles, polyaniline nanofiber and glucose oxidase (Chen et al., 2015). The use of polyaniline nanofibers and nanoparticles resulted in a higher electrocatalytic activity for the enzyme (Chen et al., 2015). The authors used the composite coating membrane that included the nanosphere Nafion and the porous PVDF to limit glucose transport and improve the duration of in vivo tests (Chen et al., 2015). The biosensor for continuous glucose sensing displayed a quick reaction time of less than 30 s, low output current (less than 1 μA), and a sensitivity of 0.23 μA/mM (Chen et al., 2015). The glucose concentration detected with this method ranged from 0 to 20 mM (Chen et al., 2015). Remarkably, mouse models were used to demonstrate the applicability and biocompatibility of this sensing technique (Chen et al., 2015).

Alcoholism is a key public health challenge that can lead to traffic fatalities and some health related hazards (Mohan et al., 2017). Unhealthy alcohol consumption leads to about 2.5 million deaths annually. It is a principal risk factor for several injuries, disorders and diseases (Mohan et al., 2017). Not long ago, a microneedle-based sensor was developed for subcutaneous alcohol quantitation (Mohan et al., 2017). An injection-molding method was used to fabricate hollow MNs from a biocompatible liquid crystal polymer (Mohan et al., 2017). The polymer used for fabrication possessed excellent mechanical characteristics: a flexural modulus of 2300 MPa, a tensile modulus of 2350 MPa, and a hardness of 118 on the Rockwell scale (Mohan et al., 2017). The authors used a mill to fabricate high-carbon stainless steel mold (Mohan et al., 2017). Pyramidal MN devices integrated with platinum and silver were subsequently produced (Mohan et al., 2017). The MN lumen was created by electropolymerizing o-phenylene diamine onto the platinum wire microtransducer, with subsequent immobilization of alcohol oxidase on a middle chitosan layer and an outer Nafion layer (Mohan et al., 2017). The MN-based enzyme electrode was used to detect ethanol in an artificial interstitial fluid without jeopardizing its stability, sensitivity and response time (Mohan et al., 2017). Studies using an ex vivo mouse skin were used to substantiate the effectiveness of the biosensor for subcutaneous alcohol sensing (Mohan et al., 2017).

Adverse events linked to Type 1 and Type 2 diabetes can be alleviated by strict blood glucose control (Nicholas et al., 2018). Self-monitoring of blood glucose (SMBG) is typically utilized for the quantification of the analyte at a specific point in time while glycated hemoglobin (HbA1c) concentration is determined over a protracted period (the previous 2–3 months) (Nicholas et al., 2018). Several authors have reported that about 40%–50% of diabetes patients either do not conduct scheduled SMBG or do not adhere to guidelines (Nicholas et al., 2018). MNs can be used to quantify glucose level in the interstitial fluid as a substitute to blood sampling. Nicholas et al. fabricated hollow microneedles with sharp beveled edges (Nicholas et al., 2018). A clear resin and a printer were used for make the master mold (Nicholas et al., 2018). A photopolymer mix comprising 1-hydroxycyclohexyl phenyl ketone (0.01% w/w) in

2-methoxyethanol, triethylene glycol dimethacrylate (15.7% w/w), trimethylolpropane trimethacrylate (6.0% w/w) and glycidyl methacrylate (10% w/w) was used for MN manufacture (Nicholas et al., 2018). The authors employed a 30G hypodermic needle for MN fabrication and used a colorimetric system in combination with the obtained hollow MN for a quick and visual sensing of glucose in simulated interstitial fluid (Nicholas et al., 2018). The MN platform quickly extracted artificial interstitial fluid within 5 s and displayed a color change that was a function of glucose concentration within 30 s (Nicholas et al., 2018). Using the naked eye, the authors used this technique to show the difference between blood glucose levels in normal glycemia (4–7 mM) and hyperglycemia (>7 mM) (Nicholas et al., 2018).

Point-of-care diagnosis with highly automated, facile and low-cost technologies is imperative for personalized longitudinal monitoring or public screening (Al Sulaiman et al., 2019). Cancer and other diseases can be diagnosed and predicted by using cell-free nucleic acids (cfNAs) as biomarkers (Al Sulaiman et al., 2019). One group of non-coding RNAs that are 19–25 nucleotides in length is called microRNAs (or miRs). These molecules can be used either in combinations or as individual biomarkers (Al Sulaiman et al., 2019). Sensing technologies that can be employed to quantify specific nucleic acids in biological fluids are becoming increasingly important. Recently, Sulaiman and coworkers used hydrogel-coated MNs for sensing and sampling of nucleic acids in the dermal ISF (Al Sulaiman et al., 2019). The authors coated poly-L-Lactide MNs with alginate modified with Peptide Nucleic Acid and used this device for sequence-specific analysis of ISF for circulating nucleic acid (Al Sulaiman et al., 2019). Pyramid-shaped MNs (7 × 7) were fabricated with the height of the MNs kept at 550 μm. The authors subsequently used this device to pierce the epidermal layer of human skin and take ISF samples for analysis (Al Sulaiman et al., 2019). Sequence-specific immobilization of miRNA of interest was based on Watson-Crick base pairing (Al Sulaiman et al., 2019). This technology was characterized by large sampling capacity (~6.5 μL in 2 min) and fast sampling kinetics. Furthermore, the nucleic acids were assayed either in solution following light-mediated delivery from the hydrogel or the patch (Al Sulaiman et al., 2019).

Shotgun quantitative proteomics has been used to detect several proteins in the interstitial fluid with subsequent comparison of the proteome to matching plasma and serum (Arévalo et al., 2019). Recently, Arévalo and coworkers re-evaluated data from an an earlier proteomic study to ascertain different proportions of distinct interstitial fluid immunoglobulins in comparison to plasma and serum (Arévalo et al., 2019). The authors found that the composition of IgA, IgD, IgG and IgE in ISF was comparable to that of plasma or serum (Arévalo et al., 2019). The lower amounts of IgM, was attributed to the relatively lower concentrations of T-cell and B-cell subgroups found in the dermis (Arévalo et al., 2019). In another study, a microneedle device was employed for the collection of dermal interstitial fluid from three healthy subjects (Tran et al., 2018). The data was compared to those of matching plasma and serum samples. Employing a shotgun proteomic method, 407 proteins were assayed with at least one unique peptide, and of those, 135 proteins were expressed differently at least 2-fold (Tran et al., 2018). It was proved that interstitial fluid is analogous to both serum and plasma (Tran et al., 2018). The identity of less than 1% of proteins were specifically determined in the interstitial fluid (Tran et al., 2018).

Although interstitial fluid contains biomarkers of interest, interstitial fluid sampling for clinical applications is still beleaguered by the deficiency of facile and clinically relevant methods that can collect adequate volume of interstitial fluid (Samant & Prausnitz, 2018). Samant and Prausnitz used microneedle (MN) patches for theoretical and experimental investigation of dermal interstitial fluid (Samant & Prausnitz, 2018). The authors fabricated hydrogel MNs from a dispersion of PVA. This dispersion was subsequently poured into silicone molds and cross-linked at high temperature with overnight drying (Samant & Prausnitz, 2018). MNs were fabricated from chemically etched stainless-steel sheets (Samant & Prausnitz, 2018). To produce porous microneedles, the authors laser-cut sheets of filter paper by utilizing metal MNs as a protective mask. The paper MNs were subsequently wedged between two metal MN arrays with the aim of providing mechanical strength (Samant & Prausnitz, 2018). The authors made hollow MNs by using a rotary tool to shorten and rebevel hypodermic needles (Samant & Prausnitz, 2018). Coated MNs were produced by dipping stainless steel microneedles into maltose solution (Samant & Prausnitz, 2018). The authors collected more than 1 μL of interstitial fluid within 20 min from human subjects and porcine cadaver skin (Samant & Prausnitz, 2018). Microneedle patches with arrays of micron-scale porous, solid or hollow needles were used to pierce superficial skin layers and reach the interstitial fluid (Samant & Prausnitz, 2018). Interestingly, interstitial fluid collection from porcine skin was a function of transport mechanism according to the following rank order (diffusion < capillary action < osmosis < pressure-driven convection) (Samant & Prausnitz, 2018).

Hydrogel MN devices can be utilized for the extraction of skin interstitial fluid (ISF) via in situ hydrogel swelling (Zheng et al., 2020). Nevertheless, currently fabricated hydrogel MNs need several minutes for the collection of adequate volume (1 mL) for effective quantitation (Zheng et al., 2020). Zheng et al. used an osmolyte-powered hydrogel MN patch capable of extracting interstitial fluid 3 times faster than the currently used systems (Zheng et al., 2020). In situ quantification of the biomarkers was then carried out (Zheng et al., 2020). The MN patch was made from hydrogel (i.e., methacrylated hyaluronic acid) and osmolytes (i.e., maltose) (Zheng et al., 2020). The osmolytes normally dissolve in the ISF and provide osmotic pressure-mediated ISF diffusion from the skin to the hydrogel (Zheng et al., 2020). In animal experiments, a system with 100 MNs was used to extract 3.82 μL of ISF from mouse skin within 3 min and 7.90 μL of ISF from pig skin ex vivo. In comparison, hydrogel MNs with no osmolytes need more than 10 min to obtain comparable results (Zheng et al., 2020). Furthermore, researchers can quantify insulin or glucose in the extracted via integration with electronic glucose sensors (Zheng et al., 2020).

Lymphatic vasculature contributes to the maintenance of immune surveillance, tissue fluid homeostasis and dietary fat absorption into the intestine (Polomska et al., 2019). Changes in lymphatic function have been linked to several diseases, including chronic inflammation, defective wound healing, venous insufficiencies, lipedema and lymphedema (Polomska et al., 2019). Reproducible and reliable methods for quantifying lymphatic vessel function are important for monitoring and for diagnosis (Polomska et al., 2019). Polomska et al. developed a facile technique for the quantitative and painless evaluation of the lymphatic function of mouse skin (Polomska et al., 2019). A lymphatic-specific tracer system, that incorporates indocyanine green (a near-infrared fluorescent dye approved for clinical use) and Kolliphor HS15 (a surface-active agent) was made (Polomska et al., 2019). The authors used MicronJet600 hollow microneedles to deliver the tracer to the dermal layer of the skin, and a custom-made, portable detection device was subsequently used to quantify the rate of fluorescence signal decay (Polomska et al., 2019). The rate of fluorescence signal decay was used as a direct test of the lymphatic vessel's capacity to drain (Polomska et al., 2019). The investigators used this method to quantitatively evaluate damaged lymphatic clearance in transgenic mice with no dermal lymphatics (Polomska et al., 2019). This technique was also used to differentiate diverse lymphatic clearance patterns in pigs under manual stimulus and in several body locations (Polomska et al., 2019).

Ribet and colleagues combined the benefits of currently marketed continuous glucose monitoring (CGM) systems and hollow microneedles (HMNs) (Ribet et al., 2018). A CGM device that can be used to conduct minimally delayed and minimally invasive in-situ glucose quantification in the dermal ISF was developed (Ribet et al., 2018). The authors used conventional silicon microfabrication techniques to build a device that is based on the incorporation of an ultra-minute electrochemical sensing probe into the bore of a single HMN (Ribet et al., 2018). Quantitative and real-time assessment can be carried out based on the diffusion of molecules over a short distance. In this instance, the sensing electrodes are placed within the needle bore facing an opening toward the dermal space (Ribet et al., 2018). The device operates based on passive capillary filling of the lumen with no need for complicated fluid extraction procedures. Significantly, the transdermal part of the system is 50 times smaller in comparison to currently marketed products (Ribet et al., 2018). This painless approach reduces tissue trauma and facilitates access to the dermis. The three-electrode sensor possesses long-term stability (up to 4 days) as well as satisfactory linearity (up to 14 mM), sensitivity (1.5 nA/mM) and selectivity in-vitro (Ribet et al., 2018). The authors merged microneedles with sensor technology for dependable insertion in forearm skin. Investigations with humans demonstrated the possibility of dynamically and correctly tracking blood glucose level over time, with a lag time of about 10 min (physiologically acceptable delay) (Ribet et al., 2018). The proposed device can be used for convenient, real-time and minimally invasive CGM in diabetics (Ribet et al., 2018).

Chen and coworkers developed microneedles for convenient and painless diagnostic testing of breast cancers before the occurrence of metastasis (Chen et al., 2020). The authors collected tissue fluid with a thin film. The collected tissue fluid sample was subsequently incubated to form blue-colored products. In cancerous regions, color changes were more obvious than in healthy regions (Chen et al., 2020). In vivo experiments demonstrated that the nano-silver based colorimetric technique can be used for breast cancer assay in one week (Chen et al., 2020). Significantly, the tumors can be effectively suppressed with doxycycline. In contrast, the traditional blood test requires 2 weeks to confirm the presence of breast cancer (Chen et al., 2020). In another interesting study, immunodiagnostic microneedles (MNs) were used for the extraction and assay of a melanoma biomarker (S100B) (Totti et al., 2019). Melanoma cells (A375) were cultured in a highly porous polymeric scaffold with subsequent capture of

S100B in situ with the MN device (Totti et al., 2019). The MNs were used to detect S100B in the scaffold, while the investigators did not detect the biomarker in the adjoining supernatant (Totti et al., 2019).

Diagnostic devices for home-based continuous glucose monitoring can assist in finding timely treatments and preventing high blood pressure, nephropathy, and stroke (Z. Wang et al., 2020). The existing techniques for measuring glucose entail invasive blood sampling. Microneedle patches can be used to extract and detect metabolites or to deliver medications into the ISF (Z. Wang et al., 2020). The merging of sampling with immediate metabolite readout is capable of enhancing compliance by patients. To avoid the use of complicated in-lab apparatus, Wang et al. developed a transdermal colorimetric microneedle system that incorporates sampling and display (Z. Wang et al., 2020). This system can be used for CGS. At abnormally elevated glucose concentration, the domino enzymatic reactions of horseradish peroxidase (HRP) and glucose oxidase (GOx) leads to the coloration of 3,3′,5,5′-tetramethylbenzidine (Z. Wang et al., 2020). The authors biomineralized HRP in the top layer with calcium phosphate (CaP) shell as a pH responsive approach for protection from nonspecific reactions and enhanced sensitivity (Z. Wang et al., 2020). Minimally invasive extraction of the ISF and the quantitative conversion of glucose to a visible color change was achieved with this colorimetric sensor (Y. Wang et al., 2020). A scanned image of the microneedle was used to provide a red green and blue (RGB) information that is useful for quantitation (Z. Wang et al., 2020).

Under normal conditions, the concentration of lactate in the blood is between 0.5 and 2.0 mM (Y. Wang et al., 2020). Wang et al. developed a sensor for the electrochemical assay of lactate in body fluids (Y. Wang et al., 2020). A gold counter electrode, gold working electrode, and a built-in solid-state silver/silver chloride electrode, were connected to a MN device that can breach the human skin and extract a small volume of body fluid painlessly for in-situ lactate quantitation (Y. Wang et al., 2020). Lactate oxidase immobilized in an albumin matrix was used to functionalize the surface of the working electrode (Y. Wang et al., 2020). The authors measured the dependence of extracted mass of the compound on pressure and evaluated the efficiency of the transdermal sensor when applied to blood plasma and buffer solutions. The device was used to detect lactate in the 1–30 mM range, which is clinically important in various pathological conditions such as lactic acidosis (Wang, Ausri et al., 2020; Wang, Li et al., 2020). In a different study, Calio and coworkers used poly (ethylene glycol) diacrylate (PEGDA) doped by redox mediator, enzyme, and photoinitiator to fabricate working electrodes (Caliò et al., 2016). Standard photolithography was used to prepare the electrodes (Caliò et al., 2016). The device functions as the working electrode for the electrochemical assay of lactic acid and glucose when lactose oxidase (LOx) and glucose oxidase (GOx) are utilized as enzymes, respectively (Caliò et al., 2016). The biosensor displayed a linear response in the concentration range from 0 to 1 mM and from 0 to 4 mM, for lactic acid and glucose respectively. The limit of detection was 1 μM (Caliò et al., 2016).

Most physiologically relevant immune cell populations can be found in peripheral and barrier tissues such as the gut, skin, and other mucosal surfaces (Mandal et al., 2018). Examples are memory T cells, plasma cells, B cells, natural killer T cells, natural killer cells, and tissue-resident macrophages. Currently, immune monitoring is carried out mainly via blood analysis (Mandal et al., 2018). A prime example is the use of flow cytometry for peripheral blood sampling for immunophenotyping (Mandal et al., 2018). It is noteworthy that important cell groups cannot be reached by conventional blood draws that are utilized for immune system monitoring. A microneedle-based technology was employed for parallel monitoring of immune responses (Mandal et al., 2018). Coated MNs fabricated from a cross-linked biocompatible polymer, and capable of swelling upon skin insertion were used to create a porous matrix for local leukocyte infiltration (Mandal et al., 2018). By incorporating specific antigens into nanocapsules within the hydrogel coating as well as molecular adjuvants, antigen-specific lymphocytes can be enriched permitting a comprehensive functional and phenotypic analysis (Mandal et al., 2018). The authors used this device for experiments in mice immunized with a model protein antigen and a skin sample infected with vaccinia virus (Mandal et al., 2018). The MNs permitted tissue-resident memory T cells to be longitudinally tracked in the skin for several months following vaccination or infection. Antigen-specific T cell population declined to low or undetectable levels in the course of these studies (Mandal et al., 2018). The use of MNs did not alter the immune status of antigen-exposed or naïve mice (Mandal et al., 2018).

Electrochemical glucose sensors based on enzymes are useful because of good selectivity, low cost, and stability (Samavat et al., 2018). Glucose sensors based on tetrathiafulvalene (TTF) exhibit superior performance compared to sensors that use other materials (Samavat et al., 2018). The immobilization of the mediator both near the electrode's surface and the enzyme's redox center is crucial for electron transfer between the enzyme, the mediator and the electrode (Samavat et al., 2018). Deposition of sensing elements is usually carried out via drop casting leading to the fabrication of low-cost biosensors. Layer-by-layer and mixed drop casting of glucose oxidase (GOx) and TTF are frequently conducted (Samavat et al., 2018). However, GOx and TTF do not mix well because of poor solubility in organic and aqueous solvents respectively. Mixing GOx and TTF is challenging via the traditional drop-cast method (Samavat et al., 2018). To overcome this difficulty, the authors used a spray mixing method to mix these compounds (Samavat et al., 2018). GOx and TTF were deposited successively via a spray layer-by-layer (SL) technique and the traditional drop-cast (DL) method (Samavat et al., 2018). The two techniques were compared with the simultaneous spray mixing (SM) technique. With SM, a uniformly mixed sensing layer was deposited on to three-dimensional MN arrays for continuous glucose monitoring (Samavat et al., 2018). GOx mixes with TTF as erosol particles to generate a uniformly mixed sensing layer (Samavat et al., 2018). SM's sensors were more efficient and this was ascribed to the even distribution of GOx and TTF throughout the sensing layer with increased redox reaction and electron transfer between TTF and GOx (Samavat et al., 2018). The fabricated microneedle system for CGM only breaches the epidermis to access ISF with insignificant pain in contrast to continuous glucose monitoring devices that are currently being marketed (Samavat et al., 2018). The later class of devices can breach the skin deeper leading to significant pain (Samavat et al., 2018). Furthermore, the MN device for continuous glucose monitoring can be produced at a lower cost (Samavat et al., 2018).

Tracking biological signals via electroencephalography (EEG), electrocardiography (ECG), and electromyography (EMG) is relevant for a more valuable understanding of the physiological or pathological conditions of the human body (Forvi et al., 2012). Wet electrodes are typically used for measuring biological potentials but this approach has a number of disadvantages (a) the sensitivity to motions artifacts, a crucial element in EMG and ECG monitoring, (b) the protracted process to be applied in EEG examinations, and (iii) not being able to track biosignals in long-term applications (Forvi et al., 2012). MN-based dry electrodes were evaluated (Forvi et al., 2012) and it was demonstrated that these electrodes: (a) permit a better electro-mechanical interface with human skin, (b) possess performance that can be compared to wet electrodes in terms of the measurement of EMG, EEG, and static ECG signals but with better monitoring of ECG signal in dynamic cases, and (c) do not require protracted skin preparation for EEG applications, and it is user friendly (Forvi et al., 2012). MN-based dry electrodes can serve as effective option to conventional wet electrodes used for the monitoring of biological signals (Forvi et al., 2012).

Biosensors for the skin monitoring of molecules are interesting devices (Sharma et al., 2017). Microneedles can be used to investigate biomarkers in the ISF of the skin, but there are technical challenges that require resolution (Sharma et al., 2017). These sensors have to be convenient, reliable, and cost-effective. Inexpensive and scalable technologies can be used to fabricate transdermal these sensing devices (Sharma et al., 2017). Sharma and coworkers developed a technique for the fabrication of microneedle arrays for continuous analyte monitoring (Sharma et al., 2017). The authors characterized the microneedle arrays mechanically and it was demonstrated that the device can breach human skin under slight thumb pressure (Sharma et al., 2017). The microneedles were subsequently functionalized and assessed as lactate and glucose biosensors (Sharma et al., 2017).

Devices for monitoring circulating proteins in ISF are receiving increasing attention (Coffey et al., 2018). Minimal invasive microneedle patches can be used to selectively capture circulating proteins for diagnostic purposes. It is important to extract an adequate amount of specific proteins from the skin in a timely fashion (Coffey et al., 2018). Coffey et al. utilized a wearable immunoassay patch with a high surface area for quick and selective sampling of immunoglobulin G (IgG) from the skin (Coffey et al., 2018). The authors studied the influence of microneedle geometry (i.e. length, projection density, and array size) on IgG capture (Coffey et al., 2018). The authors demonstrated that by increasing array size (4–36 mm^2) and the projection density (5000–30,000 projections/cm^2) there was a significant improvement in biomarker capture without any change in tolerability (Coffey et al., 2018). The researchers developed a high surface area microneedle system (30,000 projections/cm^{-2}, 36 mm^2,140 μm) with a 4.5-fold enhancement in permeated surface area in comparison to a previous microneedle array (20,408 projections/cm^{-2}, 16 mm^2, 100 μm) designed by the same research group (Coffey et al., 2018). The high surface area microneedle array captured antigen-specific IgG from mice in 30 s with 100% diagnostic sensitivity compared with 10–30 min for the previously used microneedle array (Coffey et al., 2018). The authors showed that the microprojection array system may be utilized for ultra-fast (<1 min) protein capture from the skin in a timely manner (Coffey et al., 2018).

Researchers are continuously developing cost-effective, efficient and simple diagnostic methods (Bhargav et al., 2012). Some investigators have developed microneedles functionalized with bioreceptors with the subsequent use of these devices for efficient capture of molecular markers (Bhargav et al., 2012; Dervisevic et al., 2020). In a typical scenario, an antibody-modified microneedle array is first injected into the skin for the specific binding with biomarker (s) found in the interstitial fluid (Bhargav et al., 2012; Dervisevic et al., 2020) before removing the MNA for further analysis. This method is beneficial in comparison to MNA-based approaches that extract whole interstitial fluid because additional sample pretreatment steps may be needed due to the biological complexity of the interstitial fluid (Bhargav et al., 2012; Dervisevic et al., 2020). Microprojection arrays can be used to breach the skin and capture specific biomarkers (Bhargav et al., 2012). Recently, Bhargav et al. used carbodiimide chemistry techniques to immobilize biomarker-specific probes for capture and quantitation (Bhargav et al., 2012). There was significant variation across the MN surface and capture efficiencies were relatively poor. The authors studied a modified microprojection array with the aim of understanding the source of the variability and enhancing capture efficiency (Bhargav et al., 2012). The protein immobilization stage was linked to most of the variance in surface uniformity. Changing the protein immobilization conditions led to a 14-fold decline in overall variance and about 18-fold increase in the quantity of the captured biomarker (Bhargav et al., 2012). Optimizing the surface chemistry of the MN device resulted in enhanced capture of biomarkers from mouse skin's interstitial fluid (Bhargav et al., 2012).

Transdermal electrolytic conduction is imperative for user-friendly skin-patch devices (Nagamine et al., 2017). A two-dimensional and painless MN array was studied for the sampling of interstitial fluid and transdermal drug delivery (Nagamine et al., 2017). Nagamine et al. used porous microneedles for this purpose (Nagamine et al., 2017). The microneedles were 100 μm long and each needle was supported by a thicker cylindrical protrusion from a planar substrate (Nagamine et al., 2017). Since substrate and post were porous just like the needles, the needle chip was pervious for the electrolyte (Nagamine et al., 2017). Local tracking of edema was conducted with a porous and ion-conductive microneedle array interfaced with a transdermal conductometer. The porous device was used for microneedle-based transdermal electrical diagnosis and therapy (Nagamine et al., 2017).

The human body controls acid-base homeostasis via the lungs, kidneys and buffer systems in such a manner that even a slight alteration in pH can have a significant influence on the functions of many organs (Mani et al., 2017). Mani and coworkers fabricated real-time and microneedle-based pH sensors (Mani et al., 2017). The authors employed ZnO thin films on tungsten MNs and Ag/AgCl thick films as working and reference electrodes respectively (Mani et al., 2017). Different analytical techniques were utilized for the study of the structural and morphological properties of the MNs (Mani et al., 2017). A Nernstian response of -46 mV/pH was displayed by the sensor. Various conditions such as different buffer solutions, with respect to time were used to test the stability and accuracy of the sensor and this was compared with currently marketed pH electrodes (Mani et al., 2017). In addition, the MN sensor was also studied in mouse bladders and cerebrospinal fluid. The results were reproducible after several rounds of testing and the pH sensor procedure was reversible (Mani et al., 2017).

In a separate set of experiments, Wu et al. used post stretching and extrusion-based three-dimensional printing to prepare an MN patch for glucose-responsive insulin delivery (Wu et al., 2020). The authors initially studied the printability of various biological inks made up of hydroxyapatite and alginate (Wu et al., 2020). After printing the substrate and a cylindrical array of the patch, the top surface of the cylindrical microneedle array was stretched to create needle-like tips (Wu et al., 2020). The MN patch consisted of 6×6 microneedles, and only the MNs took part in the glucose-responsive release of insulin (Wu et al., 2020). The MNs were conical with a tip diameter of 24 μm, a base diameter of 601 μm, and a height of 643 μm (Wu et al., 2020). The MNs were mechanically robust enough to pierce mouse skin and responsively release insulin according to the glucose levels (Wu et al., 2020). When the authors administered the MNs only once on mouse skin, the patches brought blood glucose concentration of diabetic mice to normal levels for up to 40 h and relieved diabetic symptoms (Wu et al., 2020).

Physicians typically use blood and/or skin tests to determine the allergen(s) (Ito et al., 2017). Ito and coworkers investigated the feasibility of utilizing dissolving MNs as a skin allergy test device (Ito et al., 2017). The authors employed poly-L-arginine as the model allergen (Ito et al., 2017). Dextran was used to fabricate three types of dissolving microneedles (DMs) consisting of diverse doses of poly-L-arginine (Ito et al., 2017): and 87.4 ± 1.1 μg (high-dose DM), 42.2 ± 0.8 μg (medium-dose DMs), 17.1 ± 0.5 μg (low-dose DMs); each 1.0 cm^2 chip had 300 DMs (Ito et al., 2017). The mean diameters of the base were 301 ± 2, 299 ± 1, and 299 ± 2 μm and the average lengths of the low-, medium-, and high-dose

DM were 489 ± 3, 485 ± 3, and 492 ± 1 μm respectively (Ito et al., 2017). For the low-, medium-, and high-dose DMs, the scratching behavior per 30 min was 9.8 ± 3.4, 60.4 ± 8.3, and 95.7 ± 10.6 times while the administered doses of poly-L-arginine were 9.3 ± 1.9, 31.1 ± 1.3, and 61.9 ± 4.7 μg respectively (Ito et al., 2017). The authors observed the dose dependence of the poly-L-arginine immunoreactivity when administered as DMs, suggesting that DMs can potentially replace the devices currently used to identify allergens (Ito et al., 2017). In another interesting study, Esfandyarpour and coworkers developed a microneedle biosensor which uses changes in impedance and ionic current to measure biomolecules such as nucleic acids and proteins (Esfandyarpour et al., 2013). The authors prepared and electrically characterized the sensors. A target protein was successfully detected with the sensor. In this study, streptavidin was the analyte and biotinylated bovine serum albumin served as the receptor (Esfandyarpour et al., 2013).

Tremendous progress has been made in our understanding of genomic signatures (Ibrahim et al., 2017; Kabbarah & Chin, 2005). However, skin disease diagnosis is still based on a subjective eye examination, which is often followed by histology (Ibrahim et al., 2017). These measures are often not sensitive enough, and in many cases, biopsies are not justified, making it challenging or impractical to obtain samples (Ibrahim et al., 2017). Ibrahim and colleagues used MiNDERA's whole transcriptome microneedle device to extract adequate amounts of mRNA from ex vivo as well as in vivo human skin (Ibrahim et al., 2017). Following extraction, the biomarkers were processed using facile techniques that can be adapted to less complex environments and analyzed by next generation sequencing (NGS) and quantitative polymerase chain reaction (Ibrahim et al., 2017). The authors used this technique to quantify mRNA transcripts in the skin (Ibrahim et al., 2017). A comparison between punch biopsy and in vivo MN extraction revealed excellent agreement between the two techniques (Ibrahim et al., 2017). It was shown that the microneedle-based method can be used for non-invasive extraction of skin transcriptome with an efficiency that is similar to biopsy (Ibrahim et al., 2017). The device facilitates the monitoring of the molecular "fingerprint" of the skin via non-invasive extraction of the transcriptome (Ibrahim et al., 2017).

Another set of investigators fabricated a non-enzymatic sensor based on a platinum black-modified stainless steel microneedle array (Dervisevic et al., 2020). A technique based on wet chemical etching of stainless steel followed by 90° bend of the MNs was utilized for the preparation of the microneedle array (Dervisevic et al., 2020). The authors electrodeposited a thin gold layer electrodeposited on the stainless steel microneedle surface and passivated the array via a parylene coating approach (Dervisevic et al., 2020). The tips of the MNs were electroplated to form either a silver/silver chloride layer or a platinum black layer (Dervisevic et al., 2020). The authors subsequently coated the nanostructured platinum black-modified microneedle tips with a Nafion membrane to erase interference from other molecules (Dervisevic et al., 2020). The microneedle array was used for continuous transdermal monitoring of glucose. In another interesting project, a hollow microneedle-mediated transdermal biosensor for high-accuracy, painless and real-time quantitation of glucose in the ISF was described (Strambini et al., 2015). The transdermal biosensor employs densely-packed silicon-dioxide hollow microneedles (up to 1×10^6 needles/cm^2) and small (lumen down to 4 μm) (Strambini et al., 2015). This permits reliable, efficient, and pump-free (self-powered) uptake of interstitial fluid in a few seconds (1 μL/s) by using microneedle capillarity (Strambini et al., 2015). By combining a glucose biosensor with MNs operating under capillary-action, glucose can be quantified with high reproducibility and accuracy over the concentration range of 0−630 mg/dL in real-time (30 s) (Strambini et al., 2015).

Diabetics typically measure the concentration of glucose in the blood glucose with lancet/blood-glucose meter followed by insulin administration 2 or 3 times daily via subcutaneous injection (Hsu et al., 2020). Lipodystrophy and pain are typical common adverse events following this approach (Hsu et al., 2020). The fundamental goal of diabetes management is to attain safe and painless continuous glycemic control and instrument-free glucose-biosensing (Hsu et al., 2020). Hsu et al. designed an integrated system consisting of an insulin-delivery microneedle patch and a glucose-measuring microneedle patch (Hsu et al., 2020). When this system was implanted in the skin, ISF was transported into the patch with color change under conditions of hyperglycemia which was subsequently measured by a smartphone (Hsu et al., 2020). In the next phase, insulin-delivery microneedle patch comprising insulin-loaded nanovesicles and free insulin was applied (Hsu et al., 2020). In chemically induced type 1 diabetic rats, the blood glucose concentration was quickly quantitated by the glucose biosensor (Hsu et al., 2020). Blood glucose concentration was also reduced to and kept at normal concentration within 13 h using the insulin-delivery microneedle patch (Hsu et al., 2020). This device can be used for syringe-free diabetes management thus circumventing both hypoglycemia and hyperglycemia (Hsu et al., 2020).

Senel et al. used conductive gold ink casting to fabricate gold microneedles (Senel et al., 2019). The investigators functionalized gold microneedles with a ferrocene- and epoxy-functional polymers and used this electrochemical biosensor to covalently immobilized urease (Senel et al., 2019). The authors used the microneedle device for urea quantitation in a skin mimic and artificial interstitial fluid, both spiked with the compound (urea) (Senel et al., 2019). The analyte was quantified in the range of 50–2500 μM, with a sensitivity of 31 nA/mM and a calculated limit of detection of 2.8 μM (Senel et al., 2019). MN-based biosensors can be used for the fabrication of wearable real-time sensing devices (Senel et al., 2019).

References

Al Sulaiman, D., Chang, J. Y. H., Bennett, N. R., Topouzi, H., Higgins, C. A., Irvine, D. J., & Ladame, S. (2019). Hydrogel-coated microneedle arrays for minimally invasive sampling and sensing of specific circulating nucleic acids from skin interstitial fluid. *ACS Nano, 13*(8), 9620–9628. https://doi.org/10.1021/acsnano.9b04783

Arévalo, M. T., Rizzo, G. M., Polsky, R., Glaros, T., & Mach, P. M. (2019). Proteomic characterization of immunoglobulin content in dermal interstitial fluid. *Journal of Proteome Research, 18*(6), 2381–2384. https://doi.org/10.1021/acs.jproteome.9b00155

Babity, S., Roohnikan, M., & Brambilla, D. (2018). Advances in the design of transdermal microneedles for diagnostic and monitoring applications. *Small, 14*(49), 201803186. https://doi.org/10.1002/smll

Bhargav, A., Muller, D. A., Kendall, M. A. F., & Corrie, S. R. (2012). Surface modifications of microprojection arrays for improved biomarker capture in the skin of live mice. *ACS Applied Materials and Interfaces, 4*(5), 2483–2489. https://doi.org/10.1021/am3001727

Bollella, P., Sharma, S., Cass, A. E. G., & Antiochia, R. (2019). Microneedle-based biosensor for minimally-invasive lactate detection. *Biosensors and Bioelectronics, 123*, 152–159. https://doi.org/10.1016/j.bios.2018.08.010

Caliò, A., Dardano, P., Di Palma, V., Bevilacqua, M. F., Di Matteo, A., Iuele, H., & De Stefano, L. (2016). Polymeric microneedles based enzymatic electrodes for electrochemical biosensing of glucose and lactic acid. *Sensors and Actuators B: Chemical, 236*, 343–349. https://doi.org/10.1016/j.snb.2016.05.156

Chen, D., Wang, C., Chen, W., Chen, Y., & Zhang, J. X. J. (2015). PVDF-Nafion nanomembranes coated microneedles for in vivo transcutaneous implantable glucose sensing. *Biosensors and Bioelectronics, 74*, 1047–1052. https://doi.org/10.1016/j.bios.2015.07.036

Chen, L., Zhang, C., Xiao, J., You, J., Zhang, W., Liu, Y., Xu, L., Liu, A., Xin, H., & Wang, X. (2020). Local extraction and detection of early stage breast cancers through a microneedle and nano-Ag/MBL film based painless and blood-free strategy. *Materials Science and Engineering: C, 109*, 110402. https://doi.org/10.1016/j.msec.2019.110402

Coffey, J. W., Corrie, S. R., & Kendall, M. A. F. (2018). Rapid and selective sampling of IgG from skin in less than 1 min using a high surface area wearable immunoassay patch. *Biomaterials, 170*, 49–57. https://doi.org/10.1016/j.biomaterials.2018.03.039

Dardano, P., Rea, I., & De Stefano, L. (2019). Microneedles-based electrochemical sensors: New tools for advanced biosensing. *Current Opinion in Electrochemistry, 17*, 121–127. https://doi.org/10.1016/j.coelec.2019.05.012

Dervisevic, M., Alba, M., Prieto-Simon, B., & Voelcker, N. H. (2020). Skin in the diagnostics game: Wearable biosensor nano- and microsystems for medical diagnostics. *Nano Today, 30*, 100828. https://doi.org/10.1016/j.nantod.2019.100828

Dixon, R. V., Skaria, E., Lau, W. M., Manning, P., Birch-Machin, M. A., Moghimi, S. M., & Ng, K. W. (2021). Microneedle-based devices for point-of-care infectious disease diagnostics. *Acta Pharmaceutica Sinica B*. https://doi.org/10.1016/j.apsb.2021.02.010

Early, S., Moon, I. S., Bommakanti, K., Hunter, I., & Stankovic, K. M. (2019). A novel microneedle device for controlled and reliable liquid biopsy of the human inner ear. *Hearing Research, 381*, 107761. https://doi.org/10.1016/j.heares.2019.06.004

Esfandyarpour, R., Esfandyarpour, H., Javanmard, M., Harris, J. S., & Davis, R. W. (2013). Microneedle biosensor: A method for direct label-free real time protein detection. *Sensors and Actuators B: Chemical, 177*, 848–855. https://doi.org/10.1016/j.snb.2012.11.064

Forvi, E., Bedoni, M., Carabalona, R., Soncini, M., Mazzoleni, P., Rizzo, F., O'Mahony, C., Morasso, C., Cassarà, D. G., & Gramatica, F. (2012). Preliminary technological assessment of microneedles-based dry electrodes for biopotential monitoring in clinical examinations. *Sensors and Actuators, A: Physical, 180*, 177–186. https://doi.org/10.1016/j.sna.2012.04.019

Hegarty, C., McConville, A., McGlynn, R. J., Mariotti, D., & Davis, J. (2019). Design of composite microneedle sensor systems for the measurement of transdermal pH. *Materials Chemistry and Physics, 227*, 340–346. https://doi.org/10.1016/j.matchemphys.2019.01.052

Hsu, W. L., Huang, C. Y., Hsu, Y. P., Hwang, T. L., Chang, S. H., Wang, H. Y. J., Feng, L. Y., Tzou, S. J., Wei, K. C., & Yang, H. W. (2020). On-skin glucose-biosensing and on-demand insulin-zinc hexamers delivery using microneedles for syringe-free diabetes management. *Chemical Engineering Journal, 398*. https://doi.org/10.1016/j.cej.2020.125536

Ibrahim, S., Dickerson, T., Taft, B., Pramanick, S., Mannath, T., & Nore, P. (2017). LB968 Non-invasive skin transcriptome extraction using MiNDERA microneedle technology. *Journal of Investigative Dermatology, B7*. https://doi.org/10.1016/j.jid.2017.07.045

Ito, Y., Matsumoto, K., Osakama, N., Yoshioka, R., Kobuchi, S., Sakaeda, T., & Takada, K. (2017). Dissolving microneedles as skin allergy test device. *Biological and Pharmaceutical Bulletin, 40*(4), 531–534. https://doi.org/10.1248/bpb.b16-00768

Kabbarah, O., & Chin, L. (2005). Revealing the genomic heterogeneity of melanoma. *Cancer Cell, 8*(6), 439–441. https://doi.org/10.1016/j.ccr.2005.11.008

Liu, G.-S., Kong, Y., Wang, Y., Luo, Y., Fan, X., Xie, X., Yang, B.-R., & Wu, M. X. (2020). Microneedles for transdermal diagnostics: Recent advances and new horizons. *Biomaterials, 232*, 119740. https://doi.org/10.1016/j.biomaterials.2019.119740

Madden, J., O'Mahony, C., Thompson, M., O'Riordan, A., & Galvin, P. (2020). Biosensing in dermal interstitial fluid using microneedle based electrochemical devices. *Sensing and Bio-Sensing Research, 29*, 100348. https://doi.org/10.1016/j.sbsr.2020100348

Mandal, A., Boopathy, A. V., Lam, L. K. W., Moynihan, K. D., Welsh, M. E., Bennett, N. R., … Irvine, D. J. (2018). Cell and fluid sampling microneedle patches for monitoring skin-resident immunity. *Science Translational Medicine, 10*(467). https://doi.org/10.1126/scitranslmed. eaar2227.

Mani, G. K., Miyakoda, K., Saito, A., Yasoda, Y., Kajiwara, K., Kimura, M., & Tsuchiya, K. (2017). Microneedle pH sensor: Direct, label-free, real-time detection of cerebrospinal fluid and bladder pH. *ACS Applied Materials and Interfaces, 9*(26), 21651–21659. https://doi.org/10.1021/acsami.7b04225

Mohan, A. M. V., Windmiller, J. R., Mishra, R. K., & Wang, J. (2017). Continuous minimally-invasive alcohol monitoring using microneedle sensor arrays. *Biosensors and Bioelectronics, 91*, 574–579. https://doi.org/10.1016/j.bios.2017.01.016

Nagamine, K., Kubota, J., Kai, H., Ono, Y., & Nishizawa, M. (2017). An array of porous microneedles for transdermal monitoring of intercellular swelling. *Biomedical Microdevices, 19*(3). https://doi.org/10.1007/s10544-017-0207-y

Ng, K., Dixon, R., Lau, W., & Moghimi, S. (2020). The diagnostic potential of microneedles in infectious diseases. *Precision Nanomedicine, 3*, 629–640.

Nicholas, D., Logan, K. A., Sheng, Y., Gao, J., Farrell, S., Dixon, D., … Callan, J. F. (2018). Rapid paper based colorimetric detection of glucose using a hollow microneedle device. *International Journal of Pharmaceutics, 547*(1–2), 244–249. https://doi.org/10.1016/j.ijpharm.2018.06.002

Paliwal, S. (2013). Skin in the diagnostics game: Wearable biosensor nano- and microsystems for medical diagnostics. *European Journal of Pharmaceutical Sciences, 50*(5).

Paliwal, S., Hwang, B. H., Tsai, K. Y., & Mitragotri, S. (2013). Diagnostic opportunities based on skin biomarkers. *European Journal of Pharmaceutical Sciences, 50*(5), 546–556.

Polomska, A. K., Proulx, S. T., Brambilla, D., Fehr, D., Bonmarin, M., Brändli, S., Meboldt, M., Steuer, C., Vasileva, T., Reinke, N., Leroux, J. C., & Detmar, M. (2019). Minimally invasive method for the point-of-care quantification of lymphatic vessel function. *JCI Insight, 4*(4). https://doi.org/10.1172/jci.insight.126515

Ribet, F., Stemme, G., & Roxhed, N. (2018). Real-time intradermal continuous glucose monitoring using a minimally invasive microneedle-based system. *Biomedical Microdevices, 20*(4), 101. https://doi.org/10.1007/s10544-018-0349-6

Samant, P. P., & Prausnitz, M. R. (2018). Mechanisms of sampling interstitial fluid from skin using a microneedle patch. *Proceedings of the National Academy of Sciences of the United States of America, 115*(18), 4583–4588. https://doi.org/10.1073/pnas.1716772115

Samavat, S., Lloyd, J., O'Dea, L., Zhang, W., Preedy, E., Luzio, S., & Teng, K. S. (2018). Uniform sensing layer of immiscible enzyme-mediator compounds developed via a spray aerosol mixing technique towards low cost minimally invasive microneedle continuous glucose monitoring devices. *Biosensors and Bioelectronics, 118*, 224–230. https://doi.org/10.1016/j.bios.2018.07.054

Senel, M., Dervisevic, M., & Voelcker, N. H. (2019). Gold microneedles fabricated by casting of gold ink used for urea sensing. *Materials Letters, 243*, 50–53. https://doi.org/10.1016/j.matlet.2019.02.014

Sharma, S., Saeed, A., Johnson, C., Gadegaard, N., & Cass, A. E. (2017). Rapid, low cost prototyping of transdermal devices for personal healthcare monitoring. *Sensing and Bio-Sensing Research, 13*, 104–108. https://doi.org/10.1016/j.sbsr.2016.10.004

Strambini, L. M., Longo, A., Scarano, S., Prescimone, T., Palchetti, I., Minunni, M., Giannessi, D., & Barillaro, G. (2015). Self-powered microneedle-based biosensors for pain-free high-accuracy measurement of glycaemia in interstitial fluid. *Biosensors and Bioelectronics, 66*, 162–168. https://doi.org/10.1016/j.bios.2014.11.010

Sun, W., Araci, Z., Inayathullah, M., Manickam, S., Zhang, X., Bruce, M. A., Marinkovich, M. P., Lane, A. T., Milla, C., Rajadas, J., & Butte, M. J. (2013). Polyvinylpyrrolidone microneedles enable delivery of intact proteins for diagnostic and therapeutic applications. *Acta Biomaterialia, 9*(8), 7767–7774. https://doi.org/10.1016/j.actbio.2013.04.045

Takeuchi, K., & Kim, B. (2018). Functionalized microneedles for continuous glucose monitoring. *Nano Convergence, 5*(1). https://doi.org/10.1186/s40580-018-0161-2

Totti, S., Ng, K. W., Dale, L., Lian, G., Chen, T., & Velliou, E. G. (2019). A novel versatile animal-free 3D tool for rapid low-cost assessment of immunodiagnostic microneedles. *Sensors and Actuators B: Chemical, 296*, 126652. https://doi.org/10.1016/j.snb.2019

Tran, B. Q., Miller, P. R., Taylor, R. M., Boyd, G., Mach, P. M., Rosenzweig, C. N., Baca, J. T., Polsky, R., & Glaros, T. (2018). Proteomic characterization of dermal interstitial fluid extracted using a novel microneedle-assisted technique. *Journal of Proteome Research, 17*(1), 479–485. https://doi.org/10.1021/acs.jproteome.7b00642

Wang, Y., Ausri, I. R., Wang, Z., Derry, C., & Tang, X. S. (2020). Towards a transdermal membrane biosensor for the detection of lactate in body fluids. *Sensors and Actuators B: Chemical, 308*, 127645. https://doi.org/10.1016/j.snb.2019.127645

Wang, Z., Li, H., Wang, J., Chen, Z., Chen, G., Wen, D., Chan, A., & Gu, Z. (2020). Transdermal colorimetric patch for hyperglycemia sensing in diabetic mice. *Biomaterials, 237*, 119782. https://doi.org/10.1016/j.biomaterials.2020.119782

Wu, M., Zhang, Y., Huang, H., Li, J., Liu, H., Guo, Z., Xue, L., Liu, S., & Lei, Y. (2020). Assisted 3D printing of microneedle patches for minimally invasive glucose control in diabetes. *Materials Science and Engineering: C, 117*, 111299. https://doi.org/10.1016/j.msec.2020.111299

Zhang, X., Chen, G., Bian, F., Cai, L., & Zhao, Y. (2019). Encoded microneedle arrays for detection of skin interstitial fluid biomarkers. *Advanced Materials, 31*(37), 1902825. https://doi.org/10.1002/adma.201902825

Zheng, M., Wang, Z., Chang, H., Wang, L., Chew, S. W. T., Lio, D. C. S., Cui, M., Liu, L., Tee, B. C. K., & Xu, C. (2020). Osmosis-powered hydrogel microneedles for microliters of skin interstitial fluid extraction within minutes. *Advanced Healthcare Materials, 9*(10), 01683. https://doi.org/10.1002/adhm. 2019

CHAPTER 7

Therapeutic drug monitoring with microneedles

Introduction

Therapeutic drug monitoring (TDM) is a term used to denote the scientific discipline that deals with drug therapy optimization via the achievement of targeted drug levels in biological matrices (Kiang et al., 2017). The range of therapeutic drug concentration is chosen with the objective of enhancing therapeutic efficacy and reducing adverse events (Kiang et al., 2017). Clinicians sometimes measure the blood level of the drug and use that information to modify the drug dosing schedule (Ates et al., 2020). Several reports have demonstrated that the concentration of a medication in the blood has a better correlation with efficacy or toxicity compared to dosage (Ates et al., 2020). Alterations that affect the pharmacokinetic profile of a drug can influence exposure of the target site to the medication (Domínguez-Gil Hurlé et al., 2006). This can significantly affect the safety and efficacy of pharmacotherapy (Domínguez-Gil Hurlé et al., 2006). TDM is routinely carried out in patients who have undergone organ transplantation (Caffarel-Salvador et al., 2015). It is one of the pillars of pharmacotherapeutic control in several forms of epilepsy and it is also performed personally by most diabetic patients every day (Caffarel-Salvador et al., 2015).

TDM for antibiotic therapy entails the use of patient's plasma for drug quantitation with the aim of reducing toxicity especially when drugs with narrow therapeutic indices are involved (Mabilat et al., 2020). The target patients for therapeutic drug monitoring are those hospitalized for longer term care or those in acute care settings (Mabilat et al., 2020). TDM can also be applied to a broader spectrum of antibiotics (those with wider therapeutic indices) with the aim of minimizing toxicity and boosting efficacy (Mabilat et al., 2020). It leverages on the concept of the most effective antibiotic exposure for a favorable therapeutic outcome (Mabilat et al., 2020). For clinicians, medications for therapeutic drug monitoring

Microneedles
ISBN 978-0-323-91811-4
https://doi.org/10.1016/B978-0-323-91811-4.00001-8

must meet certain benchmarks (Ates et al., 2020). In the first place, there should not be an alternative measurable endpoint of therapeutic relevance such as blood pressure (Ates et al., 2020). Secondly, a dependable correlation must be demonstrated between the clinical effect and the concentration of the analyte. Thirdly, the medication must have a low therapeutic index, that is, the toxic and noneffective concentration values must be close to the therapeutic concentration (Ates et al., 2020). Fourthly, the relationship between the blood concentration and the drug dosing regimen must be poor, with high interindividual variability. In addition, the time for therapy should be protracted enough for TDM to become relatively advantageous. Finally, adverse outcomes should be obvious if there is overdosing or underdosing (Ates et al., 2020). The utility of therapeutic drug monitoring tightly correlates with the sensitivity and specificity of the analytical technique. In the context of modern TDM, chromatographic techniques in tandem with immunoassays or mass spectrometers are commonly utilized (Ates et al., 2020). It is often assumed that when blood is utilized for TDM, the concentration of the medication in the blood compartment correlates to that achieved in the target site, and sometimes this is incorrect (e.g., antibiotic concentration in the fat or muscle tissues can vary from that found in the blood) (Kiang et al., 2017). In addition, taking blood samples is an invasive procedure that can have dire consequences such as psychological trauma or iatrogenic infection (Kiang et al., 2017).

In the pediatric, elderly or anemic population or those with fragile veins, it might not be practical, possible or even ethical to take blood samples for the purposes of therapy (Kiang et al., 2017). Furthermore, frequency of obtaining blood from patients is often scaled down due to unease linked to infectious diseases (Ito et al., 2016). Thus, it is important to develop new and painless techniques for TDM (Ito et al., 2016). With the significant costs linked to the collection and analysis of blood samples (e.g., emolument for phlebotomists, nurses, and lab technicians), clinicians and researchers are investigating substitute biomatrices for TDM (Kiang et al., 2017). Biological matrices used for TDM as replacement for blood should not have the above-mentioned adverse features (Kiang et al., 2017).

Traditional TDM techniques possess certain practical disadvantages in the context of large-scale practice (Ates et al., 2020). These include long turnaround times, lack of standardization in workflows, and high cost of analytical equipment with complicated sample preparation steps (Ates et al., 2020). New developments in sensing technologies can be used to circumvent these drawbacks and to leverage the full potential of TDM (Ates et al., 2020). Several less-invasive and surrogate biological matrices (e.g., urine, breast milk, semen, tears, sweat, interstitial fluid, hair and oral fluids) have been studied for possible use in TDM (Kiang et al., 2017). Apart from some scenarios where a specific biological matrix (e.g., tears) is required (e.g., quantitation of drug concentrations in the case of eye diseases), interstitial fluid and saliva can serve as substitutes to blood (Kiang et al., 2017). Interstitial fluid has a composition that is comparable to plasma even though the protein concentration is lower; for that reason, the matrix is less burdensome to analyze in comparison to blood and the medications that are found in the ISF are predominantly in an active form (Kiang et al., 2017). Since the interstitial fluid is the connection between cells and the vasculature, it serves as the passage for various cellular wastes and signaling molecules (Kiang et al., 2017).

The utilization of interstitial fluid as a substitute biological matrix for therapeutic drug monitoring has generated global interest (Kiang et al., 2017). Several investigators studying therapeutic drug monitoring for chemotherapy (i.e., methotrexate, carboplatin), anticonvulsants (i.e., valproic acid), immunosuppressants (i.e., tacrolimus), antibiotics (i.e., vancomycin, gentamicin) and other medications have found strong correlations between drug concentrations in the blood/serum and the ISF (Ranamukhaarachchi et al., 2016). Due to the low amount of interstitial fluid in the skin (about 20 nL/mm^2 in the epidermis and approximately 800 nL/mm^2 in the dermis), obtaining adequate volumes of the biological fluid for assay is a key hindrance in designing biosensors (Kiang et al., 2017). MN-integrated sensors are seen as feasible solutions to this challenge (Kiang et al., 2017). Microneedle-integrated biosensors can extract interstitial fluid and carry out on-device analysis of biomolecules, without the need to transfer the fluid to a different device or equipment (Ito et al., 2016). Recently, the use of microneedles has generated interest as a substitute approach for therapeutic drug monitoring (Ito et al., 2016). Microneedles can be used for the painless extraction of dermal interstitial fluid containing biological markers without the expertize needed for or the risk linked to blood draw (Ito et al., 2016). In the next section of this chapter, we will examine specific examples of MN-assisted therapeutic drug monitoring.

Microneedle-mediated therapeutic drug monitoring

Vancomycin

Vancomycin is an antibiotic that is utilized for the management of systemic infections, notably those caused by methicillin resistant *Staphylococcus aureus*. This glycopeptide antibiotic is used for the management of hospital and community acquired pneumonia, complicated infections, intrahospital sepsis as well as infective endocarditis (Kovacevic et al., 2020). Because wrong dosing can result in increased risk of resistance as well as treatment failure, therapeutic drug monitoring (TDM) of vancomycin has become an important part of pharmacotherapy particularly in patients suffering from sepsis (Kovacevic et al., 2020). Nonetheless, traditional TDM techniques are painful. Ranamukhaarachchi and coworkers developed a minimally-invasive and painless option for TDM by employing hollow MNs for the extraction of minute volumes (less than 1 nL) of interstitial fluid (Ranamukhaarachchi et al., 2016). A microneedle bore was modified for use as a micro-reactor during sampling with no need for the transfer of samples. An optofluidic device was incorporated into this MN for quick and sensitive assay of drug analytes via a simple approach based on absorbance (Ranamukhaarachchi et al., 2016). Vancomycin is currently assayed by utilizing volumes that range from 50 μL to 100 μL with a limit of detection (LoD) of 1.35 μM (Ranamukhaarachchi et al., 2016). The developed MN-optofluidic biosensor can quantify the drug by using a sample volume of 0.6 nL and achieving a limit of detection of less than 100 nM. This has positive implications such as a decrease in the pain suffered by patients and also a reduction in healthcare costs (Ranamukhaarachchi et al., 2016).

In another research project, Ito et al. used dissolving microneedles (DMNs) for painless extraction of interstitial fluid (ISF) for the monitoring of vancomycin (Ito et al., 2016). Chondroitin glue was prepared by adding distilled water to sodium chondroitin sulfate with subsequent mixing (Ito et al., 2016). The authors then degassed the glue via decreased pressure and poured it into a mold with 300 inverted cone-shaped cavities (Ito et al., 2016). Each well had a surface diameter and depth of 300 μm and 500 μm respectively. The investigators subsequently covered mold with a 300-g steel plate, added the glue to the wells, and dried the mold (Ito et al., 2016). They fabricated a chip from a mixture of hydroxypropyl cellulose and cellulose acetate (1:10) by employing a Handtab-100 tableting machine. After removing the plate, the glue made up of 25 mL of distilled water and 15 mg of chondroitin sulfate was used to coat the chip (Ito et al., 2016). The DMNs were obtained following pressure-drying (Ito et al., 2016). The investigators administered vancomycin intravenously into Male Wistar rats via the jugular vein. Interstitial fluid was then collected from a formed pore at 0.25, 0.5, 0.75, 1, 1.25, 1.5, and 2 h following the removal of DMNs from the skin (Ito et al., 2016). Blood samples were subsequently collected from the left femoral vein for analysis (Ito et al., 2016). There was a high correlation between ISF and plasma vancomycin concentrations. The safety of DMNs as a device for sampling interstitial fluid was confirmed via microscopy (Ito et al., 2016).

In another interesting project, Rawson and coworkers used a microneedle-based β-lactam biosensor for real-time phenoxymethylpenicillin monitoring in healthy participants (Rawson et al., 2019). The subjects wore a solid microneedle β-lactam biosensor and were administered phenoxymethylpenicillin orally (Rawson et al., 2019). Extracellular fluid and blood samples were drawn during one dosing interval. The drug concentration data determined following the use of microneedles were calibrated and compared with microdialysis as well as blood data (Rawson et al., 2019). The pharmacokinetic profile of the drug following the use of each method was assessed via a non-compartmental analysis. The authors compared maximum concentration, AUC—time curve, and time to maximum concentration (Rawson et al., 2019). Microneedle and microdialysis results were comparable for the time to maximum concentration parameter, extracellular fluid maximum concentration, and AUC (Fig. 7.1) (Rawson et al., 2019). Mean drug AUCs were also comparable (Rawson et al., 2019).

Caffarel-Salvador recently utilized hydrogel-forming microneedles for in vitro and in vivo extraction and quantitation of active pharmaceutical ingredients (Caffarel-Salvador et al., 2015). Microneedles fabricated from aqueous blends of hydrolyzed poly (methyl-vinyletherco-maleic anhydride) (11.1% w/w), poly (ethyleneglycol) 10,000 Da (5.6% w/w), and crosslinked via esterification were inserted into neonatal pig skin (Caffarel-Salvador et al., 2015). Following removal, caffeine and theophylline were extracted from the microneedle and quantitated via high-performance liquid chromatography. In vitro experiments with excised neonatal pig skin demonstrated quick (5 min) uptake of the drug (Caffarel-Salvador et al., 2015). The mean concentration obtained by extraction of microneedles inserted into skin bathed with 15 μg/mL caffeine was 0.33 μg/mL while the mean concentration of 0.10 μg/mL was achieved following extraction of microneedles inserted for 5 min into

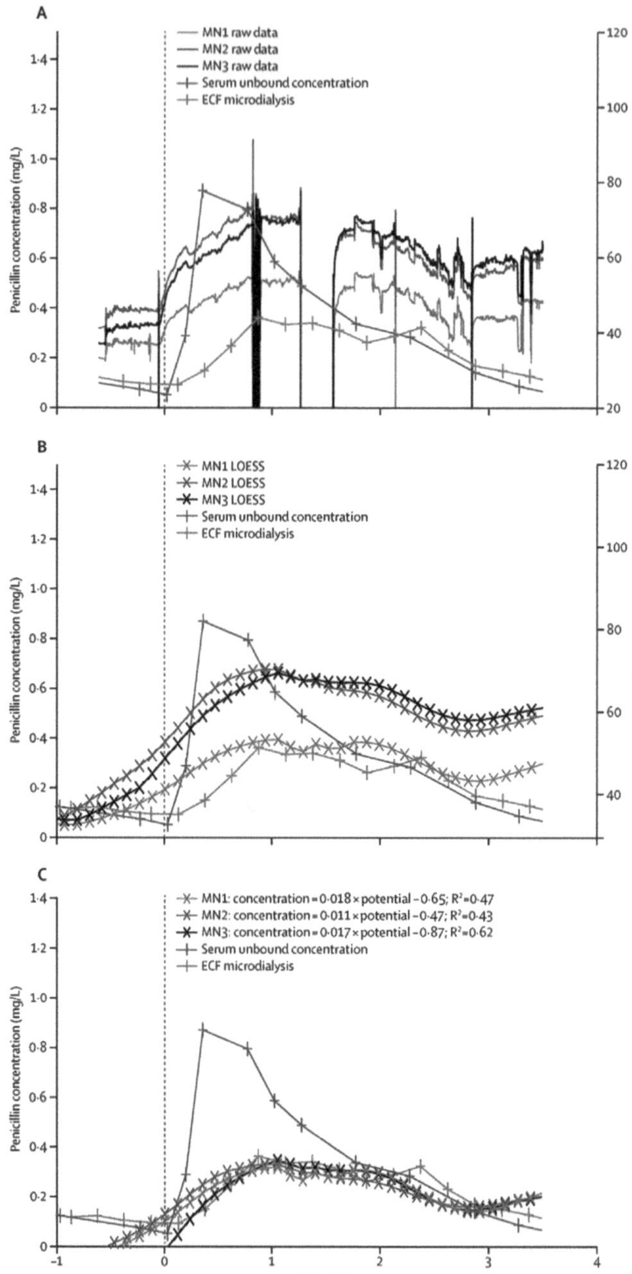

Fig. 7.1 Example of individual microneedle data calibration against blood and extracellular fluid penicillin concentrations (A) Raw sensor data (mV) with the control sensor data subtracted. (B) Locally weighted scatter plot smoothing (LOESS) analysis of sensor data (mV). (C) MN data converted to penicillin concentration (mg/L) using individual first order polynomial equations. Time = 0 min when dose of 500 mg phenoxymethylpenicillin is administered to the patient. *Ag*, silver; *Cl*, chloride; *ECF*, extracellular fluid; *MN*, microneedle. *(From Rawson, T. M., Gowers, S. A. N., Freeman, D. M. E., Wilson, R. C., Sharma, S., Gilchrist, M., MacGowan, A., Lovering, A., Bayliss, M., Kyriakides, M., Georgiou, P., Cass, A. E. G., O'Hare, D., & Holmes, A. H. (2019). Microneedle biosensors for real-time, minimally invasive drug monitoring of phenoxymethylpenicillin: a first-in-human evaluation in healthy volunteers.* The Lancet Digital Health, 1*(7), e335–e343. https://doi.org/10.1016/S2589-7500(19)30131-1.)*

skin bathed with 5 μg/mL caffeine (Caffarel-Salvador et al., 2015). Over the period following the ingestion of 100 mg Proplus1 tablets, the highest mean concentration of caffeine detected in human subjects using microneedles was 91.31 μg/mL (Caffarel-Salvador et al., 2015). The authors demonstrated, for the first time, the use of hydrogel-forming microneedles for therapeutic drug monitoring (Caffarel-Salvador et al., 2015).

References

Ates, H. C., Roberts, J. A., Lipman, J., Cass, A. E. G., Urban, G. A., & Dincer, C. (2020). On-site therapeutic drug monitoring. *Trends in Biotechnology, 38*(11), 1262–1277. https://doi.org/10.1016/j.tibtech.2020.03.001

Caffarel-Salvador, E., Brady, A. J., Eltayib, E., Meng, T., Alonso-Vicente, A., Gonzalez-Vazquez, P., Torrisi, B. M., Vicente-Perez, E. M., Mooney, K., Jones, D. S., Bell, S. E. J., McCoy, C. P., McCarthy, H. O., McElnay, J. C., Donnelly, R. F., & Chan, C. (2015). Hydrogel-forming microneedle arrays allow detection of drugs and glucose in vivo: potential for use in diagnosis and therapeutic drug monitoring. *PloS One, 10*(12), e0145644. https://doi.org/10.1371/journal.pone.0145644

Domínguez-Gil Hurlé, A., Sánchez Navarro, A., & García Sánchez, M. J. (2006). Therapeutic drug monitoring of itraconazole and the relevance of pharmacokinetic interactions. *Clinical Microbiology and Infections, 12*(7), 97–106. https://doi.org/10.1111/j.1469-0691.2006.01611.x

Ito, Y., Inagaki, Y., Kobuchi, S., Takada, K., & Sakaeda, T. (2016). Therapeutic drug monitoring of vancomycin in dermal interstitial fluid using dissolving microneedles. *International Journal of Medical Sciences, 13*(4), 271–276. https://doi.org/10.7150/ijms.13601

Kiang, T. K. L., Ranamukhaarachchi, S. A., & Ensom, M. H. H. (2017). Revolutionizing therapeutic drug monitoring with the use of interstitial fluid and microneedles technology. *Pharmaceutics, 9*(4). https://doi.org/10.3390/pharmaceutics9040043

Kovacevic, T., Miljkovic, B., Kovacevic, P., Dragic, S., Momcicevic, D., Avram, S., Jovanovic, M., & Vucicevic, K. (2020). Population pharmacokinetic model of Vancomycin based on therapeutic drug monitoring data in critically ill septic patients. *Journal of Critical Care, 55*, 116–121. https://doi.org/10.1016/j.jcrc.2019.10.012

Mabilat, C., Gros, M. F., Nicolau, D., Mouton, J. W., Textoris, J., Roberts, J. A., Cotta, M. O., van Belkum, A., & Caniaux, I. (2020). Diagnostic and medical needs for therapeutic drug monitoring of antibiotics. *European Journal of Clinical Microbiology & Infectious Diseases, 39*(5), 791–797. https://doi.org/10.1007/s10096-019-03769-8

Ranamukhaarachchi, S. A., Padeste, C., Dübner, M., Häfeli, U. O., Stoeber, B., & Cadarso, V. J. (2016). Integrated hollow microneedle-optofluidic biosensor for therapeutic drug monitoring in sub-nanoliter volumes. *Scientific Reports, 6*. https://doi.org/10.1038/srep. 29075

Rawson, T. M., Gowers, S. A. N., Freeman, D. M. E., Wilson, R. C., Sharma, S., Gilchrist, M., MacGowan, A., Lovering, A., Bayliss, M., Kyriakides, M., Georgiou, P., Cass, A. E. G., O'Hare, D., & Holmes, A. H. (2019). Microneedle biosensors for real-time, minimally invasive drug monitoring of phenoxymethylpenicillin: A first-in-human evaluation in healthy volunteers. *The Lancet Digital Health, 1*(7), e335–e343. https://doi.org/10.1016/S2589-7500(19)30131-1

CHAPTER 8

Miscellaneous routes of microneedle-assisted drug delivery

Microneedle-assisted vaccine delivery

It has been projected that the world population will be approximately 9.9 billion by 2050 (Excler et al., 2021). This represents an increase of more than 25% from the 2020 population (7.8 billion) (Excler et al., 2021). Several infectious diseases such as influenza, Middle-East respiratory syndrome (MERS), COVID-19, Ebola, severe acute respiratory syndrome (SARS), antimicrobial resistance (AMR), Zika and dengue have had deleterious impact on global health (Anser et al., 2020; Ita, 2021). With more than 100 million people infected with SARS-C0V-2 and more than 2 million dead, the addition of vaccine(s) to the existing COVID-19 countermeasures holds the best hope for pandemic control (Excler et al., 2021). Indeed, the United States Food and Drug Administration (FDA) granted Emergency Use Authorization for the Pfizer-BioNTech COVID-19 vaccine and Moderna COVID-19 vaccine (Zhao, 2021). The Medicines & Healthcare Products Regulatory Agency (MHRA) in the United Kingdom also granted a temporary authorization for emergency use to Pfizer/BioNTech for the mRNA-based vaccine (BNT162b2) (Ita & Ukaoma, 2021). On January 29th, 2021, the European Medicines Agency Committee for Medicinal Products for Human Use (EMA CHMP) granted a conditional marketing authorization to Astra-Zeneca for COVID-19 AZD1222 vaccine use (Hernández et al., 2021). AZD1222 is a monovalent vaccine consisting of a replication-deficient, single recombinant, chimpanzee adenovirus (ChAdOx1) vector altered to carry the gene encoding the of SARS CoV-2 Spike (S) glycoprotein (Hernández et al., 2021).

Microneedles
ISBN 978-0-323-91811-4
https://doi.org/10.1016/B978-0-323-91811-4.00002-X

The skin represents an excellent route for vaccination since the epidermis contains a high number of dendritic and Langerhans cells (Guillot et al., 2020). Langerhans cells (LCs) are specialized antigen-presenting cells located in the epidermal layer of the skin (Bellmann et al., 2021). Upon pathogen invasion or inflammation, LCs are activated and are subsequently transported to the lymph node where they interact with different T cells (Bellmann et al., 2021). Keratinocytes (KCs), mast cells (MCs) and fibroblasts (FBs) also contribute to the immunocompetence of the human skin (Anser et al., 2020; Ita, 2021). The delivery of vaccines in the form of injections has numerous drawbacks such as poor patient compliance resulting from needle-phobia as well as the possibility of needle stick injuries (Guillot et al., 2020). MNs can be as long as several hundred micrometers (Choi et al., 2020) and can deliver antigen and/or adjuvant intradermally in a minimally invasive fashion resulting in an immune response (Ita, 2016; Lee et al., 2021). There are numerous benefits associated with the use of MNs: the administration of MNs requires no professional training and the technique is painless (Ita, 2016). Furthermore, MNs possess broad functionality that can also be adapted for oral mucosal vaccine delivery (Creighton & Woodrow, 2019). MNs can be fabricated from diverse materials into distinct geometries possessing assorted sizes to reach specific depths in various tissues (Creighton & Woodrow, 2019).

Conventional vaccines are often unable to induce potent cytotoxic T-cell responses necessary for the prophylaxis or treatment of many infectious diseases or cancer (Erdos et al., 2020). An important focal point in vaccinology is the induction of antigen-specific cellular immunity in contrast to the more conventional viral surface proteins that are targeted by antibodies but can be highly mutable (Erdos et al., 2020). Recently, Erdos et al. fabricated dissolving MNs and used the platform to simultaneously co-deliver a toll-like receptor (TLR) 3–triggering double-stranded RNA molecule as well as adenovectors encoding the transgene for the model antigen ovalbumin (OVA) together with polyinosinic acid:polycytidylic acid (Poly [I:C]) to the skin (Erdos et al., 2020). The aim was to generate antigen-specific cellular immune responses (Erdos et al., 2020). The authors showed that the administration of Poly (I:C)-adjuvanted MNA-adenovirus vaccines resulted in significantly enhanced cytotoxic T-cell responses in comparison with adenovirus alone while generating antibody responses that are comparable to those induced by adenovirus alone (Erdos et al., 2020). In another interesting study, dissolving microneedles (DMNs) were used to deliver OVA into ex-vivo human skin (Lee et al., 2021). The DMNs were successfully fabricated via an automatic fabrication system (AFS), and hardly any antigen waste was encountered (Lee et al., 2021). Compared to the method that uses mold centrifugation, the use of AFS led to a 98.5% volume reduction of antigen/polymer solution and the production time was 24 h shorter (Lee et al., 2021). Furthermore, AFS can be scaled up and this is important from a manufacturing standpoint (Lee et al., 2021). The authors delivered 8.57 μg of ovalbumin into the skin. It was shown in a previous study that an OVA dose as low as 0.4 μg can induce IgG titers that is similar to what is achieved by 15 μg of ovalbumin delivered intramuscularly (Raphael et al., 2010).

Fig. 8.1 shows compartmental microneedle array (CMA) for the delivery of bivalent antigens of B/Yamagata and B/Victoria as a single product by single administration. Jeong et al. used a two-compartmental microneedles array to deliver two influenza vaccine strains without mixing (Jeong et al., 2020). The CMA loaded with the vaccines was administered to mice, and weight change and survival were compared with other groups of mice administered (a) two monovalent vaccines with microneedles (b) combination vaccines with microneedles, (c) a combination vaccine given intramuscularly, and (d) two monovalent vaccines

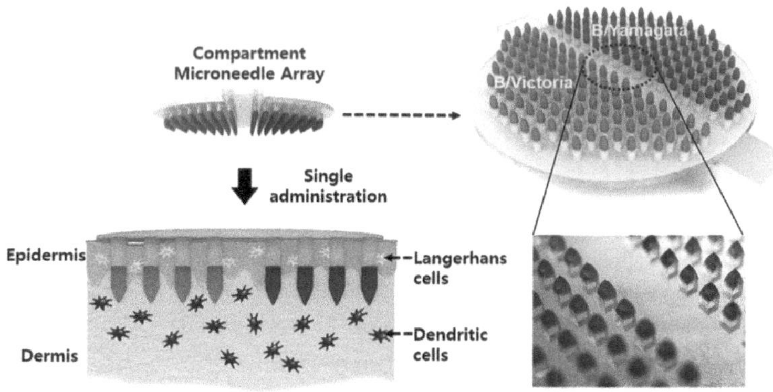

Fig. 8.1 Compartmental microneedle array (CMA) for the delivery of bivalent antigens of B/Yamagata and B/Victoria as a single product by single administration. B–Y|B-V CMA: B-Y and B-V separately in two compartments in an array of microneedles, (A) Vaccines for B/Yamagata and B/Victoria were delivered into the skin layer by CMA without mixing the two antigens (B) Two monovalent vaccines induce an immune response in the skin layer. *(From Jeong, H. R., Bae, J. Y., Park, J. H., Baek, S. K., Kim, G., Park, M.S., & Park, J.H. (n.d.). Preclinical study of influenza bivalent vaccine delivered with a two compartmental microneedle array.* Journal of Controlled Release, 324, *280–288. https://doi.org/10.1016/j.jconrel.2020.05.024.)*

given intramuscularly (Jeong et al., 2020). Plaque reduction neutralization tests were carried out by the authors to compare the CMA group with the other groups. The compartmental microneedles array exhibited similar antibody-forming efficacy in comparison to other groups (Jeong et al., 2020). The CMA mice group displayed higher weight change and survival compared to the mice that was given intramuscular (IM) injection of the combination vaccine (Jeong et al., 2020). All microneedle groups exhibited higher neutralizing antibodies compared to the intramuscular groups (Jeong et al., 2020).

Severe Acute Respiratory Syndrome Coronavirus-2 (SARS-CoV-2) and Middle East Respiratory Syndrome Coronavirus (MERS-CoV) cause COVID-19, and MERS respectively (Kim, Choi et al., 2020; Kim, Erdos, et al., 2020). There is an urgent need for efficacious and safe vaccines that can quickly engender long-lasting and potent virus-specific immune responses against coronaviruses (Kim, Choi et al., 2020; Kim, Erdos, et al., 2020). Researchers have identified the coronavirus spike (S) protein, as a key target for the development of vaccines (Kim, Choi et al., 2020; Kim, Erdos, et al., 2020). Recently, Kim et al. prepared codon optimized MERS-S1 subunit vaccines that were subsequently fused with a foldon trimerization domain (Kim, Choi et al., 2020; Kim, Erdos, et al., 2020). The authors then tethered immune stimulants (flagellin or RS09, as TLR5 or TLR4 agonists, respectively) to the trimer. Dissolving MNs (DMNs) were used by the investigators for the intradermal delivery of the subunit vaccine (Kim, Choi et al., 2020; Kim, Erdos, et al., 2020). The same approach was used to quickly develop MNs for the delivery of SARS-CoV-2 subunit vaccines. MERS-S1 subunit vaccines delivered with DMNs elicited long-lasting and strong antigen-specific antibody responses in comparison with those administered with conventional hypodermic needles (Kim, Choi et al., 2020; Kim, Erdos, et al., 2020). Significantly, SARS-CoV-2 S1 subunit vaccine administered with DMNs induced potent antigen-specific antibody responses (Kim, Choi et al., 2020; Kim, Erdos, et al., 2020).

DNA vaccines are promising for the prevention or treatment of nearly any immunogenic disease, including cancer (Cole et al., 2017). However, these vaccines possess poor immunogenicity in vivo because of lack of an optimal delivery system that can improve transfection efficiencies and protect the DNA from degradation (Cole et al., 2017). Coated MNs fabricated via layer-by-layer (LbL) deposition can be used for the delivery of DNA vaccines and adjuvant leading to the induction of type I interferons and the generation of antigen-specific antibodies by B cells (Amani et al., 2021). In addition, antigen-specific $CD8^+$ T-cells can generate interferon-gamma (IFN-γ) and facilitate cancer cell death (Amani et al., 2021). Recently, microneedles (MNs) were produced via layer-by-layer coating with negatively charged immunostimulatory adjuvant poly (I:C) and ultra-pH-responsive oligo sulfamethazine conjugated poly (β-amino ester urethane) (OSM-PEG-PAEU) (Duong et al., 2018). Interestingly, transcutaneous administration of the vaccine-loaded coated MNs in mice led to a 3-fold increase in Anti-ovalbumin immunoglobulin G1 serum antibody and 3-fold increase in cytotoxic $CD8^+$ T cell in comparison with the control soluble DNA vaccine dosage form (Duong et al., 2018).

In another interesting research project, Seok and colleagues utilized MNs coated with a polyplex containing poly lactic-*co*-glycolic acid/polyethyleneimine (PLGA/PEI) nanoparticles to deliver a pH1N1 DNA vaccine (Seok et al., 2017). Strikingly, the coated polyplex promptly dissolved in pig skin (within 5 min) and led to a higher humoral immune response compared to that induced by naked pH1N1 DNA vaccine delivery via a dry-coated MN or intramuscular polyplex delivery (Seok et al., 2017). Cole et al. also studied the complexation of DNA to RALA (a cationic delivery peptide) prior to incorporation into dissolving microneedles (Cole et al., 2017). The aim was to enhance transfection efficiency of the vaccine. The authors reported that RALA/DNA nanoparticles released from dissolving polyvinyl alcohol polymeric matrices exhibited up to 10-fold greater transfection efficiencies compared to the corresponding complexes released from polyvinylpyrrolidone matrices (Cole et al., 2017).

Hepatitis B infection can be prevented via intramuscular administration of the vaccine, but successful antibody generation by traditional vaccination requires three vaccine doses administered over two months (Kim, Erdos, et al., 2020). Kim and coworkers fabricated dissolving microneedles from polylactic acid/polyvinylpyrrolidone (PLA/PVP) and coated with carboxymethylcellulose (CMC) (Kim, Choi et al., 2020). The dissolving microneedles with dual-release rate pattern were then loaded with Hepatitis B surface antigen (HBsAg). The vaccine displayed high immunogenicity with a single administration (Kim, Choi et al., 2020). HBsAg-PLA/CMC-DMNs pierced the skin successfully. The PVA/PVP base was completely dissolved within 10 min of insertion leading to the dissolution of all MN tips into the skin (Kim, Choi et al., 2020). In the in vitro release study, the entire HBsAg in the CMC coating was liberated in less than 20 min, and the HBsAg in the PLA tips was released over more than 55 days (Kim, Choi et al., 2020). The antibody titer of one shot of HBsAg-PLA/CMC-DMNs was the same as or higher than two shots delivered by intramuscular or coated microneedle approaches (Kim, Choi et al., 2020). In another study, large-hepatitis B surface antigen (L-HBsAg) was administered via microneedles with no adjuvant (Nguyen et al., 2019). The authors compared the efficacy of intradermal immunization with that of the intramuscular route that utilized a traditional dosage form with aluminum hydroxide as an adjuvant (Nguyen et al., 2019).

Vaccination coverage in the United States of America has remained at about 42%—49% during the past eight influenza seasons even though a yearly vaccination coverage of 70% has been recommended by the US Department of Health and Human Services (Healthy People, 2020) for all persons aged $\geq$6 months (Wu et al., 2021). Microneedles can improve the acceptability and safety of the flu vaccine (Frew et al., 2020). Recently, Frew and colleagues investigated the acceptability of inactivated influenza vaccine administered via a microneedle patch compared to inactivated influenza vaccine administered via a hypodermic needle (Frew et al., 2020). The authors enrolled 100 healthy adult participants (18—49 years) for the study. At baseline, 98.6% of participants receiving microneedle-assisted vaccination reported an overall positive experience in comparison to 86.4% for participants receiving the inactivated influenza vaccine intramuscularly (Frew et al., 2020). The majority of those that took part in the study accepted vaccination via microneedles in comparison to hypodermic injection (Frew et al., 2020).

For the fabrication of MNs loaded with the process sensitive H1N1 antigen, a stabilizer is normally incorporated into the formulation for the maintenance of vaccine antigenicity (Jeong et al., 2019). Nevertheless, finding an optimal stabilizer is challenging since the incorporation of a stabilizer can restrict antigen dose due to the reduced microneedle volume (Jeong et al., 2019). Jeong et al. fabricated MNs without a stabilizer by keeping the samples and the environment at a low temperature of 10°C (Jeong et al., 2019). The polylactic acid MN array had 145 pyramid-shaped needles 370 μm in width and 800 μm in height on a circular base and 500 μm thick and 1 cm in diameter (Jeong et al., 2019). The microneedles were loaded with the H1N1 antigen and mice immunized with the microneedles displayed strong antibody responses (Jeong et al., 2019).

Multi-bolus regimens are used for most vaccines, and these can lead to financial stress for patients due to excessive costs, pain, and complex injection schedules (Tran et al., 2020). These factors can hinder widespread immunization (Tran et al., 2020) especially during periods of epidemics and global pandemics such as flu and COVID-19. There is an unmet need for novel vaccine delivery methods that require only a single administration and are injection free (Tran et al., 2020). Tran and coworkers fabricated core—shell microneedles from poly (lactic-*co*-glycolic acid) via micromolding (Tran et al., 2020). The authors used a scalable and high-throughput 3D-manufacturing technique to produce a biodegradable microneedle system. Each microneedle incorporated a core—shell microstructure that was formed via a 3D manufacturing process that combined three distinct constituents of the microneedles: a cap, a shell, and a dried vaccine core (Tran et al., 2020). The microneedles had different antigen release profiles and were utilized for the preprogrammed burst delivery of vaccines over few days to more than a month from a single administration (Tran et al., 2020).

Diverse routes of microneedle-mediated drug delivery

Several other routes of microneedle-assisted drug delivery are being explored even though the transdermal route is the most common. Gastrointestinal, perivascular, endovascular and other modes of microneedle application have been reported in the literature. In comparison with parenteral methods of drug administration, oral drug delivery has several advantages (Traverso et al., 2015). Nonetheless, this drug delivery approach is characterized by potential flaws including drug degradation and low drug absorption. This is especially relevant when introducing monoclonal antibodies, insulin, nucleic acids and other biologics into the body (Traverso et al., 2015). Extreme pH, bacteria, endonucleases, and proteases found in the gastrointestinal tract can destroy these molecules. Numerous techniques have been deployed to overcome these drawbacks and facilitate the administration of biologics through the oral route (Traverso et al., 2015). These molecules can be administered concomitantly with enzyme inhibitors and chemical modification of the medication can also be carried out. Other approaches include the utilization of nano- and microcarriers prepared from polymers, targeted nanoparticles or liposome carriers (Traverso et al., 2015). The limitation of these approaches is that reformulation of the drug is needed for the maintenance of API activity (Traverso et al., 2015). Traverso and coworkers showed in a recently published study that the use of microneedles can lead to enhanced bioavailability of macromolecules (Fig. 8.2) (Traverso et al., 2015). The authors manufactured a device from clear acrylic and fitted 25G needles manually into previously formed orifices. The system was 1 cm in diameter and 2 cm in length (Traverso et al., 2015). A central metallic core was included to boost radio-opacity for quick radiographic sensing. The authors demonstrated that the MN-based devices can be safely eliminated from the gastrointestinal tract (Traverso et al., 2015).

Veins and arteries possess three layers: the tunica interna, tunica media and the adventitia (Ita, 2016; Lee et al., 2021). Atherosclerosis, which is characterized by the buildup of plaque in arterial walls, leads to vasoconstriction. Subsequently, the formed blood clot can inhibit blood circulation leading to stroke or myocardial infarction (Lee, Goudie, et al., 2020; Lee, Lee et al., 2020). Several perivascular devices have been utilized for the delivery of antimetabolites. The aim is to block intimal hyperplasia caused by anomalous growth of smooth muscle cells in the tunica media layer (Lee et al., 2017). Be that as it may, there are conflicting challenges such as vasoconstriction, pulsatile expansion versus conformal tight device installation as well as device porosity versus local and unidirectional delivery (Lee et al., 2017). Lee et al. utilized a transfer molding technique to fabricate a microneedle array that incorporates a flexible woven surgical mesh (Lee et al., 2017). The authors investigated the mechanical

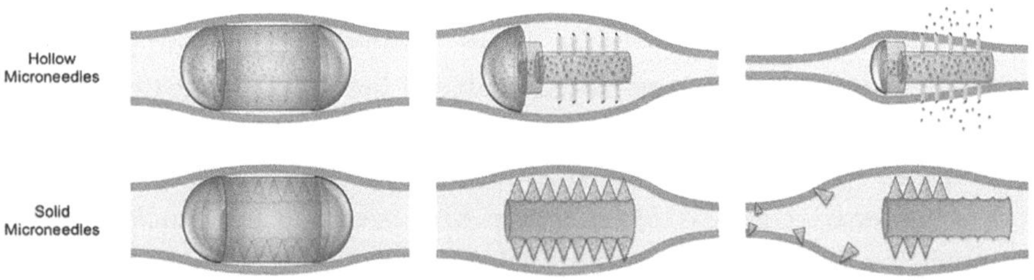

Fig. 8.2 The microneedle pill Both hollow and solid microneedles could be used. In both cases, the pill's needles are initially coated by a pH-responsive coating to aid in ingestion (left). When the pill has reached the desired location in the GI tract, the coating dissolves, revealing the microneedles (middle). In the case of hollow microneedles (top right), the drug reservoir is compressed through peristalsis, releasing the drug through the needles. In the case of solid microneedles (bottom right), the drug is formulated into the microneedles. The microneedles penetrate the tissue and break off of the pill, leaving the needle to release the drug in a controlled manner, based on the needle formulation. *(From Traverso, G., Schoellhammer, C. M., Schroeder, A., Maa, R., Lauwers, G. Y., Polat, B. E., Anderson, D. G., Blankschtein, D., & Langer, R. (2015). Micro-needles for drug delivery via the gastrointestinal tract.* Journal of Pharmaceutical Sciences, 104*(2), 362–367. https://doi.org/10.1002/jps.24182.)*

properties of the "wrappable" MN meshes and blood vessels. In vivo and ex vivo experiments showed the lessening of intimal hyperplasia, increased drug delivery efficiency, and enhanced safety of the microneedle mesh (Lee et al., 2017). Specifically, when the microneedle mesh was deployed, there was substantially lower neointimal formation (11.1%) after 4-weeks in comparison to other groups (23.7% and 22.2%) (Lee et al., 2017). Furthermore, wrappable MN meshes effectively eliminated adverse events such as the interaction of toxic medications with the adjoining healthy cells, intimal hyperplasia linked to mechanical constriction, and apoptosis (Lee et al., 2017).

Restenosis is frequently associated with vascular grafting and can lead to death (Lee et al., 2014). Drug eluting balloon (DEB)s and drug eluting stents (DESs) can be used to manage restenosis (Lee et al., 2014). Nevertheless, they are not very efficient because of drug loss and they may not be effective in restenotic regions in the presence of anastomosis sites or stents (Lee et al., 2014). In a recent study, Lee et al. reported that microneedle cuff (MNC) can be used as a perivascular delivery device for the achievement of high drug delivery efficiency to the tunica media (Lee et al., 2014). The investigators used a rabbit balloon injury model to study MN insertion and the pharmacological efficacy of MNC-mediated drug delivery in the management of neointimal hyperplasia (NH) (Lee et al., 2014). The authors optimized microneedle shape to improve insertion into the tunica media layer. Uniform distribution of paclitaxel in tunica media administered via microneedle cuff was also documented (Lee et al., 2014). There was greater drug delivery efficiency and a marked decrease in NH when MNC was used in comparison to flat type devices (Lee et al., 2014). In another research project, curved microneedles were administered to the outer surface of blood vessels with the aim of facilitating drug delivery to atherosclerotic or hyperplastic vessels (Choi et al., 2020). Post-annealing processes and spatially discrete thermal drawing were used to prepare microneedles on a curved surface. The MNs were inserted into arteries in vivo and ex vivo, and their drug-delivery capabilities as well as mechanical properties were evaluated (Choi et al., 2012).

Long-term administration of anti-platelet medications, in-stent restenosis as well as stent thrombosis can be avoided when drug eluting balloons (DEBs) are used (Lee, Goudie, et al., 2020; Lee, Lee et al., 2020). Despite their advantages, the use of DEBs leads to low drug delivery efficiency because of short balloon inflation times (30–60 s) that hinders passive drug transport from the surface of the balloon to the luminal lesion

(Lee, Goudie, et al., 2020; Lee, Lee et al., 2020). To boost drug delivery efficiency, a MN DEB was fabricated via a conformal transfer molding procedure with a thin polydimethylsiloxane mold that incorporates a negative array of 200 μm MN (Lee, Goudie, et al., 2020; Lee, Lee et al., 2020). A microneedle array comprising UV curable resin was applied to the DEB surface with subsequent drug coating. The authors investigated the mechanical properties of the microneedles and the penetration of MNs into the luminal vasculature was proved in animal experiments (Lee, Goudie, et al., 2020; Lee, Lee et al., 2020). Using a rabbit aorta for experiments, the investigators also showed an improvement in the efficiency of drug delivery in comparison to a conventional DEB. Subsequently, an atherosclerosis rabbit model was used to demonstrate the higher therapeutic efficacy of the microneedle drug eluting baloon (Lee, Goudie, et al., 2020; Lee, Lee et al., 2020).

Balloon angioplasty may cause an acute barotraumatic damage to a chronically inflamed artery leading to the recruitment of macrophages and neutrophils into the injury site (Owens et al., 2014). The degree of inflammatory response may have significant impact on subsequent restenosis in the femoropopliteal artery and this in an indication that pharmacological interventions, which alleviate the initial inflammatory cascade, might enhance patency (Owens et al., 2014). The Bullfrog Micro-Infusion Catheter is a wire-guided, rapid-exchange catheter with a balloon sheathed 35-gauge (140 mm diameter), 0.9-mm-long needle capable of transporting drug infusion to the perivascular and adventitial tissues (Owens et al., 2014). It is cleared by the FDA as a 510(k) device for use in peripheral and coronary arteries. It can be used to manage vessels that are 3–6 mm in diameter and it is advanced via a 6F sheath over a 0.014-inch wire (Owens et al., 2014). The correct orientation of the needle is carried out through three radio-opaque catheter markers. The needle is usually exposed via balloon inflation using regular angioplasty equipment (Owens et al., 2014). When the balloon is in contact with the arterial wall, contact pressure compels the needle to penetrate across the arterial wall into the perivascular and adventitia tissues (Owens et al., 2014). The limit of the contact pressure of the balloon against the artery wall is 2 atm and this forestalls injury to the artery wall. A mixture of a contrast agent and the infusate (1:4) is subsequently administered into the adventitia via fluoroscopy (Owens et al., 2014).

Sun et al. recently developed a cardiomyocytes (CMs)-integrated conductive microneedle system for the management of myocardial infarction (Sun et al., 2021). The CMs were obtained from induced pluripotent stem cells (iPSCs). The cardiac patches consisted of a methacrylated gelatin (GelMA) hydrogel scaffold upper layer, carbon nanotube (CNT) conductive middle layer and drug-encapsulated microneedle bottom layer (Sun et al., 2021). The anisotropic architecture of the microneedle patches led to the induction of directional alignment of CMs, while its conductive element supplied a scaffold for cell interaction (Sun et al., 2021). The cardiac patches were used for in vivo experiments following the induction of iPSCs to CMs in contrast to direct stem cell therapeutic patches (Sun et al., 2021). The microneedle system was tethered to the heart and released the encapsulated medications when used for the management of myocardial infarction (Sun et al., 2021). Furthermore, the presence of aligned CNT layer ensured that the cells maintained synergies with the heart in vivo. The use of the microneedle system also led to concurrent contraction of the CMs present on the patch (Sun et al., 2021).

It is challenging treating recurring oral cancers as standard surgical techniques may not be feasible for these disorders (Ma et al., 2015). Furthermore, oral cancer patients respond poorly to systemic chemotherapy (Ma et al., 2015). An attractive alternative treatment option entails the localized intratumoral injection of anti-cancer medications. A limitation of this approach, however, is that traditional hypodermic injections can lead to drug leakage, pain to the patient as well as poor distribution of the medication in the tumor (Ma et al., 2015). Ma et al. used coated microneedles for minimally invasive and direct intratumoral delivery of anti-cancer medications (Ma et al., 2015). Doxorubicin-loaded poly (lactic-*co*-glycolic) acid (DOX-PLGA) nanoparticles were used to coat in-plane microneedles (Ma et al., 2015). Microscopic assessment of pig cadaver buccal tissues and three-dimensional tissue phantoms that were treated with the in-plane microneedles coated with DOX-PLGA nanoparticles showed that the drug diffused both vertically and laterally into the tissues resulting in a cytotoxic effect (Ma et al., 2015). There was uniform distribution of the drug in a pig buccal tissue up to a depth greater than 3 mm following the use of out of plane microneedle arrays (1 cm × 1 cm) coated with free DOX (Ma et al., 2015). The authors demonstrated that drug coated microneedles can be used for effective drug delivery to oral cancers (Ma et al., 2015).

References

Amani, H., Shahbazi, M. A., D'Amico, C., Fontana, F., Abbaszadeh, S., & Santos, H. A. (2021). Microneedles for painless transdermal immunotherapeutic applications. *Journal of Controlled Release, 330*, 185–217. https://doi.org/10.1016/j.jconrel.2020.12.019

Anser, M. K., Islam, T., Khan, M. A., Zaman, K., Nassani, A. A., Askar, S. E., Abro, M. M. Q., & Kabbani, A. (2020). Identifying the potential causes, consequences, and prevention of communicable diseases (including COVID-19). *BioMed Research International, 2020*. https://doi.org/10.1155/2020/8894006

Bellmann, L., Zelle-Rieser, C., Milne, P., Resteu, A., Tripp, C. H., Hermann-Kleiter, N., Zaderer, V., Wilflingseder, D., Hörtnagl, P., Theochari, M., Schulze, J., Rentzsch, M., Del Frari, B., Collin, M., Rademacher, C., Romani, N., & Stoitzner, P. (2021). Notch-mediated generation of monocyte-derived Langerhans cells: Phenotype and function. *Journal of Investigative Dermatology, 141*(1), 84–94.e6. https://doi.org/10.1016/j.jid.2020.05.098

Choi, C. K., Kim, J. B., Jang, E. H., Youn, Y. N., & Ryu, W. H. (2012). Curved biodegradable microneedles for vascular drug delivery. *Small, 8*(16), 2483–2488. https://doi.org/10.1002/smll.201200441

Choi, I. J., Na, W., Kang, A., Ahn, M. H., Yeom, M., Kim, H. O., Lim, J. W., Choi, S. O., Baek, S. K., Song, D., & Park, J. H. (2020). Patchless administration of canine influenza vaccine on dog's ear using insertion-responsive microneedles (IRMN) without removal of hair and its in vivo efficacy evaluation. *European Journal of Pharmaceutics and Biopharmaceutics, 153*, 150–157. https://doi.org/10.1016/j.ejpb.2020.06.006

Cole, G., McCaffrey, J., Ali, A. A., McBride, J. W., McCrudden, C. M., Vincente-Perez, E. M., Donnelly, R. F., & McCarthy, H. O. (2017). Dissolving microneedles for DNA vaccination: Improving functionality via polymer characterization and RALA complexation. *Human Vaccines and Immunotherapeutics, 13*(1), 50–62. https://doi.org/10.1080/21645515.2016.1248008

Creighton, R. L., & Woodrow, K. A. (2019). Microneedle-mediated vaccine delivery to the oral mucosa. *Advanced Healthcare Materials, 8*(4). https://doi.org/10.1002/adhm.201801180

Duong, H. T. T., Yin, Y., Thambi, T., Nguyen, T. L., Giang Phan, V. H., Lee, M. S., Lee, J. E., Kim, J., Jeong, J. H., & Lee, D. S. (2018). Smart vaccine delivery based on microneedle arrays decorated with ultra-pH-responsive copolymers for cancer immunotherapy. *Biomaterials, 185*, 13–24. https://doi.org/10.1016/j.biomaterials.2018.09.008

Erdos, G., Balmert, S. C., Carey, C. D., Falo, G. D., Patel, N. A., Zhang, J., Gambotto, A., Korkmaz, E., & Falo, L. D. (2020). Improved cutaneous genetic immunization by microneedle array delivery of an adjuvanted adenovirus vaccine. *Journal of Investigative Dermatology, 140*(12), 2528–2531.e2. https://doi.org/10.1016/j.jid.2020.03.966

Excler, J. L., Saville, M., Berkley, S., & Kim, J. H. (2021). Vaccine development for emerging infectious diseases. *Nature Medicine, 27*(4), 591–600. https://doi.org/10.1038/s41591-021-01301-0

Frew, P. M., Paine, M. B., Rouphael, N., Schamel, J., Chung, Y., Mulligan, M. J., & Prausnitz, M. R. (2020). Acceptability of an inactivated influenza vaccine delivered by microneedle patch: Results from a phase I clinical trial of safety, reactogenicity, and immunogenicity. *Vaccine, 38*(45), 7175–7181. https://doi.org/10.1016/j.vaccine.2020.07.064

Guillot, A. J., Cordeiro, A. S., Donnelly, R. F., Montesinos, M. C., Garrigues, T. M., & Melero, A. (2020). Microneedle-based delivery: An overview of current applications and trends. *Pharmaceutics, 569*. https://doi.org/10.3390/pharmaceutics12060569

Healthy People. (2020). https://www.healthypeople.gov/2020/. Accessed on October 9, 2021.

Hernández, A. F., Calina, D., Poulas, K., Docea, A. O., & Tsatsakis, A. M. (2021). Safety of COVID-19 vaccines administered in the EU: Should we be concerned? *Toxicology Reports, 8*, 871–879. https://doi.org/10.1016/j.toxrep.2021.04.003

Ita, K. (2016). Transdermal delivery of vaccines – recent progress and critical issues. *Biomedicine and Pharmacotherapy, 83*, 1080–1088. https://doi.org/10.1016/j.biopha.2016.08.026

Ita, K. (2021). Coronavirus disease (COVID-19): Current status and prospects for drug and vaccine development. *Archives of Medical Research, 52*(1), 15–24. https://doi.org/10.1016/j.arcmed.2020.09.010

Ita, K., & Ukaoma, M. (2021). COVID-19 vaccines: New developments and the road ahead. *Archives of Medical Research, 52*(4), 454–455. https://doi.org/10.1016/j.arcmed.2020.12.009

Jeong, H.-R., Bae, J.-Y., Park, J.-H., Baek, S.-K., Kim, G., Park, M.-S., & Park, J.-H. (2020). Preclinical study of influenza bivalent vaccine delivered with a two compartmental microneedle array. *Journal of Controlled Release, 324*, 280–288. https://doi.org/10.1016/j.jconrel.2020.05.024

Jeong, H. R., Park, S., Park, J. H., Bae, J. Y., Kim, G.y., Baek, S. K., Park, M. S., & Park, J. H. (2019). Preparation of H1N1 microneedles by a low-temperature process without a stabilizer. *European Journal of Pharmaceutics and Biopharmaceutics, 143*, 1–7. https://doi.org/10.1016/j.ejpb.2019.08.005

Kim, J. S., Choi, J., Kim, J. C., Park, H., Yang, E., Park, J. S., Song, M., & Park, J.-H. (2020). Microneedles with dual release pattern for improved immunological efficacy of Hepatitis B vaccine. *International Journal of Pharmaceutics, 591*, 119928. https://doi.org/10.1016/j.ijpharm.2020.119928

Kim, E., Erdos, G., Huang, S., Kenniston, T. W., Balmert, S. C., Carey, C. D., Raj, V. S., Epperly, M. W., Klimstra, W. B., Haagmans, B. L., Korkmaz, E., Falo, L. D., & Gambotto, A. (2020). Microneedle array delivered recombinant coronavirus vaccines: Immunogenicity and rapid translational development. *EBioMedicine, 55*, 102743. https://doi.org/10.1016/j.ebiom.2020.102743

Lee, K. J., Goudie, M. J., Tebon, P., Sun, W., Luo, Z., Lee, J., Zhang, S., Fetah, K., Kim, H. J., Xue, Y., Darabi, M. A., Ahadian, S., Sarikhani, E., Ryu, W. H., Gu, Z., Weiss, P. S., Dokmeci, M. R., Ashammakhi, N., & Khademhosseini, A. (2020). Non-transdermal microneedles for advanced drug delivery. *Advanced Drug Delivery Reviews, 165–166*, 41–59. https://doi.org/10.1016/j.addr.2019.11.010

Lee, J. Y., Kim, D. H., Lee, K. J., Seo, I. H., Park, S. H., Jang, E. H., Park, Y., Youn, Y. N., & Ryu, W. H. (2017). Transfer-molded wrappable microneedle meshes for perivascular drug delivery. *Journal of Controlled Release, 268*, 237–246. https://doi.org/10.1016/j.jconrel.2017.10.007

Lee, K. J., Lee, J. Y., Lee, S. G., Park, S. H., Yang, D. S., Lee, J. J., Khademhosseini, A., Kim, J. S., & Ryu, W. H. (2020). Microneedle drug eluting balloon for enhanced drug delivery to vascular tissue. *Journal of Controlled Release, 321*, 174–183. https://doi.org/10.1016/j.jconrel.2020.02.012

Lee, K. J., Park, S. H., Lee, J. Y., Joo, H. C., Jang, E. H., Youn, Y. N., & Ryu, W. (2014). Perivascular biodegradable microneedle cuff for reduction of neointima formation after vascular injury. *Journal of Controlled Release, 192*, 174–181. https://doi.org/10.1016/j.jconrel.2014.07.007

Lee, J., van der Maaden, K., Gooris, G., O'Mahony, C., Jiskoot, W., & Bouwstra, J. (2021). Engineering of an automated nano-droplet dispensing system for fabrication of antigen-loaded dissolving microneedle arrays. *International Journal of Pharmaceutics, 600*, 120473. https://doi.org/10.1016/j.ijpharm.2021.120473

Ma, Y., Boese, S. E., Luo, Z., Nitin, N., & Gill, H. S. (2015). Drug coated microneedles for minimally-invasive treatment of oral carcinomas: Development and in vitro evaluation. *Biomedical Microdevices, 17*(2). https://doi.org/10.1007/s10544-015-9944-y

Nguyen, T. T., Choi, J. A., Kim, J. S., Park, H., Yang, E., Lee, W. J., Baek, S. K., Song, M., & Park, J. H. (2019). Skin immunization with third-generation hepatitis B surface antigen using microneedles. *Vaccine, 37*(40), 5954–5961. https://doi.org/10.1016/j.vaccine.2019.08.036

Owens, C. D., Gasper, W. J., Walker, J. P., Alley, H. F., Conte, M. S., & Grenon, S. M. (2014). Safety and feasibility of adjunctive dexamethasone infusion into the adventitia of the femoropopliteal artery following endovascular revascularization. *Journal of Vascular Surgery, 59*(4), 1016–1024. https://doi.org/10.1016/j.jvs.2013.10.051

Raphael, A. P., Prow, T. W., Crichton, M. L., Chen, X., Fernando, G. J. P., & Kendall, M. A. F. (2010). Targeted, needle-free vaccinations in skin using multilayered, densely-packed dissolving microprojection arrays. *Small, 6*(16), 1785–1793. https://doi.org/10.1002/smll.201000326

Seok, H. Y., Noh, J. Y., Lee, D. Y., Kim, S. J., Song, C. S., & Kim, Y. C. (2017). Effective humoral immune response from a H1N1 DNA vaccine delivered to the skin by microneedles coated with PLGA-based cationic nanoparticles. *Journal of Controlled Release, 265*, 66–74. https://doi.org/10.1016/j.jconrel.2017.04.027

Sun, L., Zhu, X., Zhang, X., Chen, G., Bian, F., Wang, J., Zhou, Q., Wang, D., & Zhao, Y. (2021). Induced cardiomyocytes-integrated conductive microneedle patch for treating myocardial infarction. *Chemical Engineering Journal, 414*, 128723. https://doi.org/10.1016/j.cej.2021.128723

Tran, K. T. M., Gavitt, T. D., Farrell, N. J., Curry, E. J., Mara, A. B., Patel, A., Brown, L., Kilpatrick, S., Piotrowska, R., Mishra, N., Szczepanek, S. M., & Nguyen, T. D. (2020). Transdermal microneedles for the programmable burst release of multiple vaccine payloads. *Nature Biomedical Engineering*. https://doi.org/10.1038/s41551-020-00650-4

Traverso, G., Schoellhammer, C. M., Schroeder, A., Maa, R., Lauwers, G. Y., Polat, B. E., Anderson, D. G., Blankschtein, D., & Langer, R. (2015). Microneedles for drug delivery via the gastrointestinal tract. *Journal of Pharmaceutical Sciences, 104*(2), 362–367. https://doi.org/10.1002/jps.24182

Wu, M. J., Chung, J. R., Kim, S. S., Jackson, M. L., Jackson, L. A., Belongia, E. A., McLean, H. Q., Gaglani, M., Reis, M., Beeram, M., Martin, E. T., Monto, A. S., Nowalk, M. P., Zimmerman, R., Santibanez, T. A., Singleton, J. A., Patel, M., & Flannery, B. (2021). Influenza vaccination coverage among persons seeking outpatient medical care for acute respiratory illness in five states in the United States, 2011–2012 through 2018–2019. *Vaccine, 39*(12), 1788–1796. https://doi.org/10.1016/j.vaccine.2021.01.065

Zhao, H. (2021). Low rates of urologic side effects following COVID-19 vaccination: An analysis of the FDA vaccine adverse event reporting system. *Urology, 153*, 11–13.

CHAPTER 9

Mathematical modeling of drug delivery from microneedles

Introduction

Mathematical modeling has been used by pharmaceutical scientists to predict the properties of microneedles in the early phase of fabrication and this approach can be used to enhance the efficacy of these devices (Zoudani & Soltani, 2020). Mathematical models can also serve as an approximation of drug transport across microneedle-treated skin. Numerous reports have shown a correlation between skin permeability and the geometrical parameters of microneedles (Zoudani & Soltani, 2020). An optimal mechanistic model can be developed. Mathematical modeling of microneedle-based transdermal drug delivery normally entails the simulation of three coupled sub-domains i.e., microneedles, skin, and the blood circulation (Yadav et al., 2020). Real or conceptual microneedle designs are converted into a computational domain before undertaking numerical simulations (Yadav et al., 2020). Finite difference, boundary element, finite element and discrete element methods can be used for numerical modeling (Li, 2017) of drug delivery from microneedles. Finite difference technique entails the discretization of a domain into small two-dimensional elements that are subsequently intertwined with their nodes (Chaulya & Prasad, 2016).

Differential equation of equilibrium can be approximated over each element. A system of simultaneous equations is thus created, and these equations can be solved using iteration techniques (Chaulya & Prasad, 2016). Boundary element method refers to the conversion of a governing differential equation into boundary integral equation based on Green's formula with the subsequent use of Kelvin's method to solve the problem (Zhou et al., 2020). The finite element method is based on the subdivision of a physical space into tetrahedral (in 3-D) or triangular (in 2-D) elements (Blazek, 2005a,b, pp. 29—75) with a specified number of nodes (Owen et al., 2020). The displacement of these nodes due to the effects of applied

Microneedles
ISBN 978-0-323-91811-4
https://doi.org/10.1016/B978-0-323-91811-4.00012-2

loads is controlled by a shape function (Owen et al., 2020). The finite element method also entails the transformation of governing differential equations into an equivalent integral form (Blazek, 2005a,b, pp. 29–75). Discrete element method is normally used to model the transport of granular materials via calculations that track individual particles that make up the granular material (Tanaka, Nishida, Kunimochi, & Tagaki, 2001). If contact forces acting on every particle are tracked at each time step, Newton's second law can then used to deduce the equation of motion of each particle (Tanaka, Nishida, Kunimochi, & Tagaki, 2001). From the standpoint of modeling, critical features of the microneedles and microneedle-treated skin are first examined. Mechanical properties of the microneedles (Poisson ratio, Young's modulus, ultimate tensile strength) as well as the skin (thickness, Young's modulus, porosity) are relevant for the development of mathematical models (Yadav et al., 2020).

Microneedle-assisted drug delivery and mathematical models

Recently, a mechanistic model was used to predict the release of sumatriptan succinate from dissolving microneedles and its permeation across the epidermal skin layers (Ronnander et al., 2020). This comprehensive mathematical model incorporates diffusion and can be used to predict in vitro diffusion of drugs from dissolving microneedles (Ronnander et al., 2020). A dissolving pyramid-shaped microneedle (Fig. 9.1) is applied

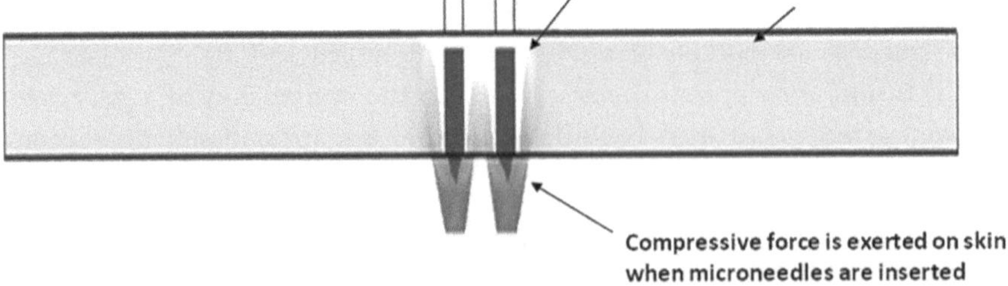

Fig. 9.1 A schematic showing compressive force exerted by microneedles when inserted into the skin. Gray regions represent areas compressed due to microneedles insertion. *(From Olatunji, O., Das, D. B, & Nassehi, V. (2012). Modeling transdermal drug delivery using microneedles: Effect of geometry on drug transport behavior.* Journal of Pharmaceutical Sciences, 101*(1), 164–175. https://doi.org/10.1002/jps.22736).*

directly onto the skin; the needles pierce the stratum corneum and permeate the upper dermal layers; in the epidermis, the needles are "wetted" by the moisture in the skin, after which they quickly dissolve, releasing the drug molecules which traverse the dermis before being absorbed into the systemic circulation (Ronnander et al., 2020). To create the model, a rectangular control volume was selected with the length and width equivalent to the needle pitch width, pw (Ronnander et al., 2020). The height is equal to the skin depth, ds, as depicted in Fig. 1, Ronnander et al. (2020). The needle is fully inserted into the skin (i.e. control volume) and has a square base of width, w, and height, h. The dissolving microneedle arrays were considered to be isotropic and dissolve evenly across the exterior surfaces of the needles (Ronnander et al., 2020). As such, the half angle (or ratio of needle height to base length) remains constant throughout the dissolution (Ronnander et al., 2020):

$$\tan\phi = \frac{w(t)}{h(t)} \tag{9.1}$$

The volume of the pyramid needle was defined in terms of the height (Ronnander et al., 2020)

$$v_c = \frac{4}{3} h^3(t) \tan^2\phi \tag{9.2}$$

The microneedle size is connected to variations in the height. After performing a mass balance on a cross-section of the needle and applying the Nernst and Brunner model, a governing equation was obtained to describe the change in microneedle height over time (Ronnander et al., 2020),

$$\frac{dh(t)}{dt} \tag{9.3}$$

In a second step, the Fick's second law of diffusion was used to explain how the concentration varies in the skin layer. The authors calculated the drug mass balance (Ronnander et al., 2020):

$$\frac{\partial c}{\partial t} = D\frac{\partial^2 t}{\partial y^2} + 4\left(\frac{k_D \tan(\phi)}{\rho \cos\phi}\right) h^2 \left(\frac{\omega\rho - c}{v_0 + v_{c,0} - v_0}\right)\left(c_s - \left(\frac{1-\omega}{\omega}\right)c\right) \tag{9.4}$$

where c denotes the concentration of the drug in the skin; D, the coefficient of diffusion (cm^2/h); kD, the dissolution rate constant (cm/h); ϕ, half the apex angle of the pyramid needle (degree); ρ, the density of a microneedle (g/cm^3); ω, the mass fraction of drug in a microneedle; v, the volume of the skin layer (cm^3); v_c, is the volume of the microneedle (cm^3) and c_s is the matrix polymer solubility in the solvent (g/cm^3) (Ronnander et al., 2020). Material balance equations were written to describe drug transport followed by absorption into the systemic circulation (Ronnander et al., 2020).

Eq. (9.5) was used to describe the cumulative quantity of the medication that was released (Ronnander et al., 2020):

$$Q(t) = p_w^2 \int_0^t J(\tau)\delta\tau \tag{9.5}$$

The percentage cumulative drug released $M(t)$ (Ronnander et al., 2020),

$$M(t) = \frac{p_w^2 \int_0^t J(\tau)\delta\tau}{\omega\rho v_{c,0}} \tag{9.6}$$

By modifying the governing equations to include a microneedle baseplate (Ronnander et al., 2020),

$$M(t) = \frac{p_w^2 \int_0^t J(\tau)\delta\tau}{\omega\rho v_{c,0} + m_0} \tag{9.7}$$

the model was utilized for the estimation of short- and long-term release profiles plotted from in vitro experiments (Ronnander et al., 2020). The baseplate is denoted by m_0. A two-dimensional (2D) problem of vertical transport of soluble medications through a porous medium with point sources mimicking dissolving microneedles (DMNs) has also been described and solved (Lyashko et al., 2018). It was shown that, by specifying the coordinates and regulating the intensity of the sources, optimal control of transdermal drug delivery can be mathematically derived and solved (Lyashko et al., 2018). An accurate and predictable distribution of drugs in epidermis can then be obtained (Lyashko et al., 2018). The authors used a two-step symmetrizable algorithm and finite-difference techniques to solve initial-boundary value problems (Lyashko et al., 2018). In another approach, Zoudani and Soltani used numerical simulation to investigate DMN dissolution in a porous medium (Zoudani & Soltani, 2020). This mathematical model was validated with an experimental investigation of DMNs used for drug transport to the cornea. The authors studied the effects of center to center spacing, pitch size and initial drug concentration on the magnitude and rate of drug concentration in the tissue layers (Zoudani & Soltani, 2020).

In an interesting set of reports, it was assumed that the dissolution process of the cone occurs uniformly. As a result, the ratio of the base radius (r) of the cone to the height (h) is constant over time (Kim et al., 2015; Zoudani & Soltani, 2020):

$$\tan\theta = \frac{r_0}{h_0} = \frac{r(t)}{h(t)} \tag{9.8}$$

θ is the half angle at the apex and denotes the initial state (Kim et al., 2015).

A specific quantity of the polymer material is transported from the MN over time during dissolution, according to this equation (Kim et al., 2015; Zoudani & Soltani, 2020):

$$J = K_D\left(C_s - \left(\frac{1-\beta}{\beta}\right)C\right) \tag{9.9}$$

where K_D is the mass transfer coefficient, C is the drug concentration, C_S is the concentration difference and β is the mass fraction of drug in the MN (Zoudani & Soltani, 2020). The authors divided the model into two distinct regions as a cone and an array of 32 hemispheres on its surface. This was done with the aim of modeling the dissolution of the modified MN shape (Zoudani & Soltani, 2020). The total volume of the MN with this new shape, is the sum of 32 hemispheres and the volume of the cone (Zoudani & Soltani, 2020):

$$V_{tot} = V_c + 32\ V_h \tag{9.10}$$

the total volume changes can be calculated based on the sum rule in differentiation (Zoudani & Soltani, 2020):

$$\frac{dV_{tot}}{dt} = \frac{dV_c}{dt} = 32\left(\frac{dV_h}{dt}\right) \tag{9.11}$$

The authors wrote the complete form of volume changes of hemisphere as follows (Zoudani & Soltani, 2020):

$$\frac{dV_h}{dt} = \frac{2\pi r_{hK_D}^2}{\rho}\left(C_s - \left(\frac{1-\beta}{\beta}\right)C\right) \tag{9.12}$$

The transport of the drug is controlled by diffusion (Zoudani & Soltani, 2020):

$$\frac{\partial C}{\partial t} = D\nabla^2 C \tag{9.13}$$

Effective diffusion coefficient can be calculated by taking into consideration the porosity factor which can be obtained from the following equation (Machekposhti et al., 2017; Olatunji et al., 2012; Zoudani & Soltani, 2020):

$$D_{eff} = \varepsilon D_r \tag{9.14}$$

where ε is the porosity of the skin (porous media), D_{eff} is effective diffusion coefficient defined in porous media, and D is the diffusion coefficient between two different fluids, D_r is the drug diffusion coefficient through porous medium, which depend on tissue properties, drug, and geometrical parameters of the MN, as described in Eqs. (9.15)–(9.18) (Olatunji et al., 2012; Zoudani & Soltani, 2020):

$$D_r = DH(\lambda) \tag{9.15}$$

$$H(\lambda) = (1 - \lambda)^2\left(1 - 2.104\lambda + 2.09\lambda^3 - 0.095\lambda^5\right) \tag{9.16}$$

$$\lambda = \left(\frac{r_p}{\sqrt{\dfrac{v_{pore} - nv_{needle}}{\pi n_p x}}}\right) \tag{9.17}$$

$$x = \tau h \tag{9.18}$$

where x represents diffusion path length, which is defined in Eq. (9.18), h is the tissue thickness, τ is the tortuosity, r_p is the hydrodynamic radius, n is the number of microneedles and n_p is the number of pores, $H(\lambda)$ is the hindrance factor, D is the drug diffusivity at infinite dilution, v_{pore} is the volume of pore in the tissue and v_{needle} is the volume of each needle (Zoudani & Soltani, 2020). The authors reported that higher drug concentration and quicker MN dissolution can be obtained by raising the initial drug concentration in the MN (Zoudani & Soltani, 2020). Even though reducing the pitch size can result in greater drug concentration, it does not significantly impact dissolution time. A new method based on array in array theory, that utilizes a cone that incorporates hemispherical convexities was suggested (Zoudani & Soltani, 2020). This approach reduces the risk of transporting inadequate quantity of the medication when partial insertion takes place, thus improving drug delivery, decreasing drug waste, and enhancing drug flux (Zoudani & Soltani, 2020).

Ronnander et al. developed a mathematical model for the prediction of sumatriptan transport across Göttingen minipig skin with subsequent absorption into systemic circulation (Ronnander et al., 2018). Sumatriptan was encapsulated in pyramidal DMN formulated from polyvinylpyrrolidone (Ronnander et al., 2018). The authors developed mass balance equations to model the transport and dissolution of the drug (Ronnander et al., 2018). A rectangular control volume that incorporates a pyramid was utilized for the development of the model. The needle possesses a side length, l and height, h. The control volume has a length on each side which is equivalent to the pitch width, pw and height that is equal to the skin depth, dS (Ronnander et al., 2018). Following needle dissolution, its volume, v_c, decreases as the volume of the rectangle skin layer compartment, v, increases (Ronnander et al., 2018). It is assumed that the MNs dissolve uniformly over time across the surfaces of the pyramidal MNs. Thus, the ratio of the microneedle height to the base length of the needle remains constant. The half angle at the apex of the MN, θ corresponds to this ratio and is defined below (Ronnander et al., 2018)

$$\tan\phi = \frac{l(t)}{2h(t)} = \frac{l_0}{2h_0} \tag{9.19}$$

The cumulative amount of medication released $Q(t)$

$$Q(t) = \beta\rho v_{c,0} - \beta\rho v_c(t) + c(t)(v_0 + v_{c,0} + v_{c,}(t)) \tag{9.20}$$

A plate, in the form of a rectangular prism, is present at the base of the DMN. The mass of sumatriptan in the baseplate is m_0

$$Q(t) = (\beta\rho v_{(c,0)} + m_0) - d_s p_w^2 (c_S D(t) + c(t)) \tag{9.21}$$

Following DMN dissolution, sumatriptan concentration in the solid and liquid phases was determined and the cumulative percentage of drug released was obtained by dividing by $Q(t)$ by $\beta\rho\, v_{c,0} + m_0$ (Chavoshi et al., 2019):

$$M(t) = 1 - \frac{d_s p_w^2 (c_{SD}(t) + c(t))}{\beta\rho v_{c,0} + m_0} \tag{9.22}$$

The authors proposed a one-dimensional system of equations with mass transfer taking place in the negative y-direction toward the base of the skin layer (that is the bloodstream). Mathematical modeling was used to evaluate and predict the influence of major parameters on sumatriptan release (Ronnander et al., 2018). Mathematica® was used for solving the equations to obtain the theoretical profiles of $c(t)$ and $h(t)$. $C(t)$ and $h\ (t)$ were used to assess the influence of design parameters (e.g., microneedle height, loading dose, microneedle pitch) on $c(t)$ (Ronnander et al., 2018). The concentration of the drug in the skin was enhanced with the height of the microneedle and the loading dose (Ronnander et al., 2018). The authors described the inverse relationship between the pitch width and the quantity of medication released in the dermal layer. In-vitro diffusion experiments with Gottingen minipig skin were used to validate the mathematical model (Ronnander et al., 2018). The developed mathematical model can be used to describe and predict in-vitro transport of three therapeutic systems loaded with sumatriptan (Ronnander et al., 2018).

Mathematical models were recently used to characterize the drug release profile and degradation rate of conical MNs containing sumatriptan (Chavoshi et al., 2019). It was assumed that the microneedles are separate, and that the sum of individual drug releases from each MN was equivalent to the overall amount of the drug released from the microneedle array (Chavoshi et al., 2019). The conical microneedles were assumed to be cylinders (with equivalent volume) for model simplification without the use of two-dimensional equations (Chavoshi et al., 2019). The following equations were used for model development:

$$v_{cone} = \frac{1}{3}\left(\frac{\pi D^2}{4}\right)L \tag{9.23}$$

$$v_e = \left(\frac{\pi D_e^2}{4}\right)L_e \tag{9.24}$$

$$v_{cone} = v_e \tag{9.25}$$

$$\text{if} \rightarrow L = L_e \Rightarrow D_e = \frac{\sqrt{3}}{3}D \tag{9.26}$$

The mass conservation for the medication, which accounts for the drug concentration alterations due to diffusion to the outer portions of the control volume, is represented as follows (Chavoshi et al., 2019):

$$\frac{\partial C_d}{\partial t} = \frac{D_d}{r}\frac{\partial}{\partial r}\left(r\left(\frac{\partial C_d}{\partial r}\right)\right) \tag{9.27}$$

Initially, the medication—with a certain concentration—is distributed evenly in the polymer. At the center line of the cylinder, the authors theoretically presumed that there are no alterations in concentration changes due to symmetry (Chavoshi et al., 2019). At the surface of the cylinder, due to the sink condition, the concentration of the drug is equal to zero (Chavoshi et al., 2019). The diffusion coefficient of the medication is not constant. As drug degradation takes place, this parameter increases. With the degradation of the polymer, there is decline in the polymer molecular weight, and there is proportionality between the diffusion coefficient of the drug and the inverse of polymer molecular weight (Chavoshi et al., 2019):

$$\frac{D_d(t)}{D_d^o} = \frac{M_n^0}{M_n(t)} \tag{9.28}$$

The authors used Runge–Kutta and finite difference techniques to solve ordinary and partial differential equations respectively (Chavoshi et al., 2019). The correlation between polymer degradation rate and molecular weight as well as molecular weight changes versus time were demonstrated (Chavoshi et al., 2019). In addition, the relationship between drug diffusivity and the molecular weight of the polymer was also shown. The authors were able to predict cumulative drug release as well as molecular weight reduction. Data series of aspirin (180.16 Da)- and albumin (66,000 Da)-loaded polylactic-*co*-glycolic acid (1:1 M ratio) hydrolysis were used to validate and evaluate the mathematical model (Chavoshi et al., 2019). The model was used to confirm experimental data from the literature. Drug release profile was predicted taking into consideration autocatalytic effects and diffusion. Significantly, the authors were able to use the mathematical model to compute drug release rates before microneedle fabrication (Chavoshi et al., 2019).

Microneedles are usually fabricated based on the relationship between certain benchmarks (e.g., microneedle material) and skin properties (e.g., skin thickness, drug permeability in skin, etc.) (Al-Qallaf & Das, 2008). The influence of various factors e.g., surface area of the patch, number of the microneedles, skin permeability should be precisely evaluated for optimum design of the microneedles (Al-Qallaf & Das, 2008). Skin permeability refers to the path length of a solute over a time period across a certain skin thickness (Al-Qallaf & Das, 2008). Geometric parameters such as surface area of the microneedles and microneedle radius were evaluated with the aim of optimization (Al-Qallaf & Das, 2008). An optimal mathematical model for the percutaneous transport of a high molecular weight medication from microneedles was also documented (Al-Qallaf & Das, 2008). Skin permeability is calculated according to this equation (Al-Qallaf & Das, 2008):

$$K = f\frac{D}{L_h} \tag{9.29}$$

where f is the fractional skin area following microneedle insertion, L_h is the length of the hole in skin, K is skin permeability of drugs, and D is the effective diffusion coefficient of drugs (Al-Qallaf & Das, 2008). The following assumptions were made by the authors (Al-Qallaf & Das, 2008):

1. Viable epidermis is a homogenous medium
2. One-dimensional diffusion of injected medication
3. Microneedle-assisted drug is delivered into two skin layers. The first layer is the epidermis and the second is the dermis from where the drug is absorbed into systemic circulation The sink condition (100% absorption) takes place at the interface between the dermis and the epidermis (Al-Qallaf & Das, 2008). Following microneedle insertion, a cylindrical shaped hole with radius (R) is formed, which refers to the microneedle radius (R), and the annular gap width (W) (Al-Qallaf & Das, 2008). The shape of the microneedle pore in the skin is conical if viewed laterally. The fractional skin area following MN insertion is given as (Al-Qallaf & Das, 2008):

$$f = n^2\pi\frac{(R+W)^2 - R^2}{A} \tag{9.30}$$

where n is the total number of microneedles, R is the microneedle radius, W is the annular gap width (W) and A is the surface area of the patch. Annular gap width (W) was linked to MN radius (R) (Al-Qallaf & Das, 2008):

$$W = \varepsilon R \tag{9.31}$$

where ε is a dimensionless parameter (Al-Qallaf & Das, 2008). The new constraint for optimization was calculated as (Al-Qallaf & Das, 2008):

$$\frac{\sqrt{A}}{n} \geq \alpha R \tag{9.32}$$

The equations for drug transport were described. The transitory drug movement in the skin was expressed via Fick's second law as (Al-Qallaf & Das, 2008):

$$\frac{\partial C}{\partial t} = D\frac{\partial^2 C}{\partial x^2} \tag{9.33}$$

where D is the diffusion coefficient, C is the drug concentration, t is time and x is the distance in a given skin layer (Al-Qallaf & Das, 2008). The concentration of the drug in the blood following transdermal drug administration was given by one-compartmental pharmacokinetic model (Al-Qallaf & Das, 2008):

$$V_b\frac{dC_b}{dt} = \left(\frac{dQ}{dt}\right)A - K_e C_b V_b \tag{9.34}$$

where K_e is the elimination rate constant, $\frac{dQ}{dt}$ is the penetration rate of the medication across the skin, A is the surface area of the microneedle patch, V_b is the volume of distribution in the blood, C_b is the concentration of the drug in the blood (Al-Qallaf & Das, 2008). The authors used the model to determine the optimum microneedle design. Data from this method was employed for the prediction of skin permeability of a high molecular weight drug injected into the skin from microneedles (Al-Qallaf & Das, 2008). The influence of microneedles on the concentration of the medication in the blood was evaluated (Al-Qallaf & Das, 2008).

A quantifiable relationship between skin diffusion properties and microneedle design parameters is important. A mathematical framework was presented by Olatunji et al. to evaluate the influence of design parameters on transdermal drug transport (Fig. 9.2) (Olatunji et al., 2012). The influence of compressive strain on the skin following microneedle insertion was taken into consideration. The authors treated the skin as a homogeneous layer since the only skin layer described in the model was the viable epidermis (VEs) (Olatunji et al., 2012). The exerted pressure was assumed to be uniformly distributed over the area of the skin under the microneedles. This assumption was considered valid since the skin is a nonrigid surface and microneedles are micrometer-sized objects (Olatunji et al., 2012). Thus, for tightly packed microneedles, the stress was distributed uniformly. The force on the skin led to a reduction in the pore size in the viable epidermis. It was assumed that the total reduction in pore size was equal to the total volume of the microneedles (Olatunji et al., 2012).

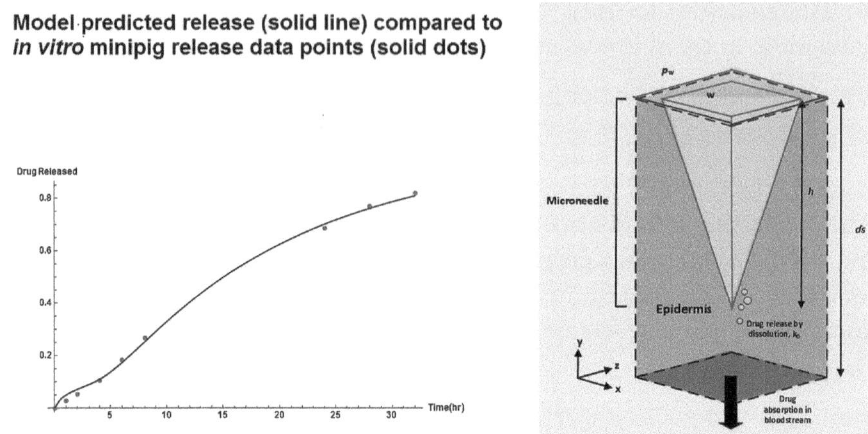

Fig. 9.2 A dissolving pyramid-shaped microneedle.

Drug permeability across the skin was mathematically described as (Olatunji et al., 2012):

$$K = \frac{\varepsilon D_r}{\tau h} \tag{9.35}$$

where ε is the skin porosity, h is the thickness of the viable epidermis, D_r is the diffusion coefficient of the drug through the water-filled pores within the viable epidermis, D_r links the drug and skin properties and τ is the tortuosity (Olatunji et al., 2012):

$$D_r = DH(\lambda) \tag{9.36}$$

where $H(\lambda)$ is a hindrance factor, D is the diffusion coefficient of the drug at infinite dilution, λ is the ratio of the hydrodynamic radius of the drug (r_p) and the radius of the hypothesized pores (r_s) in the viable epidermis (Olatunji et al., 2012):

$$\lambda = \frac{r_p}{r_s} \tag{9.37}$$

The diffusion coefficient following microneedle insertion (D_r) was then calculated (Olatunji et al., 2012):

$$D_r = DH\left(\frac{r_p}{\sqrt{\dfrac{v_{pore} - nv_{needle}}{\pi n_p x}}}\right) \tag{9.38}$$

where r_p is the hydrodynamic radius of the drug, H is a hindrance factor, n_p is the number of pores in the skin area, D is the diffusion coefficient of the drug at infinite dilution, v_{needle} is the volume of each MN, v_{pore} is the volume of pores in the skin prior to MN insertion and x is the path length of diffusion (Olatunji et al., 2012). Using this model, the authors linked microneedle geometry to drug permeability in the skin (Eqs. 9.35 and 9.38). The model was then employed for the analysis of different practical scenarios. The predicted steady-state flux was found to be lower when the influence of the microneedle strain on diffusion coefficient was considered (Olatunji et al., 2012).

Römgens et al. developed a computational model that was used to predict optimal microneedle design and arrangement (Römgens et al., 2016). The authors used a three-dimensional finite element mathematical model to describe kinetics and diffusion of microneedle-assisted antigen delivery. An assumption was made that the array comprised several conical MNs placed in a normal grid with a center-to-center spacing S (Römgens et al., 2016). The representative volume element (RVE) was a quarter of a microneedle within a rectangular skin sample (Römgens et al., 2016). The authors combined several RVEs which resulted in a microneedle array. The microneedle geometry was characterized by radius at the base, r_b and length, l_{MN}. Coated microneedles were also modeled with a coating layer thickness, t_c, of 0.02 mm (Römgens et al., 2016). In the mathematical model, the skin was assumed to comprise two layers, namely, the epidermis with a thickness, t_e, of 0.2 mm and the dermis with a thickness, t_d, of 1.8 mm (Römgens et al., 2016). Following microneedle penetration antigen was released, which will diffuse into the skin (Römgens et al., 2016). Antigen release, diffusion, and internalization by cells were described by a system of partial differential equations (Römgens et al., 2016):

$$\frac{\partial C_{MN}}{\partial t} = \nabla(D_{MN}\nabla C_{MN}), \tag{9.39}$$

$$\frac{\partial C_{ECM,e}}{\partial t} = \nabla(D_e\nabla C_{ECM,e} - k_a C_{ECM,e}), \tag{9.40}$$

$$(R_{tot,e} - C_{rec} - C_{cells}) + k_d C_{rec},$$

$$\frac{\partial C_{ECM,d}}{\partial t} = \nabla(D_e\nabla C_{ECM,d} - k_a C_{ECM,d}), \tag{9.41}$$

$$(R_{tot,d} - C_{rec} - C_{cells}) + k_d C_{rec} - k_c C_{ECM,d},$$

$$\frac{\partial C_{rec}}{\partial t} = k_a C_{ECM,i}(R_{tot,i} - C_{rec} - C_{cells}). \qquad (9.42)$$

$$-(k_d + k_{int}) C_{rec},$$

$$\frac{\partial C_{circ}}{\partial t} = k_c C_{ECM,d}, \qquad (9.43)$$

$$\frac{\partial C_{cells}}{\partial t} = k_{int} C_{rec}, \qquad (9.44)$$

where $C_{ECM,i}$ is the antigen concentration in the extracellular matrix (*ECM*) with $i = e$ and d for the epidermis and dermis, C_{MN} is the antigen concentration in the microneedle coating, C_{cells} is the antigen concentration taken up by the cells, C_{rec} is the concentration of antigen bound to receptors, and C_{circ} is the antigen concentration taken up by the blood and the lymph. These concentrations depend on time t and the x-, y-, and z-coordinates. D_j is the diffusion coefficient with $j = MN$, *e*, and *d*, representing the coating of the microneedle, epidermis, and dermis, respectively while $R_{tot,i}$ is the initial receptor concentration (Römgens et al., 2016). The skin kinetics was described with the finite element model via the subroutine *hetval* (Römgens et al., 2016), in which concentration change was computed at each integration point for every increment (Römgens et al., 2016). The results showed an optimal distance between MNs based on the number of activated antigen presenting cells, which was assumed to be linked to the generated immune response (Römgens et al., 2016). The length of the microneedles influenced the number of activated cells (Römgens et al., 2016). On the other hand, the release rate and the radius at the base of the microneedle only had minimal impact on the number of activated cells (Römgens et al., 2016). The mathematical model showed the effect of geometric parameters on generated immune response. Several parameters were modified to optimize the model (Römgens et al., 2016).

Kim et al. developed a mathematical model that can be used to estimate the quantity of medication delivered into the skin following microneedle dissolution (Kim et al., 2015). This mathematical approach takes into consideration a biological membrane (Kim et al., 2015) and describes the dissolution of conical needles that are usually fabricated from water-soluble materials (example, sucrose) premixed with medications (Kim et al., 2015). The transport of a medication from MNs is predicted based on an assumption that MNs perform as though they are in a similar milieu

(Kim et al., 2015). A single MN model is made up of a conical MN immersed in a rectangular chamber. The top and the bottom of the compartment represent the stratum corneum and the blood circulation, respectively (Kim et al., 2015). The width is equivalent to pitch p_w while the height is equivalent to the depth of the skin layer d_s. There is no mass transport through the four sides (Kim et al., 2015). Compartment volume $\upsilon(t)$ $\left(\sim p_w^2\, d_s\right)$ increases as the space previously occupied by the microneedle approaches that of the compartment as a result of the dissolution process (Kim et al., 2015). The model does not take into consideration the influence of needle geometry and pitch on skin deformation. The pitch of a microneedle array has significant effect on its function (Kim et al., 2015).

The use of a microneedle array with a small pitch MN requires more force for skin insertion. This parameter also impacts skin indentation, which may lead to leakage of the medication, inadequate insertion, and diminished diffusivity (Kim et al., 2015). The rectangular compartment assumption is invalid in certain situations. However, integrating those factors into this model may unnecessarily increase the complexity of the analysis (Kim et al., 2015). When a microneedle pierces the skin, there is a gap width between the skin layer and the surface of the microneedle (Kim et al., 2015). The gap is formed by several factors (Zoudani & Soltani, 2020): the elasticity of the skin causes the microneedle to be slightly pulled back once the insertion force is removed (Yadav et al., 2020) the microneedle array is not inserted straight into the skin (Li, 2017) as the MN dissolves, an empty space is created and it is not filed by a body fluid (Kim et al., 2015). In Kim-Ita-Simon model (Kim et al., 2015), the gap width is part of the volume of the compartment $v(t)$. The volume of the compartment changes as the microneedle dissolves. As the MN degrades, the ratio of the needle height $h(t)$ to the base radius $r(t)$ does not change. The half angle at the apex h is constant and is given by Eq. (9.8) (Kim et al., 2015). Drug concentration is even in the compartment and is only a function of time, i.e., $c(t)$. The authors parameterized the model by writing the mathematical equations in dimensionless forms (Kim et al., 2015):

$$\frac{dC}{d\tau} = -KC + 3H^2\left(\frac{\Sigma - C}{\Omega + 1 - H^3}\right)(1 - C), \tag{9.45}$$

$$\frac{dH}{d\tau} = -(1 - C), \tag{9.46}$$

$$C(\tau = 0) = 0, \tag{9.47}$$

and

$$H(\tau = 0) = 1, \tag{9.48}$$

The dimensionless variables and parameters are defined as (Kim et al., 2015):

$$C = \frac{c}{\left(\frac{\beta}{1-\beta}\right)c_s} \tag{9.49}$$

$$H = \frac{h}{h_0} \tag{9.50}$$

$$\tau = \frac{t}{\tau_0} \tag{9.51}$$

$$\Omega = \frac{v_0}{v_{c,0}} \tag{9.52}$$

$$\Sigma = \frac{(1-\beta)\rho}{c_s} \tag{9.53}$$

$$\tau_0 = \frac{\rho h_0 \sin\theta}{k_D c_s} \tag{9.54}$$

and

$$K = k_L \tau_0 \tag{9.55}$$

where the characteristic time, τ_0 is directly proportional (by a factor of 3) to the time elapsed for total dissolution of the MN (the initial mass if it were to take place at an initial rate); β is the mass fraction of drug in the MN; K denotes the ratio of the first-order drug elimination rate k_L by absorption to the drug release rate by dissolution τ_0^{-1}; Σ is the dimensionless density of matrix material in the MN; and Ω is the ratio of the initial compartment volume to that of the needle (Kim et al., 2015). These equations can be used to predict the fate of the active ingredient in the skin layer and the needle height following the use of the microneedle (Kim et al., 2015). The authors utilized the fourth order Runge–Kutta method to integrate Eqs. (9.45) and (9.46) (Kim et al., 2015) while obeying the initial conditions (Kim et al., 2015). The authors used an in-house algorithm with Mathematica and the following inputs (a profile of the needle height, $h(t)$, obtained from an experiment to compute the rate constant k_L). A Newton–Raphson's technique was used for calculation (Kim et al., 2015):

$$k_L^{(i+1)} = k_L^{(i)} - \frac{f'\left(k_L^{(i)}\right)}{f''\left(k_L^{(i)}\right)} \tag{9.56}$$

where the function f is the sum of the measured needle height and the squared error between the numerical solution of Eq. (9.46) with $k_L^{(i)}$ (Kim et al., 2015). The index of the iteration is denoted by the superscript (i). The last term of Eq. (9.56), which incorporates the first and the second derivatives, is shown via the central finite difference formula (Kim et al., 2015):

$$\frac{f'\left(k_L^{(i)}\right)}{f''\left(k_L^{(i)}\right)} = \frac{\Delta k_L\left(f\left(k_L^{(i)} + \Delta k_L\right) - f\left(k_L^{(i)} - \Delta k_L\right)\right)}{2\left(f\left(k_L^{(i)} + \Delta k_L\right) - 2f\left(k_L^{(i)}\right) + f\left(k_L^{(i)} - \Delta k_L\right)\right)} \tag{9.57}$$

where Δk_L is an arbitrary quantity. The calculation of Eq. (9.56) is repeated until the absolute value of Eq. (9.57) is less than 10^{-4} (Kim et al., 2015) Simulations with the drug fentanyl showed that the concentration of the medication was proportional to its mass fraction in the microneedle and the influence of the pitch on skin permeation was slightly nonlinear (Kim et al., 2015). A larger quantity of the drug was transported from MN arrays with lower pitch size. Microneedle dissolution did not depend on changes in the elimination rate constant (Kim et al., 2015). The authors utilized an optimization algorithm to obtain this parameter from the needle height − time data (Kim et al., 2015).

Several pharmacokinetic models have been developed for microneedle-assisted drug delivery. Rajoli et al. created a novel physiologically-based intradermal model to simulate the systemic pharmacokinetics of microneedle array patches loaded with rilpivirine and cabotegravir (Rajoli et al., 2019). The physiologically-based pharmacokinetic models were evaluated using pharmacokinetic data following intradermal and intramuscular administration of rilpivirine to rats (Rajoli et al., 2019). Rilpivirine and cabotegravir were also administered intramuscularly to healthy adult subjects (Rajoli et al., 2019). Mathematical models were subsequently used to estimate nanoformulation dose, release rates and doses for the maintenance of therapeutic plasma concentrations rationally selected (Rajoli et al., 2019). Weekly dosing was also studied with the aim of determining the potential drug delivery from a smaller patch. Self-administration of a microneedle patch that is placed on the skin for a short duration makes it unnecessary to use longer time periods in the mathematical model (Rajoli et al., 2019). The model was utilized for the estimation of optimal dose and release rates for cabotegravir and rilpivirine microneedle array patches (Rajoli et al., 2019). This method can be used to support oral antiretroviral drug administration (Rajoli et al., 2019).

The authors made the following assumptions regarding the skin compartment (Rajoli et al., 2019): (1) unidirectional drug flow from the stratum corneum to systemic circulation; (2) hair follicles cover 0.1% of the skin (3) nanoparticle release rates as well as partition and diffusion co-efficients across different skin layers were kept constant during the kinetic process (4) only the free medication released from the nanoparticles diffuse across skin layers to reach systemic circulation (Rajoli et al., 2019). The drug present at the microneedle depot in the stratum corneum was modeled as follows (Rajoli et al., 2019):

$$\frac{\partial D_{SC}}{\partial t} = -2 \times K_{TD} \cdot D_{SC} \tag{9.58}$$

$$\frac{\partial D_{SC}}{\partial t} = K_{TD} \cdot D_{SC} - \frac{2}{3} \cdot D_{SCD} \cdot K_{SC} - \frac{1}{3} \cdot D_{SCD} \cdot K_{SC} - K_{TD} \cdot D_{SCD} \tag{9.59}$$

where K_{TD} is the drug release rate from the encapsulated formulation, D_{SC} is the total drug present at the microneedle depot in the stratum corneum, K_{SC} is the permeability rate constant to the stratum corneum, D_{SCD} is the total free drug present at the microneedle depot in the stratum corneum at time t.

Drug present at the microneedle depot in the viable epidermis is (Rajoli et al., 2019):

$$\frac{\partial D_{VE}}{\partial t} = K_{TD} \cdot D_{SC} - 2 \times K_{TD} \cdot D_{VE} \tag{9.60}$$

$$\frac{\partial D_{VE}}{\partial t} = K_{TD} \cdot D_{VE} + K_{TD} \cdot D_{SC} - \frac{2}{3} \cdot D_{VED} \cdot K_{VE} - \frac{1}{3} \cdot D_{VED} \cdot K_{SC} - K_{TD} \cdot D_{VE} \tag{9.61}$$

where, D_{VED} is the total free drug present at the microneedle depot in the viable epidermis at time t, D_{VE} is the total drug present at the microneedle depot in the viable epidermis at time t, $P\,C_{S/W}$ is the partition coefficient between skin and water, K_S – permeability rate constant to the skin/hair follicles, K_{VE} — permeability rate constant to the viable epidermis. Drug present at the microneedle depot in the dermis is (Rajoli et al., 2019):

$$\frac{\partial D_{DE}}{\partial t} = K_{TD} \cdot D_{VE} + K_{TD} \cdot D_{VED} - K_{TD} \cdot D_{DE} \tag{9.62}$$

where D_{DE} is the total drug amount present at the microneedle depot in the dermis at time t. Drug transport from the microneedle to the adjoining skin layers is given by (Rajoli et al., 2019):

$$-\frac{\partial M}{\partial t} = A \cdot P_e \cdot C_b \tag{9.63}$$

where M is the quantity of medication crossing the skin, A is the surface area of drug in contact with the skin, P_e is the effective permeability and C_b is the drug concentration at the depot site. Drug crossing the stratum corneum is (Rajoli et al., 2019):

$$\frac{\partial A_{SC}}{\partial t} = \frac{2}{3} \cdot D_{SCD} \cdot K_{SC} - PC_{SC/VE} \cdot A_{SC} \tag{9.64}$$

where A_{SC} is the amount present in the stratum corneum at time t, $P\,C_{SC/VE}$, — partition coefficient between stratum corneum and viable epidermis. Drug traversing through the viable epidermis can be calculated from Eq.(9.65) (Rajoli et al., 2019):

$$\frac{\partial A_{VE}}{\partial t} = \frac{2}{3} \cdot D_{VED} \cdot K_{VE} + K_{SC/VE} \cdot A_{SC} - A_{VE}/PC_{S/W} \tag{9.65}$$

where A_{VE} is the amount present in the viable epidermis at time t, D_{VED} — total free drug present at the microneedle depot in the viable epidermis at time, $P\,C_{S/W}$ — partition coefficient between skin and water. Drug traversing through the hair follicles is (Rajoli et al., 2019):

$$\frac{\partial AH_F}{\partial t} = \frac{1}{3} \cdot D_{SCD} \cdot K_S + \frac{1}{3} \cdot D_{VED} \cdot K_S - A_{HF}/PC_{S/W} \tag{6.66}$$

where A_{HF} is the amount present in the hair follicles at time t (Rajoli et al., 2019).

Lateral transport from the microneedle reservoir was based on the drug permeability rate and the diffusion across the skin was a function of the drug partition coefficient between the layers (Rajoli et al., 2019). Quantitative structure—property relationships were used to derive the permeability rate constants (K_S, K_{SC}, K_{VE}) and drug partition coefficients $\left(PC_{SC/VE}, PC_{S/W}\right)$ (Rajoli et al., 2019).

References

Al-Qallaf, B., & Das, D. B. (2008). Optimization of square microneedle arrays for increasing drug permeability in skin. *Chemical Engineering Science, 63*(9), 2523–2535. https://doi.org/10.1016/j.ces.2008.02.007

Blazek, J. (2005a). *Principles of solution of the governing equations* (pp. 29–75). Elsevier BV. https://doi.org/10.1016/b978-008044506-9/50005-9

Blazek, J. (2005). Chapter 3 - principles of solution of the governing equations. In J. Blazek (Ed.), *Computational fluid dynamics: Principles and applications* (2nd ed., pp. 29–75). Oxford: Elsevier Science.

Chaulya, S., & Prasad, G. M. (2016). Slope failure mechanism and monitoring techniques. In *Sensing and monitoring technologies for mines and hazardous areas* (pp. 1–86).

Chavoshi, S., Rabiee, M., Rafizadeh, M., Rabiee, N., Shamsabadi, A. S., Bagherzadeh, M., Salarian, R., Tahriri, M., & Tayebi, L. (2019). Mathematical modeling of drug release from biodegradable polymeric microneedles. *Bio-Design and Manufacturing, 2*(2), 96–107. https://doi.org/10.1007/s42242-019-00041-y

Kim, K. S., Ita, K., & Simon, L. (2015). Modelling of dissolving microneedles for transdermal drug delivery: Theoretical and experimental aspects. *European Journal of Pharmaceutical Sciences, 68*, 137–143. https://doi.org/10.1016/j.ejps.2014.12.008

Li, C. C. (2017). Chapter Eight - numerical modeling. In C. C. Li (Ed.), *Rockbolting*. Butterworth-Heinemann.

Lyashko, S. I., Klyushin, D. A., Onotskyi, V. V., & Lyashko, N. I. (2018). Optimal control of drug delivery from microneedle systems*. *Cybernetics and Systems Analysis, 54*(3), 357–365. https://doi.org/10.1007/s10559-018-0037-9

Machekposhti, S., Soltani, M., Najafizadeh, P., Ebrahimi, S. A., & Chen, P. (2017). Biocompatible polymer microneedle for topical/dermal delivery of tranexamic acid. *Journal of Controlled Release, 261*, 87–92. https://doi.org/10.1016/j.jconrel.2017.06.016

Olatunji, O., Das, D. B., & Nassehi, V. (2012). Modelling transdermal drug delivery using microneedles: Effect of geometry on drug transport behaviour. *Journal of Pharmaceutical Sciences, 101*(1), 164–175. https://doi.org/10.1002/jps.22736

Owen, B., Nasar, A. M. A., Harwood, A. R. G., Hewitt, S., Bojdo, N., Keavney, B., Rogers, B. D., & Revell, A. (2020). Vector-based discrete element method for solid elastic materials. *Computer Physics Communications, 254*, 107353. https://doi.org/10.1016/j.cpc.2020.107353

Rajoli, R. K. R., Flexner, C., Chiong, J., Owen, A., Donnelly, R. F., Larrañeta, E., & Siccardi, M. (2019). Modelling the intradermal delivery of microneedle array patches for long-acting antiretrovirals using PBPK. *European Journal of Pharmaceutics and Biopharmaceutics, 144*, 101–109. https://doi.org/10.1016/j.ejpb.2019.09.011

Römgens, A. M., Bader, D. L., Bouwstra, J. A., & Oomens, C. W. J. (2016). Predicting the optimal geometry of microneedles and their array for dermal vaccination using a computational model. *Computer Methods in Biomechanics and Biomedical Engineering, 19*(15), 1599–1609. https://doi.org/10.1080/10255842.2016.1173684

Ronnander, P., Simon, L., & Koch, A. (2020). Experimental and mathematical study of the transdermal delivery of sumatriptan succinate from polyvinylpyrrolidone-based microneedles. *European Journal of Pharmaceutics and Biopharmaceutics, 146*, 32–40. https://doi.org/10.1016/j.ejpb.2019.11.007

Ronnander, P., Simon, L., Spilgies, H., & Koch, A. (2018). Modelling the in-vitro dissolution and release of sumatriptan succinate from polyvinylpyrrolidone-based microneedles. *European Journal of Pharmaceutical Sciences, 125*, 54–63. https://doi.org/10.1016/j.ejps.2018.09.010

Tanaka, K., Nishida, M., Kunimochi, T., & Tagaki, T. (2001). Numerical and experimental studies for the impact of projectiles on granular materials. In A. Levy, & H. Kalman (Eds.), *In Handbook of powder technology* (pp. 263–270). Elsevier Science B.V.

Yadav, P. R., Han, T., Olatunji, O., Pattanayek, S. K., & Das, D. B. (2020). Mathematical modelling, simulation and optimisation of microneedles for transdermal drug delivery: Trends and progress. *Pharmaceutics, 12*(8), 1–31. https://doi.org/10.3390/pharmaceutics12080693

Zhou, W., Liu, B., Wang, Q., Chang, X., & Chen, X. (2020). Formulations of displacement discontinuity method for crack problems based on boundary element method. *Engineering Analysis with Boundary Elements, 115*, 86–95. https://doi.org/10.1016/j.enganabound.2020.03.007

Zoudani, E. L., & Soltani, M. (2020). A new computational method of modeling and evaluation of dissolving microneedle for drug delivery applications: Extension to theoretical modeling of a novel design of microneedle (array in array) for efficient drug delivery. *European Journal of Pharmaceutical Sciences, 150*. https://doi.org/10.1016/j.ejps.2020.105339

CHAPTER 10

Solid microneedles

Skin anatomy and physiology

The human skin (Fig. 10.1) prevents water loss and protects the body against pathogens, different forms of trauma, including chemical, thermal, and ultraviolet radiation (Abdo et al., 2020). The skin is made up of the epidermis (outer layer), dermis (central layer) and the subcutaneous tissue (inner layer) (Haque & Talukder, 2018; Reicherter et al., 2015). The epidermis encompasses three main cell populations: keratinocytes, which make up 99% of the cells, melanocytes, and highly specialized immune cells (Langerhans cells) (Ain et al., 2021). The principal constituents of the epidermis are sheets of keratinocytes but there are also non-epithelial cells (Merkel cells, antigen-presenting dendritic Langerhans cells and melanocytes) (Abdo et al., 2020). The skin is the body's primary defense against injury and infection; it is also important for vitamin production and thermoregulation (Gravitz, 2018). From the standpoint of drug delivery, MNs can be used to overcome the skin barrier. Skin barrier function is provided by numerous skin elements: the stratum corneum, immunological barrier, chemical barrier, tight junctions and the microbiome (Bäsler et al., 2016). From the drug delivery and skin absorption perspectives, the mechanical barriers – tight junctions and the stratum corneum – play significant roles (Bäsler et al., 2016). A substantial part of the drug delivery barrier is ascribed to the outermost layer of the skin (the stratum corneum).

The stratum corneum (SC) is made up of protein-rich flattened dead cells (corneocytes) embedded in a lipid matrix (Ita & Ashong, 2020). The cells do not possess a conventional membrane but rather an inner envelope of cross-linked proteins, the cornified envelope (CE) (Schmitt & Neubert, 2018). Covalently linked to the CE are ω-hydroxy ceramides forming the lipid envelope (LE) (Schmitt & Neubert, 2018). Ceramides (CER) are classified based on a letter code, which portrays the chemical modifications of the different CER classes (Schmitt & Neubert, 2018). The first letter refers to fatty acid modification [A] reflects an α-hydroxylated fatty

Microneedles
ISBN 978-0-323-91811-4
https://doi.org/10.1016/B978-0-323-91811-4.00013-4

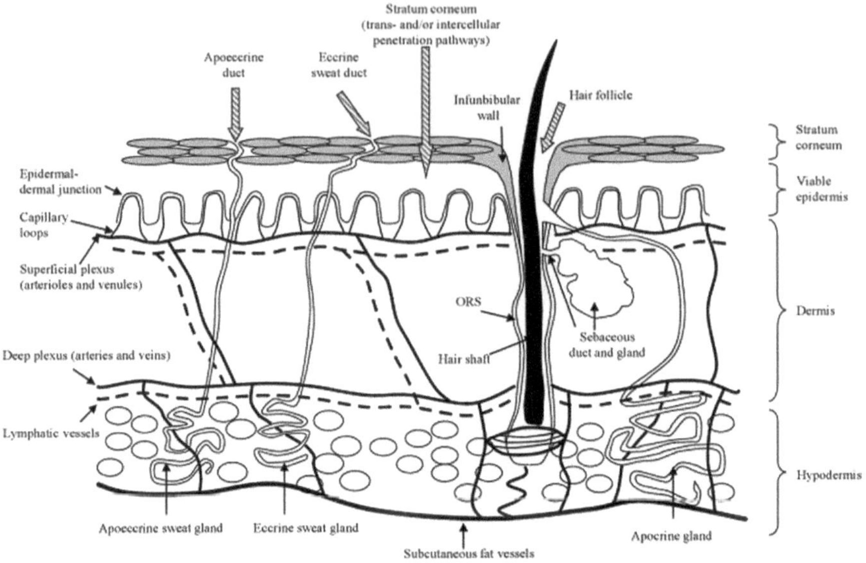

Fig. 10.1 Schematic overview of the skin layers, appendages, blood and lymphatic vessels. The stratum corneum, viable epidermis, dermis and the subcutis, lymphatic vessels, blood and appendages are shown. *(From Jepps, O. G., Dancik, Y., Anissimov, Y. G., & Roberts, M. S. (2013). Modeling the human skin barrier — toward a better understanding of dermal absorption.* Advanced Drug Delivery Reviews, *65(2), 152–168. https://doi.org/10.1016/j.addr.2012.04.003).*

acid, [N] marks a non-hydroxylated fatty acid, while [O] represents an ω-hydroxy fatty acid (Schmitt & Neubert, 2018). CER constitutes approximately 40 wt% of the dry mass of the skin's lipids. The remaining compounds are about 25 wt% cholesterol, 25 wt% free fatty acids (FFAs), 6 wt% cholesterol sulfate and minute quantities of other lipids (Schmitt & Neubert, 2018). The viable epidermis comprises the stratum basale (SB), stratum spinosum (SS) and the stratum granulosum (SG) (Bäsler et al., 2016). The terminal differentiation of keratinocytes takes place at the SG-SC interface, from which viable keratinocytes turn into corneocytes (Sahle et al., 2015). After migrating from the stratum basale, a multi-step process of differentiation takes the keratinocytes continuously in an outward direction across diverse layers of the SS and this has been demonstrated via different protein expression (involucrin as well as keratin 1 and 10) (Bäsler et al., 2016).

During the movement of keratinocytes from the stratum basale to the SG, the cells undergo several alterations. Precursor lipids are synthesized in the SB, SS and SG, assembled in the lamellar bodies in the SS and SG and the contents released via exocytosis at the SG-SC interface (Sahle et al., 2015). Keratohyalin granules loaded with lamellar bodies that contain lipids and enzymes as well as keratin intermediate filaments and pro-filaggrin are usually formed in the stratum granulosum (Bäsler et al., 2016). Lamellar bodies (LBs) are created in the stratum granulosum (SG) and in the upper stratum spinosum (Moner et al., 2016). Their contents are extruded at the SGSC interface to produce the lipid lamellar structure of this tissue (Moner et al., 2016). LBs refer to ovoid, lamellate membrane-delimited organelles that are 200–300 nm in diameter. These organelles release precursors of protease inhibitors, SC lipids, antimicrobial peptides, and hydrolases into the inter-corneocyte space of the SC (Moner et al., 2016).

Even though cells of the stratum basale remain attached to a matrix, several daughter keratinocytes penetrate the stratum spinosum via asymmetric mitoses subsequently exiting the cell cycle, growing bigger and forming strong intercellular bonds (Simpson et al., 2011). Stratum granulosum cells tend to flatten and are enclosed in a cornified envelope that is impermeable to water (Simpson et al., 2011). Although the SG is only a few cells thick, it is a vital component of the epidermis because the major changes that result in the creation of the SC occur in the SG (Wickett & Visscher, 2006). Two classes of granules are formed in the SG: lamellar bodies, which contain lipids and keratohyalin granules, which typically contain proteins (Wickett & Visscher, 2006). The stratum spinosum, which is also called the prickle cell layer is made up of polyhedral and irregular cells with "spines," that expand outward and interact with adjacent cells via desmosomes (Yousef & Sharma, 2018). Dendritic cells are sometimes present in this layer (Yousef & Sharma, 2018). The stratum spinosum consists of different cells with a variety of shape, structure and subcellular features. Suprabasal spinous cells are polyhedral and have a rounded nucleus, while cells of the upper portion of the stratum spinosum tend to have larger sizes, become flatter as they migrate toward the surface of the skin (Kolarsick et al., 2011).

The stratum germinativum, which is also called the stratum basale is made up of columnar or cuboidal cells that are attached to a basement membrane (Kolarsick et al., 2011; Lechler & Fuchs, 2005). Cells of the basal layer contain elongated or oval nuclei and there is also melanin pigment that is transported from adjacent melanocytes (Kolarsick et al., 2011). Keratin 14 is usually expressed in the basal layer, and the daughter keratinocytes are subjected to a typical basal-to-suprabasal switch from keratin 5/14 to keratin 1/10 as they move up the layers (Roger et al., 2019).

From time to time, dividing basal cells detach from an underlying basement membrane of extracellular matrix, migrate toward the skin surface, and eventually die (Lechler & Fuchs, 2005). The basement membrane (BM) connects the dermis to the epidermis (Segal et al., 2008). This membrane also controls the growth, survival, differentiation, and morphogenesis of the human skin (Segal et al., 2008). The basal membrane is linked to the papillary layer of the dermis by anchoring fibrils that enter into and are strongly bound to collagen fibrils (Behrens et al., 2012). The epidermal basement membrane thus functions as an essential interface between the connective tissues and cells of epidermis (Woodley & McNutt, 1992, pp. 19–36). Furthermore, several nutrients and signals must permeate this interface, transporting messages from the blood to the epidermis and other messages from the epidermis to other structures (Woodley & McNutt, 1992, pp. 19–36).

The dermis reinforces the epidermis (Abdo et al., 2020). Neural, vascular, lymphatic and adnexa of the skin are part of the dermis. This tissue is a durable base that prevents shear by absorbing mechanical forces (Abdo et al., 2020). It consists of a superficial papillary zone, made up of relatively thin collagen fibers and the reticular dermis: a compact layer of thick collagen fibers (Abdo et al., 2020). The major structural constituents of the dermis are type-1 collagen and elastin. Collagen provides mechanical resistance to extension, while elastin partakes in skin deformation (Yang et al., 2015). From a morphological viewpoint, the human dermis can be categorized into two parts (Korosec et al., 2019): (a) the papillary dermis with loose connective tissue and a high cell density (ii) the reticular dermis possessing low cell density and a high density of collagens and other connective tissue proteins (Korosec et al., 2019). In the papillary layer, collagen fibers are constituted into thin bundles of less than 10 μm in diameter whereas in the deep reticular dermis, an entangled mass of collagen bundles (more than 50 μm) is formed (Ventre et al., 2009). The lower dermis, comprising entirely of reticular dermis, is formed by an interwoven mass of wavy collagen bundles (Ventre et al., 2009). The reticular dermis contains hair roots, receptors, blood vessels, sensory receptors, sweat glands, and a vast web of nerves (Ye & De, 2017). Fibroblast is the primary cell type in the dermis (Abdo et al., 2020). Dermal fibroblasts synthesize and secrete structural proteins such as collagen fibers, elastin, and proteoglycans that together form the extracellular matrix, in which immune cells such as macrophages, mast cells, and dendritic cells are embedded (Ain et al., 2021). Fibroblasts can also be reprogrammed into induced pluripotent stem cells and are implicated in aging (Korosec et al., 2019). The human skin is made up of at least two fibroblast lineages with clearly defined structure and functions that range from different tissue remodeling abilities and proliferation rates to the expression of a variety of cytokine or matrix genes and modifying enzymes (Korosec et al., 2019).

The reticular and papillary strata of the human skin comprise connective tissue as well as the extracellular matrix ECM (Machaliński et al., 2016). The ECM is made up of protein fibers (reticular, collagen and elastic) and ground substance that occupy a significant portion of the extracellular space (Machaliński et al., 2016). The subcutaneous tissue (mainly fat) is frequently described as part of the skin because it is the link between the skin and the underlying muscular and bony tissues (Ye & De, 2017). It has been recently reported that the adipocyte layer in the skin below the dermis is not inert but a controlled lipid layer (Alexander et al., 2015). Subcutaneous depots appear to be ultrastructurally and structurally homogenous. In reality, they form a very polymorphous and distinct tissue subclasses due to the variety of roles played by each single tissue subcategory (Sbarbati et al., 2010). In humans, different schools of thought have identified varied number of layers. Two anatomically and histologically distinct layers of adipose tissue under the reticular dermis were named in one study (Driskell et al., 2014). In contrast, Sbarbati et al. identified three types of adipose tissues based on their ultrastructural and structural properties (Sbarbati et al., 2010). The distinct types of tissues can be categorized based on vasculo-stromal and adipocyte components. Ranging from type one to three, a gradual rise in stroma (with diverse properties) was documented (Sbarbati et al., 2010). The authors categorized white adipose tissue (WAT) on the basis of structural and ultrastructural features: structural WAT (sWAT), deposit WAT (dWAT) and fibrous WAT (fWAT) (Sbarbati et al., 2010). sWAT can be found in some areas in the hips as well as in some parts in the limbs and it is more stromal (Sbarbati et al., 2010). dWAT is defined by cells that are connected by a weak net of isolated collagen fibers and are tightly packed. Collagenic components are very poor, cells are large with few blood vessels (Sbarbati et al., 2010). The fWAT has a marked fibrous constituent and occupies regions where severe mechanic stress frequently occurs (Sbarbati et al., 2010). Sweat glands, hair follicles, and sebaceous glands constitute skin appendages (Weng et al., 2020). Hair follicles are located in the dermal layer of the skin and consist of hair papillae, hair matrix, hair bulges and root sheath (Weng et al., 2020). Taking into consideration the formidable nature of the barrier (especially the stratum corneum), several methods have been developed for enhancing drug delivery. In the next section, the effect of solid microneedles on transdermal drug delivery will be examined.

Transdermal drug delivery with solid microneedles

Solid microneedles (Fig. 10.2) constitute an important group of miniature devices for percutaneous drug transport (Abiandu & Ita, 2019). Investigators have used solid microneedles to pre-treat the skin followed by the application of an ointment, cream (Vos et al., 2020) or solution. Several techniques are available for the fabrication and evaluation of solid microneedles. Abidin et al. fabricated concave conic-shaped microneedles (Abidin, 2020). The authors applied different loads (50−800 g) perpendicular to the microneedle tip with the aim of predicting structure deformation (stress and buckling) (Abidin, 2020). Microneedles measuring (60−100 μm in height and 1−4 μm in diameter tip) were prepared using an etching method with a combination of nitric acid, hydrofluoric acid, and acetic acid. Three types of microneedles were made using acidic etching (Abidin, 2020). Stress and buckling studies of the MNs were conducted on rat skin. Stress analysis carried out with COMSOL Multiphysics simulation showed that the critical load factor for buckling analysis was 2.3359 while the highest stress

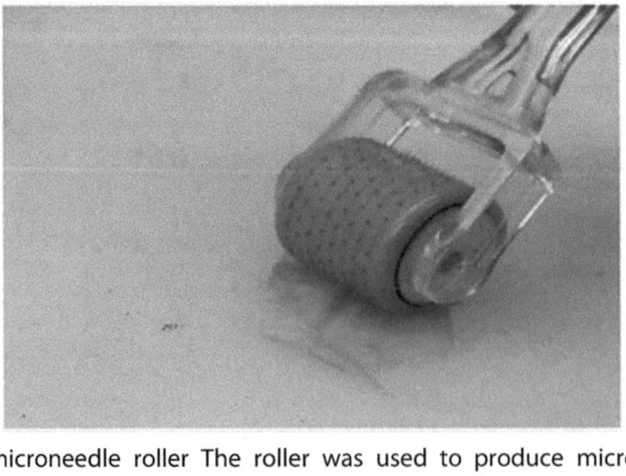

Fig. 10.2 A microneedle roller The roller was used to produce micropores in the skin. *(From Abiandu, I., & Ita, K. (2019). Transdermal delivery of potassium chloride with solid microneedles.* Journal of Drug Delivery Science and Technology, 53, *101216. https://doi.org/10.1016/j.jddst.2019.101216).*

at the microneedle tip was 6.28 × 108 Nm^{-2} (Abidin, 2020). Surface buckling deformation of a single MN did not take place with load increase. Nevertheless, the authors observed decrease in height following load increment on the investigated microneedles and this was ascribed to high microneedle tip compression (Abidin, 2020).

Solid microneedles are used principally for the delivery of potent and low dosed compounds, such as specific hormones, vaccines and analgesic drugs because of the poor drug loading/release capacity of solid microneedles (Vos et al., 2020). Occasionally, solid MNs are used to deliver drugs in the low milligram range (Vos et al., 2020). The "poke and patch" approach involves the use of solid microneedles for skin pre-treatment. Initially, the microneedles or a microneedle roller (Fig. 10.2) are applied onto the skin, removed and a patch loaded with a medication is administered leading to drug diffusion through the formed micro-conduits into systemic circulation (Abiandu & Ita, 2019; Van Der Maaden et al., 2012). Experiments with rat skin have shown that microchannels created by microneedles remain open for at least three days following microneedle treatment under occlusive conditions (Nagarkar et al., 2020). A modification of the conventional solid microneedle approach is the "scrape and patch" method that is characterized by the generation of micro-abrasions following the use of microprojections, microneedles, or microblades for skin scraping (Nagarkar et al., 2020). Drug solutions are subsequently applied. Microneedle rollers are used to pierce the skin multiple times as the roller spins (Nagarkar et al., 2020). Marketed rollers are based on this approach and are utilized to enhance transdermal drug delivery via the formation of micropores (Nagarkar et al., 2020).

One of the limitations of the "poke and patch" approach is low drug loading. To overcome the disadvantage of low loading, nanoporous microneedle arrays have been created (de Groot et al., 2017). Nanoporous microneedle arrays (npMNAs) incorporate nanofluidic structure thus making available a large reservoir for medications (Vos et al., 2020). Following the breaching of the skin, there is drug diffusion from the microneedle system (Vos et al., 2020). An interesting benefit associated with the use of this system is that the reservoir can be expanded substantially by linking the system to an external drug source resulting in the percutaneous delivery of higher drug doses (Vos et al., 2020).

Recently, Schepens et al. studied the feasibility of using npMNAs for influenza vaccination (Schepens et al., 2019). Hemagglutinin (HA) loaded in the npMNAs was efficiently transported into mouse ear skin. Intradermal

HA vaccination with npMNAs led to the generation of significantly higher specific antibodies in comparison to intramuscular immunization (Schepens et al., 2019). Remarkably, mice that were immunized with pH1N1 hemagglutinin -loaded microneedles were not infected (Schepens et al., 2019). In another study, Vos and coworkers combined npMNAs with drug storage (Vos et al., 2020). The authors investigated parameters that affected memantine transport from the reservoir (Vos et al., 2020). The authors used a mathematical model to predict the effect of reservoir properties (drug concentration and volume) as well as npMNA properties (surface area and backplate thickness) on the transport of low-molecular-weight medications (Vos et al., 2020). The authors used the model to provide a forecast for low-molecular-weight medication release from a drug reservoir via the MNA into a receptor solution (Vos et al., 2020). There was a good correlation between the model output and the transport of memantine across minipig skin. Interestingly, drug release rates from the npMNAs can be modulated to achieve zero and first order release (Vos et al., 2020).

In a recent study, de Groot and colleagues investigated ceramic npMNAs for the intradermal delivery of vaccines (de Groot et al., 2017). The authors investigated the piercing property of the ceramic (alumina) npMNAs (a length of 475 μm, 100 microneedles per array, and a mean pore size of 80 nm) (de Groot et al., 2017). The loading of DT (diphtheria toxoid), tetanus toxoid (TT), and imiquimod into, and subsequent release from the npMNAs were also evaluated in vitro. The authors loaded TT and DT into the tips of the npMNAs, and the toxoids were successfully released into ex vivo murine skin (de Groot et al., 2017). Significantly, the use of these TT- and DT-loaded npMNAs led to the induction of antigen-specific antibodies, with titers similar to those achieved with subcutaneous immunization (de Groot et al., 2017).

In another interesting study, Ilić and colleagues used combined solid microneedles with nanoemulsion for the transdermal delivery of aceclofenac (Ilić et al., 2018). The lecithin-based nanoemulsion was co-stabilized by sucrose esters. The authors pre-treated rat skin with microneedles and this led to enhanced drug transport into/across the skin from the investigated dosage forms (1.4–2.1-fold increase in bioavailability) (Ilić et al., 2018). There was also a 1.2–1.7-fold increase in the quantity of the drug retained in the skin 24 h following microneedle application (Ilić et al., 2018). Importantly, drug concentration following the use of these drug delivery systems (lecithin/polysorbate 80 vs. lecithin/sucrose palmitate) transported through MN-treated skin (259.23 ± 73.18 ng/mL vs. 173.37 ± 40.50 ng/mL vs.)

was markedly greater than what was achieved via administration through untreated skin (105.69 ± 19.53 ng/mL vs. 88.38 ± 14.46 ng/mL) (Ilić et al., 2018). The authors concluded that the combined use of sucrose palmitate-co-stabilized nanoemulsion with microneedles can result in greater concentration in the skin, whereas the combination of polysorbate 80-costabilized nanoemulsion with microneedles can be used to improve drug transport into the bloodstream (Ilić et al., 2018).

Kaur et al. investigated the influence of solid microneedles on the transdermal delivery of verapamil hydrochloride and amlodipine mesylate (Kaur et al., 2014). In vitro studies were carried out with microneedle-treated pig ear skin using vertical static Franz diffusion cells (Kaur et al., 2014). Pores created in pig skin following the use of solid microneedles are shown in Fig. 10.3. Passive diffusion across untreated porcine skin served as the control (Kaur et al., 2014). The transdermal flux of verapamil increased markedly from 8.75 $\mu g/cm^2/h$ to 49.96 $\mu g/cm^2/h$ across pig skin following the use of microneedles. Percutaneous flux of amlodipine besylate following the use of stainless steel microneedles was 22.39 $\mu g/cm^2/h$ while passive permeation rate for the medication was 1.57 $\mu g/cm^2/h$ (Kaur et al., 2014). When microneedle roller was applied to pig skin, the transdermal flux of amlodipine was 1.05 $\mu g/cm^2/h$ in comparison with a passive diffusion flux of 0.19 $\mu g/cm^2/h$ (Kaur et al., 2014).

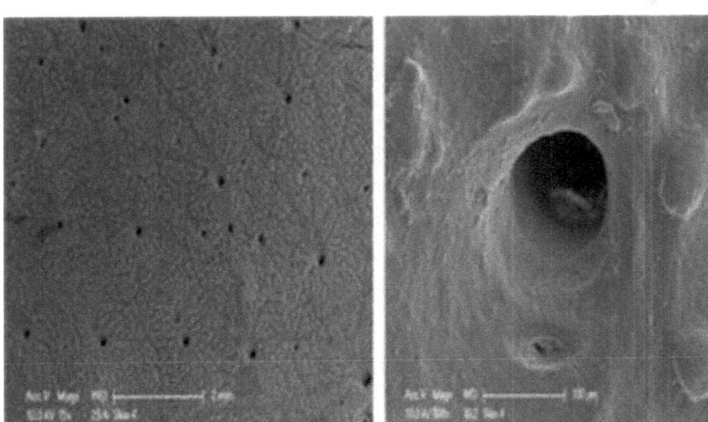

Fig. 10.3 A single micropore generated by a microneedle Microneedles were used to create micro-conduits in pig skin. *(From Kaur, M., Ita, K. B., Popova, I. E., Parikh, S. J., & Bair, D. A. (2014). Microneedle-assisted delivery of verapamil hydrochloride and amlodipine besylate.* Theme Issue: Dermal and Transdermal Drug Delivery, *86(2), 284–291. https://doi.org/10.1016/j.ejpb.2013.10.007).*

Naloxone hydrochloride (NL) is used for the reversal of respiratory depression caused by opioid overdose (Teeter et al., 2021). However, the duration of action of the medication is short. Puri et al. investigated the influence of microneedles on the skin transport of NL (Puri et al., 2021). Dr. Pen™ Ultima A6 solid microneedles were used for the experiments. The authors studied the impact of application duration, microneedle length and donor concentration on NL diffusion (Puri et al., 2021). In-vitro in-vivo correlation of the transport data was also conducted with the aim of predicting the plasma concentration kinetics of the drug. The authors reported 56- and 37-fold permeation increase following the application of 500 and 250 μm needles applied for 1 min, respectively (Puri et al., 2021). The use of 500 μm microneedles for 2 min markedly decreased the lag time to approximately 8 min and increasing the donor concentration for the same treatment group doubled the drug transport rate across pig skin (Puri et al., 2021). The pharmacokinetic profile of the drug following the use of microneedles was similar to those achieved with intranasal and intramuscular devices (Puri et al., 2021).

Hu and coworkers used a scalable manufacturing technique to fabricate microneedles via thermoplastic drawing of metallic glasses (Hu et al., 2020). By controlling the fracture and rheology of metallic glass, the investigators were able to produce MNs with tunable tips and lengths. Thermoplastic drawing can also be used to create hollow and solid microneedles via the modulation of metallic glass thickness (Hu et al., 2020). Viscous buckling of metallic liquid was used to describe the transition mechanism from solid to hollow MNs (Hu et al., 2020). In vitro experiments showed that both hollow and solid metallic glass microneedles can pierce pig skin and deliver model medications (Hu et al., 2020). In another interesting study, a touch-actuated MN array was used for the percutaneous delivery of macromolecules (Yang et al., 2015). Touch-actuated MN array patch (TMAP) combines a solid microneedle array with a conventional transdermal patch (Yang et al., 2018). The drug reservoir of TMAPs can be used to store large amounts of liquid dosage forms, particularly those of thermolabile compounds (Yang et al., 2018). In addition, TMAP can easily pierce the skin and be pulled back from it to form microconduits via touch-actuated "press and release" actions. This is normally the approach adopted for the passive transport of drug solutions (Yang et al., 2018). The authors compared subcutaneous injection, dissolvable microneedle array, solid microneedles, and traditional transdermal patch. It was evident that the use of insulin loaded TMAP resulted in the greatest hypoglycemic action on type 1 diabetic rats (Yang et al., 2018). The investigators also used a "closed-loop" system for on-demand insulin transport based on the blood glucose level feedback (Yang et al., 2018). The authors utilized a TMAP loaded with insulin to maintain type 1 diabetic rats in a normal blood glucose range for about 11.63 h (Yang et al., 2018). This was the longest therapeutic duration when comparing the different transdermal patches that were studied (Yang et al., 2018).

Even though microneedle-mediated transdermal drug transport has been widely investigated, the effort to quantitatively link in vitro and in vivo transdermal penetration data is still inadequate (Milewski et al., 2013). In a set of experiments by Milewski et al. the "poke and patch" microneedle technique was applied to Yucatan minipig to obtain pharmacokinetic data for naltrexone hydrochloride (Milewski et al., 2013). The pharmacokinetic profile showed a prompt attainment of maximum drug concentration in the plasma followed by a relatively sudden decline (Milewski et al., 2013). A diffusion-compartmental mathematical model was used to correlate the animal data with previously published in vitro steady-state transdermal penetration findings (Milewski et al., 2013). The model incorporates two parallel permeation routes, a pharmacokinetic module, microchannel closure kinetics and barrier-thickness-dependent diffusional resistance (Milewski et al., 2013). Analysis of the data showed good correlation with estimated in vitro permeation data and microchannel closure rate (Milewski et al., 2013). Data from full-thickness skin in vitro transport studies showed the best agreement with the in vivo data. A new mechanistic insight was provided following data analysis and pharmacokinetic predictions were made following the use of this approach (Milewski et al., 2013).

Ghimirey and Ita used vitro diffusion experiments to evaluate the transdermal transport of magnesium sulfate across porcine skin (Ghimirey & Ita, 2020). The authors used inductively coupled plasma-optical emission spectroscopy to measure the concentration of the drug. Transdermal permeation rates of the drug across microneedle-treated and intact pig skin were calculated from the cumulative drug amount versus time curve (Ghimirey & Ita, 2020). The permeation rate of the drug across microneedle-treated pig skin was 134.19 ± 2.4 μg/cm^2/h and the flux value following drug transport across intact pig skin was 4.64 ± 0.05 μg/cm^2/h (Ghimirey & Ita, 2020). Confocal microscopy was used to visualize the microconduits generated by the microneedle roller (with 500 μm needles) (Ghimirey & Ita, 2020). The increase in percutaneous permeation rate across pierced porcine skin was statistically significant compared to the permeation rate across intact skin (Ghimirey & Ita, 2020). Following the use of microneedles, the permeation rate (flux) of the drug over 12 h was about 33-fold greater compared to passive transport across intact pig skin (Ghimirey & Ita, 2020).

In an interesting study, Chen et al. applied MN patches on the forearm, forehead, and abdomen of healthy human subjects (Chen et al., 2019). The aim was to study the factors that influence user acceptance to MNs (density, length and size of microneedle patches) (Chen et al., 2019). Several insertion experiments showed that solid microneedles with a density of 100 MN/cm^2, a length of 400 μm, and an array size of 10 × 10 were the most beneficial for pretreating the forehead t (Chen et al., 2019). Solid microneedles that were 49 and 100 MN/cm^2 in density, 400 μm in height, with an array size of 5 × 5, 10 × 10, and 10 × 20 arrays were well-tolerated for insertion into the abdomen and were not considered painful (Chen et al., 2019). In contrast, the use of microneedles with densities of 49 and 100 MN/cm^2, lengths of 400 and 600 μm, and an array size of 10 × 10 resulted in insignificant skin irritation (forearm) and less pain (Chen et al., 2019). The safety of solid microneedles and patient compliance can be improved by optimizing microneedle dimensions as a function of application sites (Chen et al., 2019).

A microneedle array measuring 200 μm in length with a high density (576 microneedles on a 5× 5-mm area) was administered onto ex vivo human abdominal skin (600 μm in thickness) (van der Maaden et al., 2014). The investigators used different forces (3.43−22.1 N) via weight rods (350−2250 g) to apply the microneedles to the skin for 10 s (van der Maaden et al., 2014). The aim of this set of experiments was to study the parameters that have impact on the permeation efficiency of the microneedles (van der Maaden et al., 2014). The authors also investigated the influence of the time of application (5−60 s) and force (7.36 N) on microneedle penetration. An impact-insertion applicator or a manual application device operating at a velocity of 3 m/s was used for the application of a microneedle array three times onto the skin of 15 volunteers (5 females, 10 males; 21−57 years) (van der Maaden et al., 2014). Two parameters (time and application force), that can impact the permeation efficiency of the microneedles were studied (van der Maaden et al., 2014). It was shown that an increase in the applied force of up to 7.36 N (at 10 s) significantly enhanced the permeation efficiency (van der Maaden et al., 2014). An additional increase in the application time (at a constant force of 7.36 N) or increase in the application force caused only a marginal increase in the penetration ability of the microneedles (van der Maaden et al., 2014). An improvement in reproducibility was also recorded under these conditions (van der Maaden et al., 2014). Two types of application devices were compared by the authors. Following the use of the impact-insertion applicator, the skin was pierced with substantially lower intra and inter individual variability (van der Maaden et al., 2014). However, applying the device manually led to lower penetration efficiency (56%) and relatively higher inter and intra individual variability and this was ascribed to variations in the applied force (van der Maaden et al., 2014). The importance of an applicator for the efficient and reproducible penetration of the skin by microneedles was demonstrated (van der Maaden et al., 2014).

Naloxone poorly penetrates the skin due to its hydrophilic nature (Gao & Brogden, 2019). Microneedles (MNs) are capable of painlessly forming microconduits in the skin for increased naloxone absorption (Gao & Brogden, 2019). Gao and Brogden evaluated the correlation between ionization and naloxone transport across microneedle-treated skin (Gao & Brogden, 2019). In-line MN arrays fabricated from stainless steel with needles that were either 500 μm or 750 μm long were used for the experiments (Gao & Brogden, 2019). Hydrogels loaded with 0.2%, 0.5%, or 1% of the drug were prepared from 1% cross-linked polyacrylic acid (polymer) and the pH was adjusted to 5, 6.5, or 7.4 (Gao & Brogden, 2019). Pig skin was pre-treated with microneedles and the drug-containing gel with subsequent in vitro transport experiments. Rheological experiments were also conducted to study the structural properties of the gels (Gao & Brogden, 2019). The investigated gels possessed adequate spreadability and viscoelastic properties. Naloxone penetration across untreated skin was greatest from pH 7.4 gels when the drug is unionized, and this was different compared to the use of a gel at pH 5 with 100% ionization and negligible drug concentration (Gao & Brogden, 2019). When the gel (pH 5) was combined with MN treatment, the authors were able to significantly increase drug permeation leading to the achievement of therapeutic concentration (Gao & Brogden, 2019). The authors also provided evidence that confirmed that naloxone gel concentration and MN length had marked influence on absorption lag time (Gao & Brogden, 2019). In contrast, polymer concentration did not exert significant impact on drug permeability. The authors showed that transdermal delivery of naloxone across microneedle-treated skin may be a viable option for the management of opioid induced pruritus (Gao & Brogden, 2019).

Li et al. fabricated stable and mechanically robust solid microneedles from polylactic acid (PLA) (Li et al., 2017). The authors studied the effect of drug concentration, MN dimensions, drug formulation viscosity and application time on drug permeation across pretreated skin (Li et al., 2017). MNs (600 μm long) were mechanically robust (Li et al., 2017). Microneedles measuring 800 μm in height and possessing a density of 256 MNs per cm^2 had the greatest impact on drug penetration (Li et al., 2017). Increasing drug concentration enhanced the amount of drug that permeated the skin but did not have marked influence on the drug permeation rate (Li et al., 2017). When drug viscosity was increased, the amount of the medication that crossed the skin decreased. The authors utilized the microneedles for transdermal insulin flux enhancement in vivo (Li et al., 2017). The concentration of glucose in the blood of diabetic mice decreased to 29% of initial value at 5 h because of enhanced transport of insulin following microneedle application (Li et al., 2017). The biodegradable PLA solid microneedles painlessly pierced the skin and enhanced the transdermal transport of insulin (Li et al., 2017).

Takahashi and colleagues developed a highly scalable method of fabricating microneedles (Takahashi et al., 2016). The authors subjected negative photoresist to inclined/rotated ultraviolet lithography (Takahashi et al., 2016). By regulating the exposure dose and inclination angles of ultraviolet radiation, the authors were able to modify the photo-polymerized part of the photoresist and vary the contours as well as heights of the fabricated microneedles via the same mask patterns (Takahashi et al., 2016). It was shown that the dimensions of the microneedles correlated well with the theoretical predictions for different exposure doses and inclination angles (Takahashi et al., 2016). The results confirmed that the theoretical approach used by the authors can be used to produce MNs with on-demand geometry (Takahashi et al., 2016). The MNs can be utilized as molds for dissolving microneedles or as solid microneedles. The authors reported that the technique can reduce production cost and time as well as improve fabrication accuracy (Takahashi et al., 2016).

Rajabi et al. fabricated stretchable and flexible microneedle patches that combine sharp and hard stainless-steel microneedles with flexible and soft base substrates (Rajabi et al., 2016). A polymer base (elastomer) permits conformal connection between the complex topography as well as the texture of the underlying skin and the microneedle patch, while sharp and robust stainless-steel MNs pierce the skin (Rajabi et al., 2016). The authors fabricated the flexible microneedle patches by magnetically incorporating short MNs into a flexible supporting base (polymer) (Rajabi et al., 2016). The MN patches were administered onto human skin and there was adequate patch adaptation to the deformations and wrinkles of the skin (Rajabi et al., 2016). The microneedles penetrated the skin surface (Rajabi et al., 2016). The stretchable and adaptable MN patches can be used for bioelectric treatments, transdermal drug delivery, and wearable bio-electronics (Rajabi et al., 2016).

Conical microneedle arrays were fabricated from stainless steel via a micro milling process (García-López et al., 2018). Geometrical parameters of the needles were easily modulated via a dry cutting technique relative to a wet regime. The investigators achieved a maximum needle tip diameter of 188 μm via dry cutting versus values less than 30 μm following the use of a wet technique (García-López et al., 2018). Surface roughness increased following the fabrication via machining of 36 and 9 needles using a wet cutting method and a dry cutting regime, respectively (García-López et al., 2018). Nevertheless, tool breakage was documented following the milling of 12 needles via a dry cutting mode. The authors conducted cost analysis which showed that the micro milling method is competitive for fabricating microneedle arrays relative to other techniques (García-López et al., 2018).

Narayanan and Raghavan prepared microneedles from silicon (Narayanan & Raghavan, 2017). The needles possessed high aspect ratio and sharp tips (Narayanan & Raghavan, 2017). Etching parameters were optimized and utilized to produce tapered and long microneedles (Narayanan & Raghavan, 2017). The MNs were used to deliver drugs across the skin. The factors for optimization were the etching time, concentration of tetramethylammonium Hydroxide (TMAH), temperature, window size of the optical mask, and etching rates (Narayanan & Raghavan, 2017). It was revealed that increasing the window size led to an increase in the etch rates in both lateral and vertical directions. In contrast, when the temperature was more than 90°C, etching became fast and with less control (Narayanan & Raghavan, 2017). Sample placement and TMAH concentration were tuned to achieve a rate of etching in the vertical compared to the lateral direction (Narayanan & Raghavan, 2017). This was carried out with the aim of producing microneedles with high aspect ratio (Narayanan & Raghavan, 2017). MNs with tip diameter of 0.40 μm, aspect ratio of 1.43, an average height of 158 μm, tip angle of 19.4° and base width of 110.5 μm were prepared (Narayanan & Raghavan, 2017). The microneedles possessed a microhardness value of 44.4 (HRC) which was 52.2 times greater than the maximum tensile strength of the skin which makes it safer and simpler to insert microneedles into the skin (Narayanan & Raghavan, 2017).

Metallic microneedles are used for biosensing and drug delivery because of tissue compatibility and mechanical strength (Cahill et al., 2018). Be that as it may, contemporary microneedle designs do not have adequate surface area. Furthermore, there are manufacturing constraints that are associated with low scalability (Cahill et al., 2018). Cahill used a scalable manufacturing method to fabricate stainless steel microneedles (Cahill et al., 2018). These microneedles can be employed for biosensing and for drug delivery. The investigators embossed stainless steel-based feedstock, sintered it at 1100 C and then carried out electropolishing (Cahill et al., 2018). The microneedles were mechanically robust with adequate porosity (36%) and high surface area (Cahill et al., 2018). In addition, loading efficiency and capillarity were optimized by creating small pores (average diameter 2.22 μm) (Cahill et al., 2018). The porous network possessed a theoretical wicking rate of 4.7 μL/s and 27 ± 5 μL/s of fluid can be wicked up via capillary action (Cahill et al., 2018). When the investigators applied the microneedles to a skin model loaded with a metabolite, the device processed 17 ± 3 μL of the metabolite solution (Cahill et al., 2018). The efficiency of the MNs (22.4 ± 4.9 $\mu g/cm^2$) was three times greater than that of topical application (7.1 ± 4.3 $\mu g/cm^2$) (Cahill et al., 2018). The porous metallic MN patches pierced pig skin under a 19 N load (Cahill et al., 2018).

Globally, basal cell carcinoma (BCC) is recognized as the most prevalent type of cancer (Cameron et al., 2019; Gellatly et al., 2020). Imiquimod is sometimes used as a topical formulation for the management of BCC (Al-Mayahy et al., 2019). This is a cost-effective and non-invasive method of treatment. Despite the reported clinical efficacy, this drug is only approved by the Food and Drug Administration for the management of superficial BCC (Al-Mayahy et al., 2019). Al-Mayahy and coworkers evaluated imiquimod transport from Aldara™ cream (Al-Mayahy et al., 2019). Using Aldara™ cream and Franz diffusion cells for experiments, the authors showed that a large proportion of the formulation remains on the surface of the skin with more than 80% of the medication recovered from the skin wash (Al-Mayahy et al., 2019). Time-of-Flight Secondary Ion Mass Spectrometry (ToF-SIMS) of tape stripped skin samples revealed imiquimod in the upper tape strips, while the deeper strips had no drug (Al-Mayahy et al., 2019). Given the non-uniform permeation and the restricted depth of penetration at tape strips 6—18 following the application of topical imiquimod cream, the skin was pretreated with a microneedle device (Dermapen®) as a means of enhancing intradermal drug delivery (Al-Mayahy et al., 2019). The amount of the drug that was recovered in the remaining skin and the tape strips was about three times greater when Aldara™ was applied on microneedle-treated skin in comparison to untreated skin (Al-Mayahy et al., 2019). The ToF-SIMS ion images of the tape strips showed the drug in the microconduits which subsequently diffused laterally to peripheral epidermal layers (Al-Mayahy et al., 2019). ToF-SIMS analysis ws used to chemically map drug transport into the skin. In addition, enhanced intradermal transport of imiquimod following the utilization of a microneedle device was demonstrated (Al-Mayahy et al., 2019).

Photodynamic therapy (PDT) of nodular or deep skin tumors is presently hindered by the low tissue penetration of 5-aminolevulinic acid (ALA) (Donnelly et al., 2008). From the standpoint of PDT, penetration depth refers to the distance into tissue at which the intensity of the administered radiation is reduced to about 37% (Mahmoudi et al., 2019). Donnelly and coworkers used silicon microneedles to improve the skin transport of the drug in vitro and in animal experiments (Donnelly et al., 2008). The authors coated silicon wafer with nitride and oxide layer using low pressure chemical vapor deposition (Donnelly et al., 2008). The layer was subsequently patterned lithographically via plasma etching. Following patterning, the wafer was etched with 29% w/v potassium hydroxide solution at a temperature of 79°C in a water bath (Donnelly et al., 2008). The fabricated microneedles possessed an aspect ratio of 3:2 (height: base diameter). In the ultimate step of the fabrication procedure, the authors surface coated the MNs with titanium (adhesive layer) and platinum (Donnelly et al., 2008). Scanning electron microscopy was utilized for the examination of the microneedle arrays (Donnelly et al., 2008). Piercing excised murine skin with 6 × 7 arrays of microneedles (interspacing of 750 μm, 270 μm in height and a diameter of 240 μm at the base) resulted in a marked enhancement in the transdermal transport of ALA released from a bio-adhesive patch (Donnelly et al., 2008). Skin pretreatment with microneedles increased ALA transport to the upper layers of excised pig skin but, at the average depth of 1.875 mm, drug concentration was comparable to control values, and this was ascribed to the binding of ALA by the skin (Donnelly et al., 2008). Experiments with nude mice demonstrated that the use of microneedles can decrease ALA dose needed to induce photosensitizer protoporphyrin IX concentration in the skin as well as the drug application time (Donnelly et al., 2008). Shorter application times improve clinician and patient convenience and also allow the management of more patients in the same session (Donnelly et al., 2008). Decreasing the required dose of ALA is also beneficial as the drug is costly and degrades quickly through a second order reaction (Donnelly et al., 2008).

The clinical use of topical anesthetics for the management of pediatric patients is hindered by the slow onset of action (Davies et al., 2018) Poor transport of the anesthetic agent via the stratum corneum has been postulated as the principal cause for the delayed onset of anesthesia (Davies et al., 2018). MNs can painlessly pierce the skin. Davies et al. used MN treatment of ex vivo human skin to enhance the transdermal delivery of tetracaine (Davies et al., 2018) The aim of the study was to accelerate the onset of anesthesia. The authors used Franz-type diffusion cells to evaluate the transport of a marketed tetracaine product (Ametop® gel) across human epidermis (Davies et al., 2018). The 2.5 cm^2 membranes were prepared and either pre-treated with (i) an array of 30 stainless-steel microneedles (200 μm in width and 500 μm in length) which were subsequently inserted for 10 s; (ii) a single puncture using a 26-gauge hypodermic needle or (iii) left untreated (Davies et al., 2018). Following transport experiments, the epidermal membranes were visually characterized (Davies et al., 2018). At 30 min, 5.43 $\mu g/cm^2$ of the drug had crossed the intact membrane in comparison to 12.13 $\mu g/cm^2$ that permeated the microneedle-treated membrane (Davies et al., 2018). The use of a hypodermic needle resulted in the formation of a large single conduit in the epidermis (about. 4250 μm^2) whereas the pierced surface area following the use of microneedles was about 75,000 μm^2 (Davies et al., 2018). The use of microneedles for skin pretreatment significantly improved the transdermal transport of tetracaine. The onset time for anesthesia can be reduced by 50% following the application of microneedle-assisted Ametop® gel (Davies et al., 2018).

Humoral immunity can be enhanced via prolonged exposure of lymphoid tissues to vaccine antigens, but traditional bolus immunization can result in the rapid clearance of antigens (Boopathy et al., 2019). Boopathy et al. used solid pyramidal microneedles to deliver vaccine antigens (Boopathy et al., 2019). Solid pyramid-shaped MN arrays were prepared from silk fibroin tips that incorporate a stabilized human immunodeficiency virus envelope trimer immunogen and adjuvant (Boopathy et al., 2019). A dissolving polymer was used to create the microneedle base (Boopathy et al., 2019) Following microneedle administration, fibroin silk tips loaded with the vaccine are positioned in the epidermis/upper dermis where they release vaccine over a time frame defined by the silk matrix crystallinity (Boopathy et al., 2019). Microneedle-assisted immunization in mice led to the release of Env trimer over 14 days. There was a good correlation between the vaccine release and a 16-fold enhancement in bone marrow (BM) plasma cells, a ~1300-fold increase in serum IgG titers as well as increased germinal center (GC) B cell responses in comparison to bolus immunization (Boopathy et al., 2019). The authors demonstrated that implantable microneedles can be used to substantially enhance humoral immunity derived from subunit vaccines (Boopathy et al., 2019).

In an interesting study, Ogunjimi et al. investigated the synergistic influence of microneedles and microspheres on the transdermal delivery of naltrexone (Ogunjimi et al., 2020). Spray-drying was used for the microsphere formulation and the effect of microneedle skin treatment was evaluated (Ogunjimi et al., 2020). An "in-plane" array of 750-μm long stainless steel microneedles was utilized for experiments (Ogunjimi et al., 2020). The microneedle array was administered 20 times to generate 100 microconduits on a skin area of 0.64 cm^2 (analogous to the size of in-line diffusion cell diffusion area) (Ogunjimi et al., 2020). The microneedle array was initially applied 10 times in one direction, the skin was subsequently turned about 45° and then, an additional 10 microneedles were applied to create the microchannels (Ogunjimi et al., 2020). The authors treated skin samples with microneedles with subsequent gentian violet dye staining for micropore visualization (Ogunjimi et al., 2020). The transdermal penetration of the drug across microneedle-treated skin was 11.6 ± 2.2 μg/cm^2/h from chitosan-NTX microspheres, and this was substantially higher than the permeation rate across untreated skin (Ogunjimi et al., 2020).

References

Abdo, J. M., Sopko, N. A., & Milner, S. M. (2020). The applied anatomy of human skin: A model for regeneration. *Wound Medicine, 28*. https://doi.org/10.1016/j.wndm.2020.100179

Abiandu, I., & Ita, K. (2019). Transdermal delivery of potassium chloride with solid microneedles. *Journal of Drug Delivery Science and Technology, 53*, 101216. https://doi.org/10.1016/j.jddst.2019.101216

Abidin, H. E. Z. (2020). Stress and deformation of optimally shaped silicon microneedles for transdermal drug delivery. *Journal of Pharmacological Sciences, 109*(8), 2485–2492.

Ain, Q. U., Campos, E. V. R., Huynh, A., Witzigmann, D., & Hedtrich, S. (2021). Gene delivery to the skin – how far have we come? *Trends in Biotechnology, 39*(5), 474–487. https://doi.org/10.1016/j.tibtech.2020.07.012

Al-Mayahy, M. H., Sabri, A. H., Rutland, C. S., Holmes, A., McKenna, J., Marlow, M., & Scurr, D. J. (2019). Insight into imiquimod skin permeation and increased delivery using microneedle pre-treatment. *European Journal of Pharmaceutics and Biopharmaceutics, 139*, 33–43. https://doi.org/10.1016/j.ejpb.2019.02.006

Alexander, C. M., Kasza, I., Yen, C. L. E., Reeder, S. B., Hernando, D., Gallo, R. L., Jahoda, C. A. B., Horsley, V., & MacDougald, O. A. (2015). Dermal white adipose tissue: A new component of the thermogenic response. *Journal of Lipid Research, 56*(11), 2061–2069. https://doi.org/10.1194/jlr.R062893

Bäsler, K., Bergmann, S., Heisig, M., Naegel, A., Zorn-Kruppa, M., & Brandner, J. M. (2016). The role of tight junctions in skin barrier function and dermal absorption. *Journal of Controlled Release, 242*, 105–118. https://doi.org/10.1016/j.jconrel.2016.08.007

Behrens, D. T., Villone, D., Koch, M., Brunner, G., Sorokin, L., Robenek, H., Bruckner-Tuderman, L., Bruckner, P., & Hansen, U. (2012). The epidermal basement membrane is a composite of separate laminin- or collagen IV-containing networks connected by aggregated perlecan, but not by nidogens. *Journal of Biological Chemistry, 287*(22), 18700–18709. https://doi.org/10.1074/jbc.M111.336073

Boopathy, A. V., Mandal, A., Kulp, D. W., Menis, S., Bennett, N. R., Watkins, H. C., Wang, W., Martin, J. T., Thai, N. T., He, Y., Schief, W. R., Hammond, P. T., & Irvine, D. J. (2019). Enhancing humoral immunity via sustained-release implantable microneedle patch vaccination. *Proceedings of the National Academy of Sciences of the United States of America, 116*(33), 16473–16478. https://doi.org/10.1073/pnas.1902179116

Cahill, E. M., Keaveney, S., Stuettgen, V., Eberts, P., Ramos-Luna, P., Zhang, N., Dangol, M., & O'Cearbhaill, E. D. (2018). Metallic microneedles with interconnected porosity: A scalable platform for biosensing and drug delivery. *Acta Biomaterialia, 80*, 401–411. https://doi.org/10.1016/j.actbio.2018.09.007

Cameron, M. C., Lee, E., Hibler, B. P., Barker, C. A., Mori, S., Cordova, M., Nehal, K. S., & Rossi, A. M. (2019). Basal cell carcinoma: Epidemiology; pathophysiology; clinical and histological subtypes; and disease associations. *Journal of the American Academy of Dermatology, 80*(2), 303–317. https://doi.org/10.1016/j.jaad.2018.03.060

Chen, B. Z., Liu, J. L., Li, Q. Y., Wang, Z. N., Zhang, X. P., Shen, C. B., Cui, Y., & Guo, X. D. (2019). Safety evaluation of solid polymer microneedles in human volunteers at different application sites. *ACS Applied Bio Materials, 2*(12), 5616–5625. https://doi.org/10.1021/acsabm.9b00700

Davies, L. B., Gateley, C., Holland, P., Coulman, S. A., & Birchall, J. C. (2018). Accelerating topical anaesthesia using microneedles. *Skin Pharmacology and Physiology, 30*(6), 277–283. https://doi.org/10.1159/000479530

Donnelly, R. F., Morrow, D. I. J., McCarron, P. A., Woolfson, A. D., Morrissey, A., Juzenas, P., Juzeniene, A., Iani, V., McCarthy, H. O., & Moan, J. (2008). Microneedle-mediated intradermal delivery of 5-aminolevulinic acid: Potential for enhanced topical photodynamic therapy. *Journal of Controlled Release, 129*(3), 154–162. https://doi.org/10.1016/j.jconrel.2008.05.002

Driskell, R. R., Jahoda, C. A. B., Chuong, C. M., Watt, F. M., & Horsley, V. (2014). Defining dermal adipose tissue. *Experimental Dermatology, 23*(9), 629–631. https://doi.org/10.1111/exd.12450

Gao, X., & Brogden, N. K. (2019). Development of hydrogels for microneedle-assisted transdermal delivery of naloxone for opioid-induced pruritus. *Journal of Pharmaceutical Sciences, 108*(11), 3695–3703. https://doi.org/10.1016/j.xphs.2019.08.025

García-López, E., Siller, H. R., & Rodríguez, C. A. (2018). Study of the fabrication of AISI 316L microneedle arrays. In *Procedia manufacturing* (Vol. 26, pp. 117–124). Elsevier B.V. https://doi.org/10.1016/j.promfg.2018.07.014

Gellatly, M., Cruzval-O'Reilly, E., Mervak, J. E., & Mervak, B. M. (2020). Metastatic basal cell carcinoma with atypical pattern of spread. *Radiology Case Reports, 15*(12), 2641–2644. https://doi.org/10.1016/j.radcr.2020.09.054

Ghimirey, K. B., & Ita, K. (2020). Microneedle-assisted percutaneous transport of magnesium sulfate. *Current Drug Delivery, 17*(2), 140–147. https://doi.org/10.2174/1567201817666191217093936

Gravitz, L. (2018). Skin. *Nature, 563*(7732).

de Groot, A. M., Anouk, A. C., Kuijt, N., Kooten, P. J. S.v., Vos, P. J., Sijts, A. J. A. M., & van der Maaden, K. (2017). Nanoporous microneedle arrays effectively induce antibody responses against diphtheria and tetanus toxoid. *Frontiers in Immunology, 8*. https://doi.org/10.3389/fimmu.2017.01789

Haque, T., & Talukder, M. M. U. (2018). Chemical enhancer: A simplistic way to modulate barrier function of the stratum corneum. *Advanced Pharmaceutical Bulletin, 8*(2), 169–179. https://doi.org/10.15171/apb.2018.021

Hu, Z., Meduri, C., Ingrole, R., Gill, H., & Kumar, G. (2020). Solid and hollow metallic glass microneedles for transdermal drug-delivery. *Applied Physics Letters, 116*, 203703.

Ilić, T., Savić, S., Batinić, B., Marković, B., Schmidberger, M., Lunter, D., Savić, M., & Savić, S. (2018). Combined use of biocompatible nanoemulsions and solid microneedles to improve transport of a model NSAID across the skin: In vitro and in vivo studies. *European Journal of Pharmaceutical Sciences, 125*, 110–119. https://doi.org/10.1016/j.ejps.2018.09.023

Ita, K., & Ashong, S. (2020). Percutaneous delivery of antihypertensive agents: Advances and challenges. *AAPS PharmSciTech, 21*(2). https://doi.org/10.1208/s12249-019-1583-9

Jepps, O. G., Dancik, Y., Anissimov, Y. G., & Roberts, M. S. (2013). Modeling the human skin barrier — towards a better understanding of dermal absorption. *Advanced Drug Delivery Reviews, 65*(2), 152–168. https://doi.org/10.1016/j.addr.2012.04.003

Kaur, M., Ita, K. B., Popova, I. E., Parikh, S. J., & Bair, D. A. (2014). Microneedle-assisted delivery of verapamil hydrochloride and amlodipine besylate. *Theme Issue: Dermal and Transdermal Drug Delivery, 86*(2), 284–291. https://doi.org/10.1016/j.ejpb.2013.10.007

Kolarsick, P. A. J., Kolarsick, M. A., & Goodwin, C. (2011). Anatomy and physiology of the skin. *Journal of the Dermatology Nurses' Association*, 214–215. https://doi.org/10.1097/JDN.0b013e31822bdc94

Korosec, A., Frech, S., Gesslbauer, B., Vierhapper, M., Radtke, C., Petzelbauer, P., & Lichtenberger, B. M. (2019). Lineage identity and location within the dermis determine the function of papillary and reticular fibroblasts in human skin. *Journal of Investigative Dermatology, 139*(2), 342–351. https://doi.org/10.1016/j.jid.2018.07.033

Lechler, T., & Fuchs, E. (2005). Asymmetric cell divisions promote stratification and differentiation of mammalian skin. *Nature, 437*(7056), 275–280. https://doi.org/10.1038/nature03922

Li, Q. Y., Zhang, J. N., Chen, B. Z., Wang, Q. L., & Guo, X. D. (2017). A solid polymer microneedle patch pretreatment enhances the permeation of drug molecules into the skin. *RSC Advances, 7*(25), 15408–15415. https://doi.org/10.1039/c6ra26759a

van der Maaden, K., Sekerdag, E., Jiskoot, W., & Bouwstra, J. (2014). Impact-insertion applicator improves reliability of skin penetration by solid microneedle arrays. *The AAPS Journal, 16*(4), 681–684. https://doi.org/10.1208/s12248-014-9606-7

Machaliński, B., Brodkiewicz, A., Szumilas, K., Rogińska, D., Kawa, M. P., Stecewicz, I., Trybek, G., Marchlewicz, M., & Wiszniewska, B. (2016). Morphologic changes in the dermis after the single administration of autologous fibroblastic cells: A preliminary study. *Transplantation Proceedings, 48*(8), 2833–2839. https://doi.org/10.1016/j.transproceed.2016.05.016

Mahmoudi, K., Garvey, K. L., Bouras, A., Cramer, G., Stepp, H., Jesu Raj, J. G., Bozec, D., Busch, T. M., & Hadjipanayis, C. G. (2019). 5-aminolevulinic acid photodynamic therapy for the treatment of high-grade gliomas. *Journal of Neuro-Oncology, 141*(3), 595–607. https://doi.org/10.1007/s11060-019-03103-4

Milewski, M., Paudel, K. S., Brogden, N. K., Ghosh, P., Banks, S. L., Hammell, D. C., & Stinchcomb, A. L. (2013). Microneedle-assisted percutaneous delivery of naltrexone hydrochloride in yucatan minipig: In vitro-in vivo correlation. *Molecular Pharmaceutics, 10*(10), 3745–3757. https://doi.org/10.1021/mp400227e

Moner, V., Fernández, E., Rodríguez, G., Cócera, M., Barbosa-Barros, L., de la Maza, A., & López, O. (2016). Lamellar body mimetic system: An up-to-down repairing strategy of the stratum corneum lipid structure. *International Journal of Pharmaceutics, 510*(1), 135–143. https://doi.org/10.1016/j.ijpharm.2016.06.033

Nagarkar, R., Singh, M., Nguyen, H. X., & Jonnalagadda, S. (2020). A review of recent advances in microneedle technology for transdermal drug delivery. *Journal of Drug Delivery Science and Technology, 59*. https://doi.org/10.1016/j.jddst.2020.101923

Narayanan, S. P., & Raghavan, S. (2017). Solid silicon microneedles for drug delivery applications. *International Journal of Advanced Manufacturing Technology, 93*(1–4), 407–422. https://doi.org/10.1007/s00170-016-9698-6

Ogunjimi, A. T., Fiegel, J., & Brogden, N. K. (2020). Design and characterization of spray-dried chitosan-naltrexone microspheres for microneedle- assisted transdermal delivery. *Pharmaceutics, 12*(6). https://doi.org/10.3390/pharmaceutics12060496

Puri, A., Frempong, D., Mishra, D., & Dogra, P. (2021). Microneedle-mediated transdermal delivery of naloxone hydrochloride for treatment of opioid overdose. *International Journal of Pharmaceutics, 604*, 120739. https://doi.org/10.1016/j.ijpharm.2021.120739

Rajabi, M., Roxhed, N., Shafagh, R. Z., Haraldson, T., Fischer, A. C., Wijngaart, W. van der, Stemme, G., Niklaus, F., & Loh, X. J. (2016). Flexible and stretchable microneedle patches with integrated rigid stainless steel microneedles for transdermal biointerfacing. *PloS One, 11*(12), e0166330. https://doi.org/10.1371/journal.pone.0166330

Reicherter, J. M., Krapohl, D. J., & Shaw, P. K. (2015). Chapter 2 - anatomy and physiology for polygraph examiners. In *Fundamentals of polygraph practice* (pp. 29–60). Academic Press. https://doi.org/10.1016/B978-0-12-802924-4.00002-5

Roger, M., Fullard, N., Costello, L., Bradbury, S., Markiewicz, E., O'Reilly, S., Darling, N., Ritchie, P., Määttä, A., Karakesisoglou, I., Nelson, G., von Zglinicki, T., Dicolandrea, T., Isfort, R., Bascom, C., & Przyborski, S. (2019). Bioengineering the microanatomy of human skin. *Journal of Anatomy, 234*(4), 438–455. https://doi.org/10.1111/joa.12942

Sahle, F. F., Gebre-Mariam, T., Dobner, B., Wohlrab, J., & Neubert, R. H. H. (2015). Skin diseases associated with the depletion of stratum corneum lipids and stratum corneum lipid substitution therapy. *Skin Pharmacology and Physiology, 28*(1), 42–55. https://doi.org/10.1159/000360009

Sbarbati, A., Accorsi, D., Benati, D., Marchetti, L., Orsini, G., Rigotti, G., & Panettiere, P. (2010). Subcutaneous adipose tissue classification. *European Journal of Histochemistry, 54*(4), 218–222. https://doi.org/10.4081/ejh.2010.e48

Schepens, B., Vos, P. J., Saelens, X., & van der Maaden, K. (2019). Vaccination with influenza hemagglutinin-loaded ceramic nanoporous microneedle arrays induces protective immune responses. *European Journal of Pharmaceutics and Biopharmaceutics, 136*, 259–266. https://doi.org/10.1016/j.ejpb.2019.02.002

Schmitt, T., & Neubert, R. H. H. (2018). State of the art in Stratum corneum research: The biophysical properties of ceramides. *Chemistry and Physics of Lipids, 216*, 91–103. https://doi.org/10.1016/j.chemphyslip.2018.09.017

Segal, N., Andriani, F., Pfeiffer, L., Kamath, P., Lin, N., Satyamurthy, K., Egles, C., & Garlick, J. A. (2008). The basement membrane microenvironment directs the normalization and survival of bioengineered human skin equivalents. *Matrix Biology, 27*(3), 163–170. https://doi.org/10.1016/j.matbio.2007.09.002

Simpson, C. L., Patel, D. M., & Green, K. J. (2011). Deconstructing the skin: Cytoarchitectural determinants of epidermal morphogenesis. *Nature Reviews Molecular Cell Biology, 12*(9), 565–580. https://doi.org/10.1038/nrm3175

Takahashi, H., Heo, Y. J., Arakawa, N., Kan, T., Matsumoto, K., Kawano, R., & Shimoyama, I. (2016). Scalable fabrication of microneedle arrays via spatially controlled UV exposure. *Microsystems and Nanoengineering, 2*. https://doi.org/10.1038/micronano.2016.49

Teeter, B., Thannisch, M., Martin, B., Zaller, N., Jones, D., Mosley, C., & Curran, G. (2021). Opioid overdose counseling and prescribing of naloxone in rural community pharmacies: A pilot study. *Exploratory Research in Clinical and Social Pharmacy, 2*, 100019. https://doi.org/10.1016/j.rcsop.2021.100019

Van Der Maaden, K., Jiskoot, W., & Bouwstra, J. (2012). Microneedle technologies for (trans)dermal drug and vaccine delivery. *Journal of Controlled Release, 161*(2), 645–655. https://doi.org/10.1016/j.jconrel.2012.01.042

Ventre, M., Mollica, F., & Netti, P. A. (2009). The effect of composition and microstructure on the viscoelastic properties of dermis. *Journal of Biomechanics, 42*(4), 430–435. https://doi.org/10.1016/j.jbiomech.2008.12.004

Vos, P. J., Kuijt, N., Kaya, M., Rol, S., & van der Maaden, K. (2020). Nanoporous microneedle arrays seamlessly connected to a drug reservoir for tunable transdermal delivery of memantine. *European Journal of Pharmaceutical Sciences, 150*. https://doi.org/10.1016/j.ejps.2020.105331

Weng, T., Wu, P., Zhang, W., Zheng, Y., Li, Q., Jin, R., Chen, H., You, C., Guo, S., Han, C., & Wang, X. (2020). Regeneration of skin appendages and nerves: Current status and further challenges. *Journal of Translational Medicine, 18*(1). https://doi.org/10.1186/s12967-020-02248-5

Wickett, R. R., & Visscher, M. O. (2006). Structure and function of the epidermal barrier. *American Journal of Infection Control, 34*.

Woodley, D. T., & McNutt, S. (1992). *The basement membrane zone at the dermal — epidermal junction of human skin*. Springer Science and Business Media LLC. https://doi.org/10.1007/978-1-4612-2914-8_2

Yang, J., Chen, Z., Ye, R., Li, J., Lin, Y., Gao, J., Ren, L., Liu, B., & Jiang, L. (2018). Touch-actuated microneedle array patch for closed-loop transdermaldrug delivery. *Drug Delivery, 25*(1), 1728–1739. https://doi.org/10.1080/10717544.2018.1507060

Yang, W., Sherman, V. R., Gludovatz, B., Schaible, E., Stewart, P., Ritchie, R. O., & Meyers, M. A. (2015). On the tear resistance of skin. *Nature Communications, 6*. https://doi.org/10.1038/ncomms7649

Ye, H., & De, S. (2017). Thermal injury of skin and subcutaneous tissues: A review of experimental approaches and numerical models. *Burns, 43*(5), 909–932. https://doi.org/10.1016/j.burns.2016.11.014

Yousef, H., & Sharma, S. (2018). *Anatomy, skin (integument), epidermis*. In StatPearls.

CHAPTER 11

Coated microneedles

Introduction

At times, medications and vaccines are coated onto the surface of microneedles for transdermal drug delivery (Chen et al., 2017; Ita, 2018). Solid microneedles form the base of the microneedles (MNs) while dispersions/drug solutions are used to formulate the coating (Nagarkar et al., 2020). Solid microneedles are first fabricated and then coated with a formulation (Kapoor et al., 2020). The dispersion used for coating contains the active pharmaceutical ingredient(s) (Jeong, Bae, et al., 2020; Jeong, Jun, et al., 2020). It must be emphasized that controlled wetting and spreading of the drug solution on the underlying microneedle is relevant for the generation of uniform coating (Chen et al., 2017; Ita, 2018). By adding a surfactant, researchers can increase formulation viscosity, reduce surface tension and decrease contact angle resulting in improved coating thickness and wetting (Ita, 2018). Investigators have studied the effect of coated microneedles (CMNs) on transdermal drug delivery (Ita, 2018). The feasibility of creating a device with an optimized formulation as well as the simplicity of production are benefits of coated microneedles (Ita, 2018). Formulation optimization is especially important because this approach can be used to improve drug stability, particularly for thermolabile molecules such as biologics (Ita, 2018). Coated microneedles (Fig. 11.1) are resistant to moisture as well as high humidity (Jeong, Bae, et al., 2020; Jeong, Jun, et al., 2020). CMNs also possess superb mechanical strength (Jeong, Bae, et al., 2020; Jeong, Jun, et al., 2020).

The utilization of coated microneedles is also advantageous because it leads to prolonged shelf-life since the incorporated drugs are in a dry state (Ita, 2018). Even though coating thickness can be adjusted to avoid needle occlusion, the amount of medication delivered using this method is limited. This is because only a minute quantity of the drug can be coated onto the shafts and tips of the microneedles (Ita, 2018). The amount of drug in coated microneedles is limited (normally less than 1 mg). For that reason,

Microneedles
ISBN 978-0-323-91811-4
https://doi.org/10.1016/B978-0-323-91811-4.00007-9

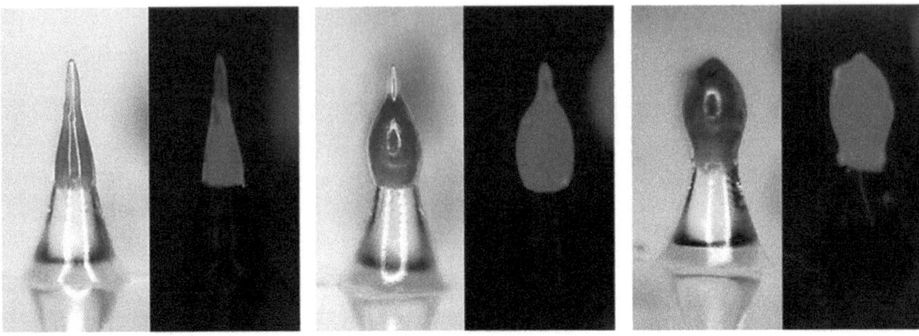

Fig. 11.1 Coated microneedles fabricated via dip-coating A coating solution containing 0.5% (w/w) sulforhodamine B was applied to microneedles. *(From Chen, Y., Chen, B. Z., Wang, Q. L., Jin, X., & Guo, X. D. (2017). Fabrication of coated polymer microneedles for transdermal drug delivery.* Challenges and Advances in Microneedle Technology, 265, *14–21. https://doi.org/10.1016/j.jconrel.2017.03.383).*

coated microneedles are seldom utilized for the transdermal delivery of high dose drugs (Ita, 2018). Coated microneedles loaded with vaccines can generate adequate immune reaction (Ita, 2018).

Drug delivery with coated microneedles

Narayanan and Raghavan used a metal sputtering method to coat pyramid-shaped silicon microneedles (Pradeep Narayanan & Raghavan, 2019). The sharp tips of the MNs were fabricated via etching with tetramethylammonium hydroxide. The gold-coated microneedles were biocompatible and mechanically robust (Pradeep Narayanan & Raghavan, 2019). The MNs had a tip angle of 24.5°, base width of 52.8 μm, diameter of 45 μm, height of 250 μm and an aspect ratio of 4.73 (Pradeep Narayanan & Raghavan, 2019). These devices were used for transdermal drug delivery over a protracted period. The authors determined the Vickers hardness value of 3800 Hv for the MNs (Pradeep Narayanan & Raghavan, 2019). The skin resistive force was significantly less than the Vickers hardness value (Pradeep Narayanan & Raghavan, 2019).

The release of medications under acidic conditions is attractive but hydrogels and polymers that combine biodegradability and acid sensitivity are few (Ullah, Khan, Choi, & Kim, 2019). Ullah et al. coated stainless-steel microneedles with a porous polymer layer via a facile technique (Ullah, Khan, Choi, & Kim, 2019). The coated microneedles were used for self-controlled drug transport. An uncomplicated but rugged procedure of coating was utilized by the authors. Lidocaine was released from the coated microneedles in a pH-responsive manner (Ullah, Khan, Choi, & Kim, 2019). To prepare the coated microneedles. Lidocaine and sodium bicarbonate (a pH-sensitive salt) were packed into the pores of a polymer layer (Ullah, Khan, Choi, & Kim, 2019). Following the exposure of the formulation to an acidic milieu, the reaction of sodium bicarbonate with protons (H^+) led to the liberation of carbon dioxide (Ullah, Khan, Choi, & Kim, 2019). The gas created pressure leading to the puncturing of the thin polymer membrane and the liberation of the medication (Ullah, Khan, Choi, & Kim, 2019). Images generated via scanning electron microscopy confirmed the presence of closed pores following the use of porous polymer-coated microneedles exposed to phosphate buffered saline (PBS) at pH 7.4 (Ullah, Khan, Choi, & Kim, 2019). When microneedles were exposed to PBS at pH 5.5, open pores were generated, and the thin membrane ruptured. The investigators showed via in vitro experiments that medication release from the microneedles was responsive to pH change (Ullah, Khan, Choi, & Kim, 2019). At the pH of 7.4, an insignificant drug release was documented for MNs. In contrast, marked release took place under an acidic microenvironment (pH 5.5) (Ullah, Khan, Choi, & Kim, 2019). Similar results were reported for drug release in vitro using pig skin and phosphate buffered saline (Ullah, Khan, Choi, & Kim, 2019). The pH-sensitive polymer-coated microneedles can be utilized for the controlled release of medications under acidic conditions (Ullah, Khan, Choi, & Kim, 2019). Stimulus-responsive polymeric carriers are also valuable because they can be used for on-demand controlled drug release, leading to low toxicity and enhanced clinical efficacy (Ullah, Khan, Choi, & Kim, 2019).

Several vaccines can sometimes be combined into a single pharmaceutical formulation (Jeong, Jun, et al., 2020). But this approach is complicated. In an interesting study, microneedles were used for the delivery of two types of influenza vaccines from a compartmental microneedle array (CMA) (Jeong, Bae, et al., 2020). The authors incorporated two chambers (compartments) into the microneedle array. To avoid mixing the two vaccines, rectangular systems were secured to each end of the microneedle device to integrate the chambers with the coating equipment (Jeong, Bae, et al., 2020). The two reservoirs were filled with the coating solution. The solution encompassed influenza vaccines for B/Victoria (B—V) and B/Yamagata (B—Y) (Jeong, Bae, et al., 2020). The authors aligned the CMA with the container by dipping the first chamber of the microneedle system into the first reservoir and the second chamber into the second reservoir (Jeong, Bae, et al., 2020). The authors used the CMA to deliver the vaccine to mice. Survival as well as weight change were then compared to other mice groups that were given (a) microneedle-mediated combination vaccines (b) two monovalent vaccines via microneedles (c) a combination vaccine intramuscularly (d) two monovalent vaccines intramuscularly (Jeong, Bae, et al., 2020). Plaque reduction neutralization tests were also conducted by the investigators to compare all the groups (Jeong, Bae, et al., 2020; Jeong, Jun, et al., 2020).

The compartmental microneedle array made by Jeong et al. displayed a relative standard error of less than 7% between samples in dose uniformity (Jeong, Bae, et al., 2020). It also demonstrated similar antibody-forming efficiency relative to other groups, particularly in studies involving rodent challenge by the B/Yamagata virus (Jeong, Bae, et al., 2020). The CMA mice group exhibited improved weight change and survival than the mice that received the combination vaccine intramuscularly (Jeong, Bae, et al., 2020). All mice in the groups that were administered vaccines via microneedles displayed greater neutralizing antibodies compared to those in the intramuscular groups (Jeong, Bae, et al., 2020). The authors administered the vaccines via a CMA without mixing and the efficacy was similar to what was achieved with intramuscular delivery of the combination vaccine. It was shown in the study that multivalent vaccines can be given (with no mixing) through a compartmental microneedle array as a single drug delivery system (Jeong, Bae, et al., 2020).

It is more challenging to use microneedles for the delivery of hydrophobic medications due to the utilization of organic solvents as well as limited dosing (Zhao et al., 2017). These conditions can have negative effects on the release and bioactivity of medications (Zhao et al., 2017). In an interesting study, Zhao et al. used coated microneedles for the transport of hydrophobic auto-antigens (Zhao et al., 2017). The peptides were used experimentally for antigen specific immunotherapy for type 1 diabetes. The authors employed three co-solvents (2-methyl-2-butanol, water, and acetic acid) as well as polyvinyl alcohol 2000 (PVA 2000) to prepare a formulation capable of dissolving both hydrophobic and hydrophilic peptide auto-antigens at high, and therapeutically significant concentrations (12.5 mg/mL or 25 mg/mL) (Zhao et al., 2017). The drug: excipient ratio was limited to 10:1 w/w to maximize dose and guarantee that the dry coating on the microneedles does not negatively influence skin piercing by the device (Zhao et al., 2017). The delivery efficiency of the peptide can be impacted by several parameters (Zhao et al., 2017). The authors demonstrated that electropolishing can enhance the efficiency of delivery from

$2.0 \pm 1.0\%$ to $59.9 \pm 6.7\%$. When the peptide mass per needle was increased from 0.37 μg to 2 μg, the coating was thicker and there was a 20% decrease in the skin delivery efficiency (Zhao et al., 2017). Other relevant parameters that can have significant effects on the performance of coated microneedles are: the intrinsic hydrophobicity of the peptide, excipients (that can enhance drug dissolution), and the type of rodent skin model utilized for experiments (Zhao et al., 2017). The results of the investigation provided guidance for optimizing the coating process, the formulation, and ultimately improving the coated microneedles (Zhao et al., 2017).

Although peanut allergies are common, there are no treatment options approved by the FDA for this class of diseases (Shakya et al., 2019). The impact of microneedles on the transdermal delivery of peanut protein extract (PE) was recently investigated. The authors also wanted to ascertain if the immune response derived from that pharmacological approach was capable of desensitizing mice who had received peanuts (Shakya et al., 2019). The rodents were divided into two groups. Mice in one group were given cutaneous immunotherapy using peanut extract-coated microneedles and those in the second group were not treated. Peanut extract was then administered to the two groups orally (Shakya et al., 2019). Following oral challenge, the signs of peanut hypersensitivity engendered by peanuts were markedly less in the MN treated mice in comparison to untreated mice (Shakya et al., 2019). The authors ascribed the pharmacological effect to down-regulation of anaphylaxis mediators (mast cell protease-1 and histamine) in the microneedles treated group (Shakya et al., 2019). The authors reported an up-regulation of Th1 cytokines (IFN-γ and IL-2) in comparison to Th2 cytokines (IL-4 and IL-5) in the splenocyte culture supernatants of the MN treated group relative to the untreated group. It was postulated that MNs enhanced immune modulation via the Th1 route (Shakya et al., 2019). In addition, the mice treated with PE-coated MNs maintained the integrity of their small bowel villi and displayed decreased infiltration of eosinophils in comparison to the untreated mice. Minimally invasive microneedle-based immunotherapy via the skin can be employed as an alternative approach for the management of peanut allergies (Shakya et al., 2019).

Ullah and coworkers used a facile technique to coat microneedles with poly (lactic acid-*co*-glycolic acid) (Ullah et al., 2018). This is a porous polymer that is capable of delivering compounds at a high rate (Ullah et al., 2018). The authors coated robust stainless-steel microneedles with a porous and thin polymer (poly (lactic acid-*co*-glycolic acid), PLGA) layer to increase calcein dye delivery rate. The surface of the microneedles was modified via plasma treatment and subsequent dip coating with polyethyleneimine to increase interfacial adhesion between the microneedles and PLGA (Ullah et al., 2018). The mean failure load (the least force adequate for the detachment of the polymer layer from the stainless steel surface) was 2.2 N for the non-modified surface and 25 N for the modified surface coating (Ullah et al., 2018). The investigators used the PLGA-coated MNs to transport calcein dye effectively into pig skin (depth of 750 μm), thus showing that the coated MNs are capable of penetrating the skin effortlessly without the microneedles being deformed (Ullah et al., 2018). The authors also delivered rhodamine B dye into gelatin gel via porous-coated and non-coated microneedles. Rhodamine administered via porous polymer-coated MNs penetrated the gel deeper in comparison with non-coated microneedles (Ullah et al., 2018). Rhodamine B dye and lidocaine were also administered via porous polymer-coated and non-coated microneedles (Ullah et al., 2018). The delivery of lidocaine by the porous polymer-coated MNs was 25 times higher in comparison with the non-coated microneedles (Ullah et al., 2018). Furthermore, the amount of rhodamine B delivered with the porous polymer-coated microneedles was 5 times more in comparison with the non-coated microneedles (Ullah et al., 2018).

It is challenging delivering DNA-based cancer vaccines to antigen presenting cells for the stimulation of cellular and humoral response (Duong et al., 2018). An optimal approach for painless and safe vaccination is difficult to achieve even though electroporation-based transfection has been used to improve vaccine efficacy (Duong et al., 2018). Duong et al. formulated a DNA vaccine delivery system by employing microneedles coated via a layer-by-layer method with ultra-pH-responsive oligo sulfamethazine conjugated poly (β-amino ester urethane) and immunostimulatory adjuvant poly (I:C) (Duong et al., 2018). The microneedle coating was loaded with nano-engineered DNA vaccine (Duong et al., 2018). Transcutaneous delivery of the coated microneedles onto mouse skin led to the release of the DNA vaccines and adjuvants (Duong et al., 2018). The vaccine was used to generate type I interferons and facilitate dendritic cell maturation resulting in the production of antigen-specific antibody (Duong et al., 2018). The vaccine induced $CD8^+$ T cell formation and antibody-dependent cell-mediated cytotoxicity (ADCC) (Duong et al., 2018). Remarkably, the transcutaneous delivery of the smart microneedle-mediated vaccine delivery system evoked a three-fold excess of cytotoxic $CD8^+$ T cell and three-fold greater frequencies of Anti-OVA IgG1 serum antibody in mice than a soluble DNA vaccine delivery system (Duong et al., 2018). The microneedle-mediated system was used against the murine B16/OVA melanoma tumors in C57BL/6 mice via the synergistic activation of cytotoxic $CD8^+$ T cells and antigen specific ADCC (Duong et al., 2018). Microneedle-mediated vaccine elicited cellular and humoral immunity with enhanced compliance, efficacy, and safety (Duong et al., 2018).

A drug delivery system that mimics the function of pancreatic beta cells and "releases" insulin in response to elevated blood glucose concentration without human intervention is desirable (Ullah et al., 2020). An uncomplicated method was used to coat stainless steel MNs with a porous polymer layer that releases insulin in a glucose-responsive manner (Ullah et al., 2020). The closed loop system was made by introducing insulin, glucose oxidase (glucose-specific enzymes [GOx]) and pH-sensitive sodium bicarbonate into the pores of a porous polymer coating (Ullah et al., 2020). Glucose was subsequently oxidized by GOx to gluconic acid, thus inducing a decline in the pH of the microenvironment (Ullah et al., 2020). The reaction of sodium bicarbonate with protons led to the liberation of carbon dioxide which created pressure inside the pores, with subsequent breaching of the thin polymer film and insulin release (Ullah et al., 2020). Images obtained via field emission scanning electron microscopy were used to demonstrate that upon exposure of the coated microneedles to elevated glucose level, the pores were opened and the thin film was punctured while the exposure of the microneedles to glucose-free phosphate buffer saline (pH 7.4) did not lead to the opening of the pores (Ullah et al., 2020). These microneedles were used to show the feasibility of in vitro (in porcine skin and PBS) and in vivo (in diabetic rats) glucose-mediated insulin release (Ullah et al., 2020).

Novel technologies that can be used to quantify cell-free nucleic acid biomarkers from liquid biopsies are important for the early diagnosis of several diseases (Al Sulaiman et al., 2019). The skin interstitial fluid has been mostly overlooked despite having molecular biomarkers that originate from cells (Al Sulaiman et al., 2019). Al Sulaiman et al. fabricated MNs that were subsequently coated with an alginate—peptide nucleic acid hybrid material (Al Sulaiman et al., 2019). The coated microneedles were used for sequence-specific sampling, isolation, and detection of nucleic acid biomarkers from the skin's interstitial fluid (Al Sulaiman et al., 2019). The microneedles can identify specific nucleic acid biomarkers either in solution after light-triggered release from the hydrogel or on the patch itself (Al Sulaiman et al., 2019). The method is characterized by large sampling capacity (~6.5 μL in 2 min) and quick sampling kinetics (Al Sulaiman et al., 2019). Techniques that can be used to identify and quantify cell-free nucleic acids in bodily fluids in a minimally invasive and automated manner have enormous potential for personalized medicine (Al Sulaiman et al., 2019). In another study, Jeong et al. coated polypropylene, polylactic acid, polyethylene, and nylon MNs with polyvinyl alcohol (Jeong, Jun, et al., 2020). Following the application of the coating formulation to the MNs, the aspect ratio decreased from 2.2, 2.5 and 3.0 for uncoated microneedles in comparison to 1.3, 1.4 and 1.6 computed for coated microneedles, respectively (Jeong, Jun, et al., 2020). In vivo studies showed that when PP MNs with a 3.0 aspect ratio were applied, the rate of puncture of uncoated MNs increased from 59% to more than 96% but went down to 13% for re-inserted MNs (Jeong, Jun, et al., 2020).

A layer-by-layer coating method was recently used to coat pH-sensitive microneedles (Fig. 11.2) (Schipper et al., 2017). The aim of this approach was to enhance microneedle sharpness and increase the surface area for antigen coating. The authors alternatingly coated pH-sensitive microneedle arrays (almost uncharged at pH 7.4 and positively charged at coating pH 5.8) with negatively charged diphtheria toxoid (DT) and positively charged N-trimethyl chitosan (a cationic adjuvant) (Schipper et al., 2017).

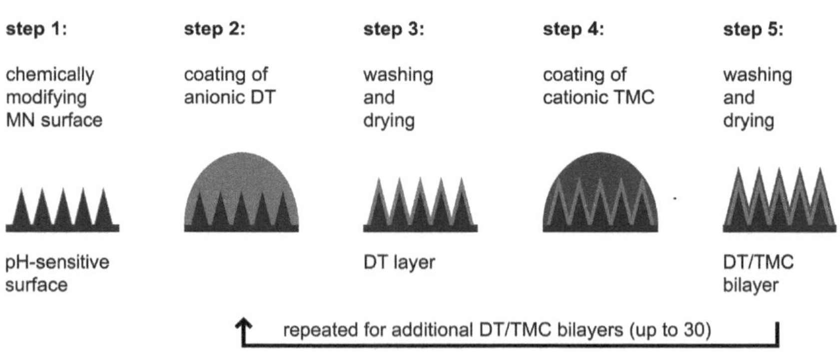

Fig. 11.2 Schematic illustration of the layer-by-layer coating approach of pH-sensitive microneedle surfaces. Fabrication steps include microneedle surface modification, coating with anionic diphtheria toxoid, washing and coating with the positively charged N-trimethyl chitosan. *(From Schipper, P., van der Maaden, K., Groeneveld, V., Ruigrok, M., Romeijn, S., Uleman, S., Oomens, C., Kersten, G., Jiskoot, W., & Bouwstra, J. (2017). Diphtheria toxoid and N-trimethyl chitosan layer-by-layer coated pH-sensitive microneedles induce potent immune responses upon dermal vaccination in mice.* Journal of Controlled Release, 262, *28–36. https://doi.org/10.1016/j.jconrel.2017.07.017).*

Vaccination with microneedles coated with 10 bilayers of diphtheria toxoid/N-trimethyl chitosan (corresponding with ± 0.6 μg DT applied intradermally) led to comparable DT-specific immune responses as subcutaneous immunization with 5 μg of diphtheria toxoid containing aluminum phosphate as an adjuvant (8-fold dose reduction) (Schipper et al., 2017).

TMC was fluorescently labeled with rhodamine B isothiocyanate (TMC-RHO) while DT was fluorescently labeled with AlexaFluor® 488 (DT-AF488) (Schipper et al., 2017). Schipper and colleagues then coated microneedle arrays with 30 bilayers of DT-AF488 and TMC-RHO. The quantity of the coated DT-AF488 and TMC-RHO increased linearly for up to 30 bilayers (Fig. 11.3) (Schipper et al., 2017).

Chinnadayyala and coworkers electrodeposited a highly porous platinum black (Pt-black) on microneedle substrate (Chinnadayyala et al., 2021). The fabrication procedure was carried out via potentiometry using a charge

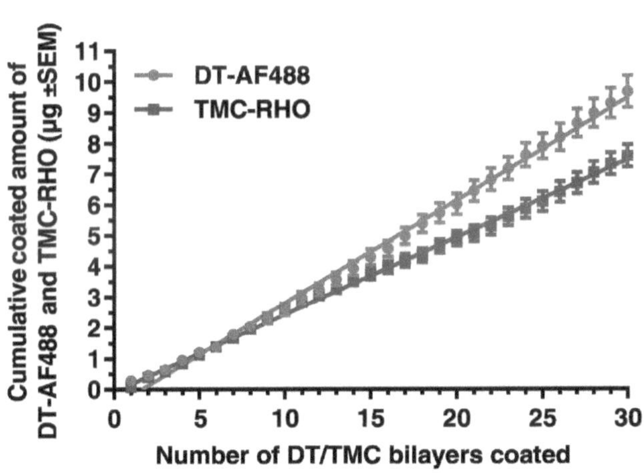

Fig. 11.3 Coating of microneedles with 30 bilayers of DT fluorescently labeled with AlexaFluor® 488 (DT-AF488) and TMC fluorescently labeled with rhodamine B isothiocyanate (TMC-RHO). Cumulative amounts of coated fluorescently labeled DT-AF488 and TMC-RHO by applying 30 DT-AF488/TMC-RHO bilayers onto microneedle arrays via a layer-by-layer coating approach. The mean cumulative amounts ± SEM of DT-AF488 (*green dots* [gray dots in print version]) and TMC-RHO (*red squares* [dark black squares in print version]) coated onto the microneedle array are presented as a function of the number of coated DT-AF488/TMC-RHO bilayers (n = 14). *(From Schipper, P., van der Maaden, K., Groeneveld, V., Ruigrok, M., Romeijn, S., Uleman, S., Oomens, C., Kersten, G., Jiskoot, W., & Bouwstra, J. (2017). Diphtheria toxoid and N-trimethyl chitosan layer-by-layer coated pH-sensitive microneedles induce potent immune responses upon dermal vaccination in mice.* Journal of Controlled Release, *262, 28–36. https://doi.org/10.1016/j.jconrel.2017.07.017).*

density of -2.5 mA/cm^2. The authors packed the microneedle surface with an ionomer (Nafion) (Chinnadayyala et al., 2021). Each microneedle was 110 μm wide, 650 μm long and 150 μm thick (Chinnadayyala et al., 2021). The coated microneedles were utilized for glucose monitoring in phosphate buffered saline (PBS) and interstitial fluid (ISF) (Chinnadayyala et al., 2021). The sensing system displayed a broad glucose dynamic range (1 − 20 mM) in artificial interstitial fluid and phosphate buffer saline in-vitro (Chinnadayyala et al., 2021). The technique was characterized by acceptable sensitivities in interstitial fluid (4.380 μA/mM cm^2) and phosphate buffered saline (5.786 μA/mM cm^2) (Chinnadayyala et al., 2021). The stability (a loss of 2% after 7 h), reproducibility, selectivity, and detection limits (PBS: 10 μM, ISF: 22.5 μM) were also satisfactory. The glucose recovery in ISF (98%—102.5%) and response time of 2 s were also adequate (Chinnadayyala et al., 2021).

In another interesting study, Liang et al. developed four different techniques (dip-coating with a roller, dam board, a limit and a fixture) for microneedle coating (Liang et al., 2020). The drug loading of one microneedle fabricated from the four methods was from 15 to 16 ng (Liang et al., 2020). In vitro experiments were carried out to assess drug transport and the results showed that microneedles fabricated using the above-mentioned techniques attained approximately 90% drug delivery efficiency (Liang et al., 2020).

Temporomandibular disorders (TMDs) refer to pathologies of the synovial temporomandibular joint (TMJ) and the masticatory muscles (Abdalla et al., 2019; Palmer & Durham, 2021). It has been estimated that 40% of the people in the United States of America suffer from TMDs (Abdalla et al., 2019). Muscle relaxants, nonsteroidal anti-inflammatory medications and pain killers are frequently used to manage the symptoms of TMJ (Abdalla et al., 2019). The use of these medications over an extended period may lead to serious cardiovascular and gastrointestinal side effects (Abdalla et al., 2019). Recently, Abdalla and coworkers fabricated coated microneedles loaded with tramadol and investigated the needles as a treatment option for TMDs (Abdalla et al., 2019). Stainless steel (grade 316) sheets were used for microneedle fabrication via a wet etch procedure (Abdalla et al., 2019). Coating of the microneedles was carried out using an in-house micro-precision dip coating device (Abdalla et al., 2019). Each MN patch was 1 cm long × 1 cm wide and consisted of 57 MNs, each 200 μm wide, 700 μm long, and 50 μm thick (Abdalla et al., 2019).

The authors used the coated microneedles or intra-TMJ injection to deliver the drug to Male Wistar rats followed by a nociceptive challenge with 1.5% formalin given 15 min later (Abdalla et al., 2019). The authors assessed the nociceptive behavior of rats and it was revealed that MNs coated with the drug provided an antinociceptive action that was comparable to that attained with tramadol injection (Abdalla et al., 2019). Significantly, there was a sustained antinociceptive effect (2 days) following the use of coated microneedles in comparison to intra-TMJ injection (less than 2 h) (Abdalla et al., 2019).

In another interesting set of experiments, Lee and colleagues fabricated coated microneedles via electropolymerization of a monomer precursor of 2,2′:5′,5″-terthiophene-3′-carboxylic acid (a conducting polymer) on gold-zinc oxide (AuZnOx) layer (Lee et al., 2020). Microneedles were coated with a gold-zinc oxide layer. The oxide was employed for the prevention of gold surface passivation by chloride ions found in biological samples (Lee et al., 2020). The authors optimized parameters capable of impacting the sensing analytical signal. These include temperature, applied potential, pH, electropolymerization cycles, and monomer concentration (Lee et al., 2020). Remarkably, the sensor was characterized by two dynamic ranges of 0.01–5.0 and 5.0–50.0 mM, a detection limit of 1.2 ± 0.026 μM and a response time within 1 s (Lee et al., 2020). The investigators used the sensor to quantify glucose concentration in different cancer cells (Saos-2, A549, HeLa, MG-62 and PC-12) and in normal (Vero) cells. A higher intracellular glucose concentration was detected in cancer cells in comparison to normal cells (Lee et al., 2020).

Homeostasis refers to a self-regulating process by which biological systems remain stable while adapting to changing external conditions (Billman, 2020). Zhang et al. used step-by-step mold replication to fabricate a dodecyl-modified chitosan (DCS)-coated bioinspired multilayer microneedle system for prompt hemostasis and tissue fixation (Zhang et al., 2021). The microneedle design was inspired by the hierarchical anatomy of the feet of insects. The dodecyl-modified chitosan can be fixed onto a cell membrane, inducing blood cell coagulation, and facilitating hemostasis (Zhang et al., 2021). The authors used rabbit experiments to show that acute tissue injuries such as kidney bleeding, spleen bleeding, and liver bleeding, can be repaired promptly by using the DCS-coated multilayer MNs (Zhang et al., 2021).

References

Abdalla, H. B., Jain, A. K., Napimoga, M. H., Clemente-Napimoga, J. T., & Gill, H. S. (2019). Microneedles coated with tramadol exhibit antinociceptive effect in a rat model of temporomandibular hypernociception. *Journal of Pharmacology and Experimental Therapeutics, 370*(3), 834–842. https://doi.org/10.1124/jpet.119.256750

Al Sulaiman, D., Chang, J. Y. H., Bennett, N. R., Topouzi, H., Higgins, C. A., Irvine, D. J., & Ladame, S. (2019). Hydrogel-coated microneedle arrays for minimally invasive sampling and sensing of specific circulating nucleic acids from skin interstitial fluid. *ACS Nano, 13*(8), 9620–9628. https://doi.org/10.1021/acsnano.9b04783

Billman, G. E. (2020). Homeostasis: The underappreciated and far too often ignored central organizing principle of physiology. *Frontiers in Physiology, 11*. https://doi.org/10.3389/fphys.2020.00200

Chen, Y., Chen, B. Z., Wang, Q. L., Jin, X., & Guo, X. D. (2017). Fabrication of coated polymer microneedles for transdermal drug delivery. *Challenges and Advances in Microneedle Technology, 265*, 14–21. https://doi.org/10.1016/j.jconrel.2017.03.383

Chinnadayyala, S. R., park, J., Satti, A. T., Kim, D., & Cho, S. (2021). Minimally invasive and continuous glucose monitoring sensor based on non-enzymatic porous platinum black-coated gold microneedles. *Electrochimica Acta, 369*, 137691. https://doi.org/10.1016/j.electacta.2020.137691

Duong, H. T. T., Yin, Y., Thambi, T., Nguyen, T. L., Giang Phan, V. H., Lee, M. S., Lee, J. E., Kim, J., Jeong, J. H., & Lee, D. S. (2018). Smart vaccine delivery based on microneedle arrays decorated with ultra-pH-responsive copolymers for cancer immunotherapy. *Biomaterials, 185*, 13–24. https://doi.org/10.1016/j.biomaterials.2018.09.008

Ita, K. (2018). Modulation of transdermal drug delivery with coated microneedles. *Journal of Drug Delivery Science and Technology, 45*, 203–212. https://doi.org/10.1016/j.jddst.2018.03.021

Jeong, H. R., Bae, J. Y., Park, J. H., Baek, S. K., Kim, G., Park, M. S., & Park, J. H. (2020). Preclinical study of influenza bivalent vaccine delivered with a two compartmental microneedle array. *Journal of Controlled Release, 324*, 280–288. https://doi.org/10.1016/j.jconrel.2020.05.024

Jeong, H. R., Jun, H., Cha, H. R., Lee, J. M., & Park, J. H. (2020). Safe coated microneedles with reduced puncture occurrence after administration. *Micromachines, 11*(8). https://doi.org/10.3390/MI11080710

Kapoor, Y., Milewski, M., Dick, L., Zhang, J., Bothe, J. R., Gehrt, M., Manser, K., Nissley, B., Petrescu, I., Johnson, P., Burton, S., Moseman, J., Hua, V., Grunewald, T., Tomai, M., & Smith, R. (2020). Coated microneedles for transdermal delivery of a potent pharmaceutical peptide. *Biomedical Microdevices, 22*(1), 7. https://doi.org/10.1007/s10544-019-0462-1

Lee, W. C., Gurudatt, N. G., Park, D. S., Kim, K. B., Choi, C. S., & Shim, Y. B. (2020). Microneedle array sensor for monitoring glucose in single cell using glucose oxidase-bonded polyterthiophene coated on AuZn oxide layer. *Sensors and Actuators B: Chemical, 320*. https://doi.org/10.1016/j.snb.2020.128416

Liang, L., Chen, Y., Zhang, B., Zhang, X., Liu, J., Shen, C., Cui, Y., & Guo, X. (2020). Optimization of dip-coating methods for the fabrication of coated microneedles for drug delivery. *Journal of Drug Delivery Science and Technology, 55*, 101464.

Nagarkar, R., Singh, M., Nguyen, H. X., & Jonnalagadda, S. (2020). A review of recent advances in microneedle technology for transdermal drug delivery. *Journal of Drug Delivery Science and Technology, 59*. https://doi.org/10.1016/j.jddst.2020. 101923

Palmer, J., & Durham, J. (2021). Temporomandibular disorders. *BJA Education, 21*(2), 44–50. https://doi.org/10.1016/j.bjae.2020.11.001

Pradeep Narayanan, S., & Raghavan, S. (2019). Fabrication and characterization of gold-coated solid silicon microneedles with improved biocompatibility. *The International Journal of Advanced Manufacturing Technology, 104*(9), 3327–3333.

Schipper, P., van der Maaden, K., Groeneveld, V., Ruigrok, M., Romeijn, S., Uleman, S., Oomens, C., Kersten, G., Jiskoot, W., & Bouwstra, J. (2017). Diphtheria toxoid and N-trimethyl chitosan layer-by-layer coated pH-sensitive microneedles induce potent immune responses upon dermal vaccination in mice. *Journal of Controlled Release, 262*, 28–36. https://doi.org/10.1016/j.jconrel.2017.07.017

Shakya, A. K., Ingrole, R. S. J., Joshi, G., Uddin, M. J., Anvari, S., Davis, C. M., & Gill, H. S. (2019). Microneedles coated with peanut allergen enable desensitization of peanut sensitized mice. *Journal of Controlled Release, 314*, 38–47. https://doi.org/10.1016/j.jconrel.2019.09.022

Ullah, A., Choi, H. J., Jang, M., An, S., & Kim, G. M. (2020). Smart microneedles with porous polymer layer for glucose-responsive insulin delivery. *Pharmaceutics, 12*(7), 1–15. https://doi.org/10.3390/pharmaceutics12070606

Ullah, A., Khan, H., Choi, H. J., Kim, G. M., et al. (2019). Smart microneedles with porous polymer coatings for pH-responsive drug delivery. *Polymers (Basel), 11*(11), 1834.

Ullah, A., Kim, C. M., & Kim, G. M. (2018). Porous polymer coatings on metal microneedles for enhanced drug delivery. *Royal Society Open Science, 5*(4). https://doi.org/10.1098/rsos. 171609

Zhang, X., Chen, G., Cai, L., Wang, Y., Sun, L., & Zhao, Y. (2021). Bioinspired pagoda-like microneedle patches with strong fixation and hemostasis capabilities. *Chemical Engineering Journal, 414*, 128905. https://doi.org/10.1016/j.cej.2021.128905

Zhao, X., Coulman, S. A., Hanna, S. J., Wong, F. S., Dayan, C. M., & Birchall, J. C. (2017). Formulation of hydrophobic peptides for skin delivery via coated microneedles. *Journal of Controlled Release, 265*, 2–13. https://doi.org/10.1016/j.jconrel.2017.03.015

CHAPTER 12

Hollow microneedles

Introduction

Hollow microneedles are designed to administer therapeutic agents continually through the bore of a needle (Ramadon et al., 2021). Pressure is used to drive a drug solution through a micrometer-sized needle (Nagarkar et al., 2020). By employing this method, therapeutically significant concentration of the medication can be transported through the skin into the systemic circulation. By using hollow microneedles, greater quantities of medications can be transported in comparison with coated and solid microneedles (Ramadon et al., 2021). Hollow microneedles have reached the pharmaceutical market as medical devices. Hollow microneedles (HMNs) can be produced from a wide range of materials including silicon, metals, glass, and certain types of polymers (Ramadon et al., 2021). Poly (imide), poly (oxymethylene) and poly (methyl methacrylate) have been used for the fabrication of hollow polymer microneedles (Mansoor et al., 2012; Moon et al., 2005; Ramadon et al., 2021; Yung et al., 2012). Nevertheless, hollow microneedles are difficult to produce because of their fragility and complex structure (Nagarkar et al., 2020). Furthermore, blockage of the hollow microneedle with skin tissue can occur upon insertion (Cárcamo-Martínez, Mallon, Anjani, et al., 2021; Cárcamo-Martínez, Mallon, Domínguez-Robles et al., 2021). This challenge can be overcome by either placing the bore on the side of the MN or retracting the MN array (Cárcamo-Martínez, Mallon, Anjani, et al., 2021; Cárcamo-Martínez, Mallon, Domínguez-Robles et al., 2021). Traditionally, hollow MNs are fabricated via techniques used for the production of microelectromechanical systems (MEMS) (Cárcamo-Martínez, Mallon, Anjani, et al., 2021; Cárcamo-Martínez, Mallon, Domínguez-Robles et al., 2021). These methods include laser cutting and ablation, micromolding, isotropic and anisotropic etching, metal electroplating and lithography (Cárcamo-Martínez, Mallon, Anjani, et al., 2021; Cárcamo-Martínez, Mallon, Domínguez-Robles et al., 2021). 3D printing, which is also called additive manufacturing, can be

Microneedles
ISBN 978-0-323-91811-4
https://doi.org/10.1016/B978-0-323-91811-4.00010-9

used for the fabrication of hollow microneedles (Cárcamo-Martínez, Mallon, Anjani, et al., 2021; Cárcamo-Martínez, Mallon, Domínguez-Robles et al., 2021).

In an interesting study, 3D-printed hollow microneedles with an inner diameter of 35 μm and an outer diameter of 100 μm were utilized for the perforation of the round window membrane (RWM) and subsequent aspiration of the perilymph (Szeto et al., 2021). This research is particularly important because the RWM is an appealing source for obtaining perilymph samples (Szeto et al., 2021). A 1 μL volume of perilymph is enough for proteomic analysis, even though the total volume of perilymph within the scala tympani of the guinea pig is less than 5 μL (Szeto et al., 2021). The authors observed no permanent damage to hearing at 72 h following perforation as evaluated by distortion product otoacoustic emissions and compound action potential (Szeto et al., 2021). Importantly, all the perforations healed within 72 h (Szeto et al., 2021). Remarkably, 620 proteins were detected in two analyzed samples. One of the proteins was cochlin, which is regarded as a perilymph marker (Szeto et al., 2021).

Skin delivery of drugs with hollow microneedles

Xenikakis and coworkers prepared hollow microneedles (Fig. 12.1) from a biocompatible resin material and used polylactic acid filament to fabricate the reservoirs (Xenikakis et al., 2021). The authors employed 3D printing technology for HMN production and used the device for the transdermal delivery of a peptide (Xenikakis et al., 2021). Leakage inspection studies and contact angle measurements were carried out to confirm the transport of liquid formulations through the bore of the needle (Xenikakis et al., 2021). The authors studied the effect of hollow microneedles on the permeation of octreotide acetate across artificial membranes (Xenikakis et al., 2021). Human keratinocytes were used to investigate the cytocompatibility of the HMNs while finite element analysis was used to conduct penetration tests and mechanical failure of this device (Xenikakis et al., 2021).

Janus Kinase inhibitors have been investigated as potential therapeutic agents for the management of inflammatory disorders of the skin (Solimani et al., 2019). These inhibitors target a family of four proteins: JAK1, JAK2, JAK3, and TYK2 (Solimani et al., 2019). Unfortunately, side effects, such as blood clots infections have been documented (Cárcamo-Martínez, Mallon, Anjani, et al., 2021; Cárcamo-Martínez, Mallon, Domínguez-Robles et al., 2021). Cárcamo-Martínez and colleagues investigated the

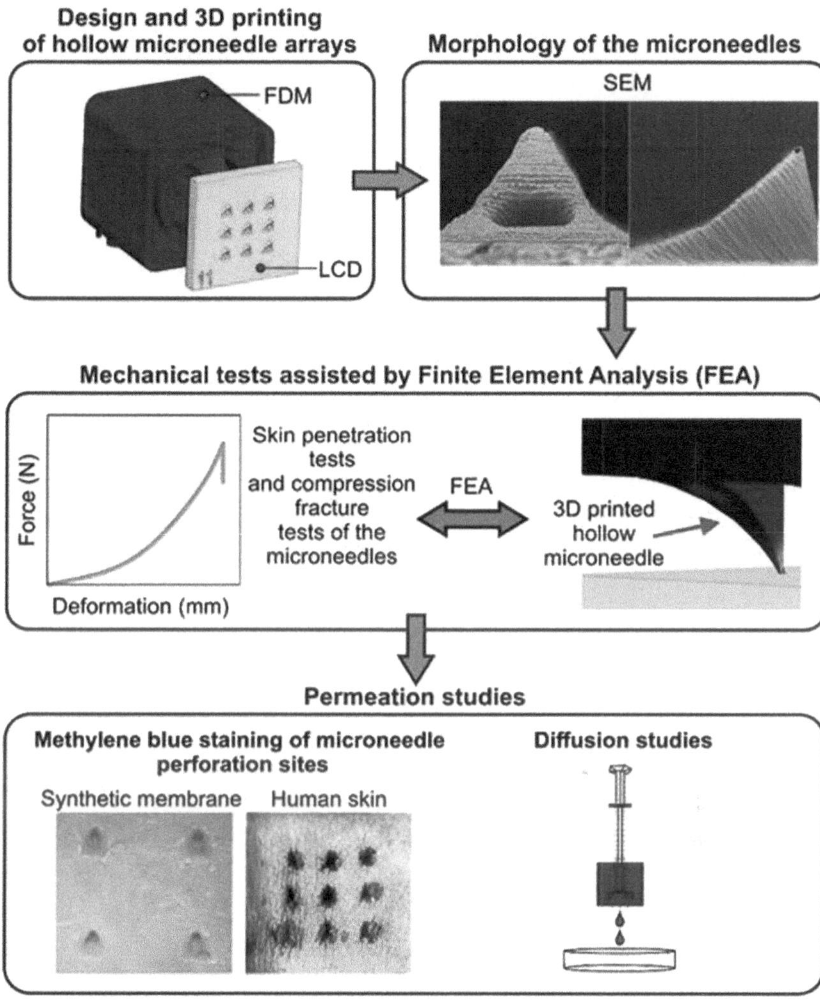

Fig. 12.1 Three-dimensional printing of hollow microneedle (HMN) devices (A) Three-dimensional CAD model of a representative HMN 3×3 array along with. STL model with supports in 3D printing platform, (B) 3D printed microneedle arrays fitted in a reservoir adapter suitable for drug administration, (C) Schematic representation of the LCD 3D printer for the fabrication of the HMN arrays and (D) Schematic representation of an FDM 3D printed for the fabrication of the reservoirs. *(From Xenikakis, I., Tsongas, K., Tzimtzimis, E. K., Zacharis, C. K., Theodoroula, N., Kalogianni, E. P., Demiri, E., Vizirianakis, I. S., Tzetzis, D., & Fatouros, D. G. (2021). Fabrication of hollow microneedles using liquid crystal display (LCD) vat polymerization 3D printing technology for transdermal macromolecular delivery.* International Journal of Pharmaceutics, 597, *120303. https://doi.org/10.1016/j.ijpharm.2021.120303.)*

effect of hollow microneedles on the intradermal delivery of tofacitinib citrate (Cárcamo-Martínez, Mallon, Anjani, et al., 2021; Cárcamo-Martínez, Mallon, Domínguez-Robles et al., 2021). The authors carried out a comparative assessment of dissolving microneedles, powder-loaded hollow microneedle arrays and a control Aqueous Cream BP (Cárcamo-Martínez, Mallon, Anjani, et al., 2021; Cárcamo-Martínez, Mallon, Domínguez-Robles et al., 2021). Hollow microneedles were fabricated and subsequently loaded with tofacitinib citrate via crosslinked hydrogels containing modifying agents (sorbitol, urea and sodium chloride) (Cárcamo-Martínez, Mallon, Anjani, et al., 2021; Cárcamo-Martínez, Mallon, Domínguez-Robles et al., 2021). Hollow microneedle arrays displayed insertion capabilities and a resistance to compression values that were similar to dissolving microneedles (Cárcamo-Martínez, Mallon, Anjani, et al., 2021; Cárcamo-Martínez, Mallon, Domínguez-Robles et al., 2021). Hollow MN arrays containing NaCl in the formulation led to slightly higher depositions of tofacitinib in epidermis and dermis of neonatal pig skin in comparison to Aqueous Cream BP, but were inferior to dissolving MN arrays from the standpoint of tofacitinib deposition (Cárcamo-Martínez, Mallon, Anjani, et al., 2021; Cárcamo-Martínez, Mallon, Domínguez-Robles et al., 2021).

Yadav et al. recently developed a 3D printed assembly of a hollow microneedle array, linked to a reservoir void, and additively fabricated via stereolithography (Yadav et al., 2021). The hollow microneedles, fabricated from a proprietary resin, were used for transdermal delivery of rifampicin (Yadav et al., 2021). To enhance its mechanical strength and the integrity of the HMNs array, the structure of the hollow microneedle array was designed with sub-apical holes present in a quarter of the needle tip (Yadav et al., 2021). In vivo evaluation of rifampicin delivery via the microneedle reservoir system in Sprague-Dawley rats showed efficient transport and enhanced bioavailability (Yadav et al., 2021).

Lee et al. fabricated hollow microneedles that were integrated with a low-temperature system for the delivery of a sol-gel transition system (Lee et al., 2017). This system was utilized for the extended release and topical delivery of phenylephrine (PE) (Lee et al., 2017). The device was employed for the management of sporadic fecal incontinence. The low temperature of the sol was maintained (Lee et al., 2017). The authors prepared assorted sol-gel formulations from hydroxypropyl methylcellulose (HPMC) and Pluronic F127 (PF-127). The flow property, gelation temperature, and delayed diffusion were documented (Lee et al., 2017). An air-charged catheter was utilized for the measurement of the resting pressure of anal sphincter as a reaction to the drug-loaded sol-gel formulation. The investigators also assessed the biocompatibility of formulations through the observation of immunological response (Lee et al., 2017). When a formulation (PF25-HPMC1-PE) prepared from HPMC 1%, PF-127 25%, and loaded with phenylephrine was administered into the peri-anal skin of a rat, the highest pressure recorded on the anal sphincter muscle took place at 6–8 h (Lee et al., 2017). The increase in anal pressure lasted twice longer in comparison to what was obtained following the administration of a phosphate-buffered saline (PBS-PE) (Lee et al., 2017). There was no statistically significant difference between the PBS-PE and the PF25-HPMC1-PE formulations as determined via the number of mast cells recorded following the administration of the formulations into rats (Lee et al., 2017). It was shown by Lee and colleagues that the combined use of an injectable sol-gel formulation and a low-pain hollow microneedle array (HMNA) increased the efficacy of sporadic fecal incontinence management (Lee et al., 2017). The low-temperature hollow microneedle system can be used in treatment options that demand local targeting, depot effect, and the control of pain (Lee et al., 2017).

In a recent report, Wang et al. documented the creation of an HMNA via a polymer-based procedure that combines replica molding and ultraviolet photolithography (Wang et al., 2009). The main attribute of the fabrication method is the use of photopatterning for the creation of a hollow cavity for microfluidic access (Wang et al., 2009). This permits the employment of a high throughput as well as a batch process (Wang et al., 2009). The authors designed, produced and described a hollow microneedle array generated from SU-8 and comprising 400 μm wide and 825 μm tall MNs with a tip diameter of 15–25 μm and a hollow channel diameter of 120 μm (Wang et al., 2009).

In another interesting study, Trautmann et al. fabricated hollow MNs by using computer-aided design as well as polymerization and microchannel creation (Trautmann et al., 2019). The authors used a laser system designed for two-photon polymerization to prepare the MNs. A Ytterbium-doped femtosecond system with a repetition rate of 55 MHz and pulse duration of 300 fs was used at 515 nm to fabricate randomly shaped MNs (Trautmann et al., 2019). The BlueCut® femtosecond fiber laser that was used in the fabrication process was combined with a second-harmonic generation device. The laser system generated light pulses at a wavelength of 515 nm with a variable repetition rate from 100 kHz up to 10 MHz and adjustable pulse duration from 345 fs to 2.8 ps

(Trautmann et al., 2019). The investigators utilized the direct writing mode that entailed using extended laser beam from one of the two lasers with a 20x objective possessing a numerical aperture of 0.5 (Trautmann et al., 2019). In an alternative mode, a 63x objective with a numerical aperture of 0.75 was focused on a photosensitive material layer made from OrmoComp® (a micro resist technology® organic-inorganic hybrid polymer for ultraviolet imprint and molding) (Trautmann et al., 2019). MNs were produced by using an underlying substrate, a previously fabricated polymethyl methacrylate chip with microconduits or a glass coverslip with laser-drilled cavities (Trautmann et al., 2019). The samples were positioned on a plate which was movable in the plane perpendicular to the incoming beam via a translation stage (Trautmann et al., 2019). A camera determined the openings and cavities of the vertical microconduits in the substrate and placed the designed MNs correctly onto the openings (Trautmann et al., 2019). The hollow MNs were incorporated into a microfluidic system which was used to deliver a rhodamine B solution (Trautmann et al., 2019).

Economidou et al. fabricated a novel system that combines 3D printing, microneedles (MNs) and microelectromechanical systems (3DMNMEMS), thus permitting a flexible approach that enables control of transdermal drug delivery by the user (Economidou et al., 2021). The authors used advanced imaging techniques that enabled them to monitor liquid distribution within the skin in real time. The hollow microneedles were designed and 3D printed via stereolithography, followed by integration into microelectromechanical systems (Economidou et al., 2021). Animal experiments showed that insulin delivery via the 3DMNMEMS led to enhanced glycemic control in diabetic animals in comparison to subcutaneous drug administration (Economidou et al., 2021). The3DMNMEMS may find utility in personalized medicine.

In another set of experiments, Vinayakumar et al. used femto second laser micromachining to fabricate hollow stainless steel MNs (Vinayakumar et al., 2016). The authors employed the fabrication approach to prepare a complex system via a single procedure devoid of masks (Vinayakumar et al., 2016). The robustness of the produced MN array was determined via transverse and axial loading. Skin histology was conducted to determine the MN piercing of rat skin (Vinayakumar et al., 2016). The authors also investigated fluid flow with various inlet pressures via the MN array. The MN array was packaged for the protection of the hollow microneedle bore from blockage (Vinayakumar et al., 2016). The authors utilized the MN array for insulin delivery to diabetic rats (Vinayakumar et al., 2016).

The results of the experiments were similar when compared to standard subcutaneous delivery at the same doses (Vinayakumar et al., 2016).

Pamornpathomkul and coworkers combined a hollow microneedle array with cationic niosomes Span 20:cholesterol:cationic lipid (N^1,N^1 - dimyristeroyloxyethyl-spermine at the molar ratio of 2.5:2.5:0.5 mM) for skin immunization with plasmid DNA-encoding ovalbumin (pOVA) (Pamornpathomkul et al., 2018). The results demonstrated that the utilization of cationic niosomes with hollow MNs for pOVA delivery elicited both cell-mediated and humoral immune responses mediated by interleukin-4 (IL-4), immunoglobulin G (IgG), and interferon gamma (IFN-γ) cytokine (Pamornpathomkul et al., 2018). The authors reported superior immune response in comparison to naked pOVA when hollow MNs with cationic niosome/pOVA complexes were used (Pamornpathomkul et al., 2018). This was ascribed to increased cytokine release and enhanced IgG antibody production (Pamornpathomkul et al., 2018). It was demonstrated that using hollow MNs with cationic niosome/pOVA complexes generated increased cytokine secretion and greater IgG immune response compared with traditional subcutaneous (SCs) injections. Strikingly, there was bleeding or skin infection in a group of mice vaccinated via hollow MNs (Pamornpathomkul et al., 2018).

Hollow MNs can be used to surmount the stratum corneum barrier and improve the transcutaneous delivery of nanoparticles (Niu et al., 2019). Niu et al. studied the effect of a hollow microneedle array on the intradermal delivery of polymeric nanoparticles (NPs) in rats (Niu et al., 2019). Transcutaneous delivery was carried out with a plastic, hollow MN array with 18 microneedles (3 M®) (Niu et al., 2019). The delivery system employed in the set of experiments was made up of a microfluidics compartment, array of microneedles, adhesive ring and tubing connection (Niu et al., 2019). A hand-held applicator was used to drive the hollow microneedle array (Niu et al., 2019). Intradermal delivery of polymeric nanoparticles mediated by hollow microneedles led to a distinctive pharmacokinetic profile, characterized by an early rapid release via lymph nodes and relatively limited systemic transport (Niu et al., 2019). This mode of administration led to the achievement of a superior pharmacokinetic profile compared to that achieved via subcutaneous and intravenous routes of administration (Niu et al., 2019). The authors studied hollow microneedles as an alternative option for vaccine delivery. High local concentration was achieved in the lymphatic system (Niu et al., 2019). Poly (D,L-lactide-coglycolide) nanoparticles were used to encapsulate a model antigen ovalbumin (OVA) as well as monophosphoryl Lipid A and imiquimod. OVA loaded nanoparticles elicited quicker antibody affinity maturation kinetics in comparison to soluble OVA-based vaccine (Niu et al., 2019). Furthermore, antigen loaded nanoparticles transported with a hollow microneedle array induced a markedly higher number of interferon (IFN)-γ secreting lymphocytes and greater IgG2a antibody response (markers of Th1 response) compared to soluble antigen delivered through HMNA or antigen loaded nanoparticles delivered via intramuscular injection (Niu et al., 2019). Overall, the results demonstrated that hollow MN assisted intradermal transport of polymeric nanoparticles has a tremendous impact on the effectiveness of vaccines (Niu et al., 2019).

Du et al. investigated the effect of hollow microneedles (Fig. 12.2) on the intradermal delivery of nanoparticulate vaccines (Du et al., 2017). An in-house hollow-microneedle applicator was used to inject nanoparticle suspensions into mouse skin at a depth of approximately 120 μm (Du et al., 2017). Ovalbumin/poly (I:C) solution and ovalbumin/poly (I:C)-loaded nanoparticles generated comparable total IgG1 and IgG responses (Du et al., 2017). Co-encapsulation of poly (I:C) and ovalbumin in nanoparticles markedly enhanced IgG2a response compared to ovalbumin/poly (I:C) solution (Du et al., 2017). It was also shown that liposomes and poly (lactic-*co*-glycolic) nanoparticles led to more potent IgG2a responses in comparison with gelatin nanoparticles and mesoporous silica nanoparticles (Du et al., 2017). Significantly, the highest $CD4^+$ and $CD8^+$ T cell

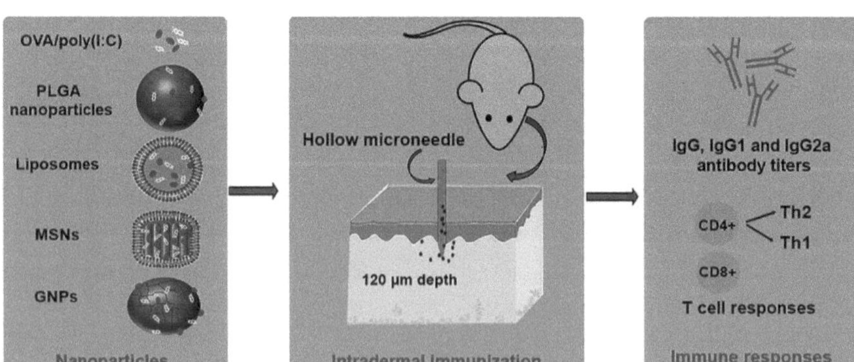

Fig. 12.2 Intradermal vaccination with hollow microneedles Nanoparticle-encapsulated protein antigen and adjuvant. *(From Du, G., Hathout, R. M., Nasr, M., Nejadnik, M. R., Tu, J., Koning, R. I., Koster, A. J., Slütter, B., Kros, A., Jiskoot, W., Bouwstra, J. A., & Mönkäre, J. (2017). Intradermal vaccination with hollow microneedles: A comparative study of various protein antigen and adjuvant encapsulated nanoparticles.* Journal of Controlled Release, 266, *109–118. https://doi.org/10.1016/j.jconrel.2017.09.021.)*

responses were engendered by ovalbumin/poly (I:C)-loaded liposomes. The authors showed that applicator controlled hollow microneedle delivery is an excellent approach for intradermal injection of nanoparticulate vaccines (Du et al., 2017).

Typically, the fabrication of hollow beveled silicon MNs is based either on the use of wet chemicals such as hydrogen fluoride or potassium hydroxide for the formation of the beveled surface (Bolton et al., 2020). Bolton et al. used deep reactive-ion etching (DRIE) technology for the production of hollow silicon microneedles (Bolton et al., 2020). The investigators were able to fabricate vertical sidewalls and a beveled tip for the MNs. The injection potential of the hollow microneedles was evaluated by delivering fluorescein isothiocyanate-labeled insulin (FITC-insulin) (Bolton et al., 2020). FITC-insulin flowed easily through the microneedle bores of three out of five microneedles. The authors attributed lack of flow of two MNs to adhesive blockage of the bore (Bolton et al., 2020). To prove that the needles can be used for injection into pig skin, the authors loaded a syringe with DIL (1,1′-dioctadecyl3,3,3′,3′-tetramethylindocarbocyanine perchlorate) fluorescent dye and attached it to a syringe pump (Bolton et al., 2020). DIL was administered into pig skin at a constant flow rate of 10 μL/min (Bolton et al., 2020). The dye permeated the porcine skin via the hollow microneedles and did not leak out through the microconduits (Bolton et al., 2020).

In another interesting set of experiments, a digitally controlled hollow microneedle injection system (DC-hMN-iSystem) with an ultra-low dead volume was used for automated micro-injections (0.25–10 μL) into human skin (van der Maaden et al., 2018). The hollow microneedles were fabricated from fused silica capillaries via hydrofluoric acid etching (van der Maaden et al., 2018). The maximum bevel length of the fabricated hollow microneedles was 97 μm, with the minimum insertion depth needed to obtain a leakage-free injection was 100 μm (van der Maaden et al., 2018). Furthermore, digitally controlled volumes as low as 1 μL were precisely and accurately injected via the microneedle system (van der Maaden et al., 2018). The DC-hMN-iSystem was subsequently used for the Injection of low volumes of a liposomal therapeutic cancer vaccine (van der Maaden et al., 2018). A synthetic long peptide derived from human papilloma virus was formulated into cationic liposomes, which was then administered intradermally via the hollow microneedle system as a therapeutic cancer vaccine (van der Maaden et al., 2018). The administration of this vaccine resulted in strong T-cell (both CD4 and CD8) responses in mice (van der Maaden et al., 2018).

References

Bolton, C. J. W., Howells, O., Blayney, G. J., Eng, P. F., Birchall, J. C., Gualeni, B., Roberts, K., Ashraf, H., & Guy, O. J. (2020). Hollow silicon microneedle fabrication using advanced plasma etch technologies for applications in transdermal drug delivery. *Lab on a Chip, 20*(15), 2788–2795. https://doi.org/10.1039/d0lc00567c

Cárcamo-Martínez, Á., Mallon, B., Anjani, Q. K., Domínguez-Robles, J., Utomo, E., Vora, L. K., Tekko, I. A., Larrañeta, E., & Donnelly, R. F. (2021). Enhancing intradermal delivery of tofacitinib citrate: Comparison between powder-loaded hollow microneedle arrays and dissolving microneedle arrays. *International Journal of Pharmaceutics, 593*, 120152. https://doi.org/10.1016/j.ijpharm.2020.120152

Cárcamo-Martínez, Á., Mallon, B., Domínguez-Robles, J., Vora, L. K., Anjani, Q. K., & Donnelly, R. F. (2021). Hollow microneedles: A perspective in biomedical applications. *International Journal of Pharmaceutics, 599*, 120455. https://doi.org/10.1016/j.ijpharm.2021.120455

Du, G., Hathout, R. M., Nasr, M., Nejadnik, M. R., Tu, J., Koning, R. I., Koster, A. J., Slütter, B., Kros, A., Jiskoot, W., Bouwstra, J. A., & Mönkäre, J. (2017). Intradermal vaccination with hollow microneedles: A comparative study of various protein antigen and adjuvant encapsulated nanoparticles. *Journal of Controlled Release, 266*, 109–118. https://doi.org/10.1016/j.jconrel.2017.09.021

Economidou, S. N., Uddin, M. J., Marques, M. J., Douroumis, D., Sow, W. T., Li, H., Reid, A., Windmill, J. F. C., & Podoleanu, A. (2021). A novel 3D printed hollow microneedle microelectromechanical system for controlled, personalized transdermal drug delivery. *Additive Manufacturing, 38*, 101815. https://doi.org/10.1016/j.addma.2020.101815

Lee, H., Park, J. H., & Park, J. H. (2017). Administration of a sol-gel formulation of phenylephrine using low-temperature hollow microneedle for treatment of intermittent fecal incontinence. *Pharmaceutical Research, 34*(12), 2809–2816. https://doi.org/10.1007/s11095-017-2261-9

van der Maaden, K., Heuts, J., Camps, M., Pontier, M., Terwisscha van Scheltinga, A., Jiskoot, W., Ossendorp, F., & Bouwstra, J. (2018). Hollow microneedle-mediated micro-injections of a liposomal HPV E743–63 synthetic long peptide vaccine for efficient induction of cytotoxic and T-helper responses. *Journal of Controlled Release, 269*, 347–354. https://doi.org/10.1016/j.jconrel.2017.11.035

Mansoor, I., Hafeli, U. O., & Stoeber, B. (2012). Hollow out-of-plane polymer microneedles made by solvent casting for transdermal drug delivery. *Journal of Microelectromechanical Systems, 21*(1), 44–52. https://doi.org/10.1109/JMEMS.2011.2174429

Moon, S. J., Lee, S. S., Lee, H. S., & Kwon, T. H. (2005). Fabrication of microneedle array using LIGA and hot embossing process. In *Microsystem technologies* (Vol. 11, pp. 311–318). https://doi.org/10.1007/s00542-004-0446-8. Issues 4–5.

Nagarkar, R., Singh, M., Nguyen, H. X., & Jonnalagadda, S. (2020). A review of recent advances in microneedle technology for transdermal drug delivery. *Journal of Drug Delivery Science and Technology, 59*. https://doi.org/10.1016/j.jddst.2020.101923

Niu, L., Chu, L. Y., Burton, S. A., Hansen, K. J., & Panyam, J. (2019). Intradermal delivery of vaccine nanoparticles using hollow microneedle array generates enhanced and balanced immune response. *Journal of Controlled Release, 294*, 268–278. https://doi.org/10.1016/j.jconrel.2018.12.026

Pamornpathomkul, B., Niyomtham, N., Yingyongnarongkul, B. E., Prasitpuriprecha, C., Rojanarata, T., Ngawhirunpat, T., & Opanasopit, P. (2018). Cationic niosomes for enhanced skin immunization of plasmid DNA-encoding ovalbumin via hollow microneedles. *AAPS PharmSciTech, 19*(1), 481–488. https://doi.org/10.1208/s12249-017-0855-5

Ramadon, D., McCrudden, M. T. C., Courtenay, A. J., & Donnelly, R. F. (2021). Enhancement strategies for transdermal drug delivery systems: Current trends and applications. *Drug Delivery and Translational Research.* https://doi.org/10.1007/s13346-021-00909-6

Solimani, F., Meier, K., & Ghoreschi, K. (2019). Emerging topical and systemic JAK inhibitors in dermatology. *Frontiers in Immunology, 10.* https://doi.org/10.3389/fimmu.2019.02847

Szeto, B., Aksit, A., Valentini, C., Yu, M., Werth, E. G., Goeta, S., Tang, C., Brown, L. M., Olson, E. S., Kysar, J. W., & Lalwani, A. K. (2021). Novel 3D-printed hollow microneedles facilitate safe, reliable, and informative sampling of perilymph from Guinea pigs. *Hearing Research, 400*, 108141. https://doi.org/10.1016/j.heares.2020.108141

Trautmann, A., Roth, G. L., Nujiqi, B., Walther, T., & Hellmann, R. (2019). Towards a versatile point-of-care system combining femtosecond laser generated microfluidic channels and direct laser written microneedle arrays. *Microsystems and Nanoengineering, 5*(1). https://doi.org/10.1038/s41378-019-0046-5

Vinayakumar, K. B., Kulkarni, P. G., Nayak, M. M., Dinesh, N. S., Hegde, G. M., Ramachandra, S. G., & Rajanna, K. (2016). A hollow stainless steel microneedle array to deliver insulin to a diabetic rat. *Journal of Micromechanics and Microengineering, 26*(6). https://doi.org/10.1088/0960-1317/26/6/065013

Wang, P. C., Wester, B. A., Rajaraman, S., Paik, S. J., Kim, S. H., & Allen, M. G. (2009). Hollow polymer microneedle array fabricated by photolithography process combined with micromolding technique. In *Proceedings of the 31st annual international conference of the IEEE engineering in medicine and biology society: Engineering the future of biomedicine, EMBC 2009* (pp. 7026–7029). IEEE Computer Society. https://doi.org/10.1109/IEMBS.2009.5333317

Xenikakis, I., Tsongas, K., Tzimtzimis, E. K., Zacharis, C. K., Theodoroula, N., Kalogianni, E. P., Demiri, E., Vizirianakis, I. S., Tzetzis, D., & Fatouros, D. G. (2021). Fabrication of hollow microneedles using liquid crystal display (LCD) vat polymerization 3D printing technology for transdermal macromolecular delivery. *International Journal of Pharmaceutics, 597*, 120303. https://doi.org/10.1016/j.ijpharm.2021.120303

Yadav, V., Sharma, P. K., Murty, U. S., Mohan, N. H., Thomas, R., Dwivedy, S. K., & Banerjee, S. (2021). 3D printed hollow microneedles array using stereolithography for efficient transdermal delivery of rifampicin. *International Journal of Pharmaceutics, 605*, 120815. https://doi.org/10.1016/j.ijpharm.2021.120815

Yung, K. L., Xu, Y., Kang, C., Liu, H., Tam, K. F., Ko, S. M., Kwan, F. Y., & Lee, T. M. H. (2012). Sharp tipped plastic hollow microneedle array by microinjection moulding. *Journal of Micromechanics and Microengineering, 22*(1), 015016. https://doi.org/10.1088/0960-1317/22/1/015016

CHAPTER 13
Hydrogel-forming microneedles

Introduction

There are several excellent reports in the literature that focus on the use of hydrogel-forming microneedles (Fig. 13.1) for transdermal drug delivery (Courtenay et al., 2018; Tekko et al., 2020). Hydrogel-forming microneedles (HFMNs) are usually prepared from swellable polymers (crosslinked hydrogels) and they possess a distinct working mechanism in comparison to other types of microneedles (Turner et al., 2021). HFMNs can penetrate the stratum corneum and draw up interstitial fluid, thus inducing the swelling of the polymeric matrix (Ita & Ita, 2020, pp. 143–181; Tekko et al., 2020). The transport of active pharmaceutical ingredients across the swollen matrix provides a route for drug delivery into the skin (Courtenay et al., 2018; Tekko et al., 2020). Therapeutic agents can be loaded into the polymeric structure during fabrication or placed into a separate reservoir that is attached to the HFMNs (Turner et al., 2021). Thus the amount of medication that can be incorporated, HFMNs is not constrained only to the needle surfaces or into the needles (Courtenay et al., 2020). On the contrary, medications can be incorporated into an accompanying reservoir (directly compressed tablet, a polymeric film, or lyophilized reservoir). This leads to a significant increase in the amount of therapeutic agent that can cross into the skin via the HFMN array.

Diverse types of polymers have been used to prepare HFMNs. Courtenay and coworkers used aqueous blends containing poly (ethylene glycol) 10,000, poly (methylvinyl ether-*co*-maleic anhydride) and 3% w/w anhydrous sodium carbonate to fabricate HFMNs (Courtenay et al., 2020). Poly (vinyl alcohol) alcohol crosslinked-gelatin, 2-hydroxyethyl methacrylate (HEMA)-crosslinked ethylene glycol dimethacrylate (EGDMA) and poly (methyl vinyl ether-*co*-maleic acid)-crosslinked pectin have also been utilized for the fabrication of HFMNs (Demir et al., 2013, 2017; Hardy et al., 2016; Ramadon et al., 2021). Physical stability and mechanical strength are the major drawbacks associated with HFMNs (Lee et al., 2019).

Microneedles
ISBN 978-0-323-91811-4
https://doi.org/10.1016/B978-0-323-91811-4.00008-0

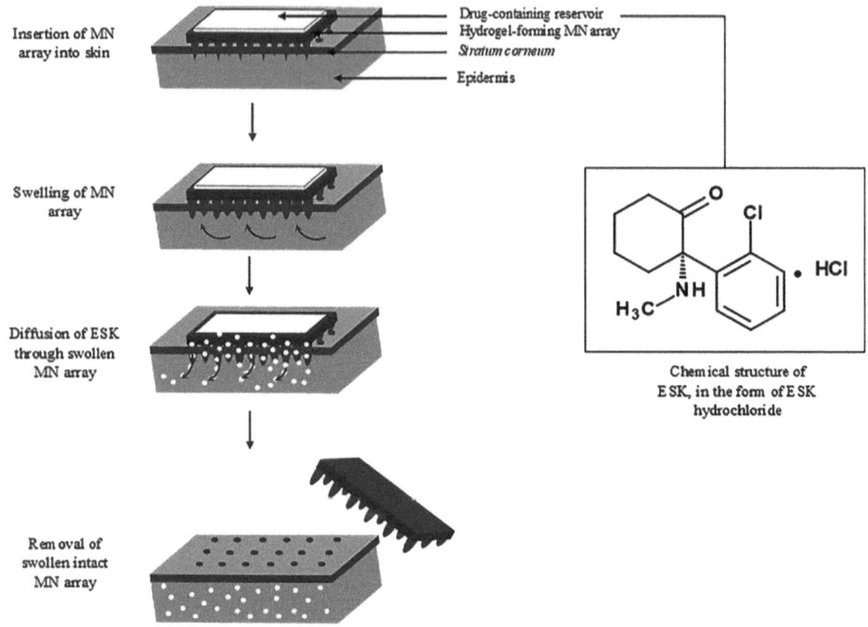

Fig. 13.1 Schematic representation of the mechanism of action of an esketamine-containing microneedle patch Esketamine-containing microneedle (MN) patches consist of hydrogel-forming MN arrays and esketamine-containing reservoir. Hydrogel-forming MN arrays take up skin interstitial fluid, inducing diffusion of esketamine from an esketamine-containing reservoir through the swollen micro-projections. *(From Courtenay, A. J., McAlister, E., McCrudden, M. T. C., Vora, L., Steiner, L., Levin, G., Levy-Nissenbaum, E., Shterman, N., Kearney, M.C., & McCarthy, H. O. (2020). Hydrogel-forming microneedle arrays as a therapeutic option for transdermal esketamine delivery.* Journal of Controlled Release, *322, 177–186.)*

These shortcomings are similar to those observed with dissolving microneedle systems (Lee et al., 2019). Materials that are robust enough to pierce the skin are used for the fabrication of HFMNs. These materials swell quickly and release the therapeutic agent following mixing with the skin's interstitial fluid (Lee et al., 2019). In contrast to dissolving microneedles, HFMNs do not remain in the skin. This is clinically beneficial (Lee et al., 2019). Another significant edge of HFMNs is that dose limitation is less problematic in comparison to coated and dissolving microneedles because HFMNs utilize a drug reservoir (Lee et al., 2019). Furthermore, the absorption profiles and drug release kinetics of HFMNs can be altered by introducing functional nanoparticles into MN system or by changing the composition of the hydrogel-forming polymers (Lee et al., 2019).

Microneedle length, tip diameter and array size are important parameters that must be considered during the design of HFMNs (Lee et al., 2019).

Diabetes mellitus represents a clinical syndrome that is characterized by hyperglycemia (Denyer et al., 2021). In patients with Type 1 diabetes, there is absolute insulin deficiency due to immune-mediated beta cell destruction, while patients with Type 2 diabetes are typically autoantibody negative and usually exhibit insulin resistance with residual beta cell function, resulting in increased serum insulin C-peptide concentration (Denyer et al., 2021). Microneedles with self-regulated drug releasing profiles may enhance the quality of care for diabetics (Shen et al., 2021). Shen and coworkers prepared a biocompatible and glucose responsive phenylboronic acid-based hydrogel microneedle patch (Shen et al., 2021). Glucose-responsive properties were accomplished via reversible phenylborate ester crosslinking points based on diol-containing units and phenylboronic acid units (Shen et al., 2021). The mechanical strength of the MNs was strengthened by using reversible crosslinking points (Shen et al., 2021).

Transdermal drug delivery from hydrogel-forming microneedles

Lutton and coworkers designed and evaluated a novel method for the fabrication of HMNs (Lutton et al., 2015). The authors prepared 14 × 14 microneedle arrays. Metal microneedle master templates were custom designed and used for the production of molds. Injection molding method was used for the production of silicone molds that were subsequently aligned on the roller base plate (Lutton et al., 2015). Four silicone molds were utilized for fabrication (Lutton et al., 2015). The roller was first rolled along the entire of the housing and then back to the previous home position. The technique of spotting four molds and rolling was repeated with the aim of subjecting the molds to a compressive force (Lutton et al., 2015). The investigators subsequently placed the formulation over the exposed molds to create the base plate (Lutton et al., 2015). Researchers typically choose the hydrogel mass to correspond with baseplate thickness but 9.5 g was used in this instance to fill the frame (Lutton et al., 2015). According to the investigators, it is easier to scale up production with this technique and this is relevant for commercialization (Lutton et al., 2015). The microneedles that were produced via the injection method were similar to those prepared via centrifugation (Lutton et al., 2015).

Both microneedle types were efficiently inserted into a skin simulant with a depth close to 60% of the needle length. The height reduction of the microneedles following insertion was about 3% in both instances (Lutton et al., 2015).

Peng et al. recently fabricated hybrid microneedle patches with hydrogel-forming microneedle bases (HFMB) and implantable poly (lactic-*co*-glycolic acid) (PLGA) tips (Peng et al., 2021). Three classes of pre-formed microneedle bases HFMB with needle heights of 600 μm, conventional dissolving baseplates, and HFMB with needle heights of 800 μm (MN8) were studied by the authors (Peng et al., 2021). It was observed that the addition of the pre-formed hydrogel-forming microneedle bases increased ex vivo drug delivery efficiency up to 80% of the loaded drug (Peng et al., 2021). In addition, the in vitro and ex vivo insertion properties of the microneedle systems were enhanced, and the implantation process was accelerated to within 60 s (Peng et al., 2021). It was easier to peel off the hydrogel baseplate from the skin in comparison to the dissolving baseplate. Significantly, the antifungal effects against *Candida albicans* were demonstrated and the release of amphotericin B from the drug loading PLGA tips was sustained over 7 days (Peng et al., 2021). The hybrid microneedle system can be used to deliver drug-eluting microimplants into the skin with the possibility of easily removing the HFMB (Peng et al., 2021).

Aung et al. prepared hydrogel forming microneedles from polyvinyl alcohol (PVA) and polyacrylic acid-*co*-maleic acid (PAMA) (4:1) (Aung et al., 2020). The authors laser-engineered silicone micromold templates and employed them for the production of α-arbutin-loaded microneedles (Aung et al., 2020). The polymer dispersion loaded with the active pharmaceutical ingredient was cast onto silicone micromold (Aung et al., 2020) and centrifuged at 3500 rpm at 25°C for 30 min to fill the mold cavities with subsequent drying for 48 h at 25°C (Aung et al., 2020). Crosslinking (90°C for 30 min) was conducted after drying. The authors carefully cut the sidewalls of the microneedles with a heated scalpel following removal of the molds (Aung et al., 2020). The microneedles were strong, sharp and capable of piercing neonatal pig skin. The investigators documented maximum swelling of the hydrogel forming microneedles at 4 h (Aung et al., 2020). In vitro transport experiments proved that the use of α-arbutin loaded hydrogel-forming microneedles provided markedly higher transport (2.8 times) of the biologically active compound in comparison with a gel and a marketed cream (Aung et al., 2020). Animal experiments also provided evidence of higher intradermal delivery of α-arbutin with hydrogel forming microneedles (1.47 μg/mL) in comparison to that achieved with a commercial cream (0.15 μg/mL) (Aung et al., 2020). Furthermore, the microneedles were stable at 25°C for 3 months while the micro-conduits generated following the HFMNs resealed within 1 h (Aung et al., 2020).

Methotrexate (MTX) is used for the management of juvenile idiopathic arthritis (JIAs) and rheumatoid arthritis (RA) (Alexeeva et al., 2019; Luedders et al., 2020). The drug is commonly administered either by subcutaneous (SCs) injection or orally (Tekko et al., 2020). Subcutaneous injection is painful and can result in high peak blood concentration of the drug. On the other hand, oral administration of the medication can lead to erratic bioavailability, vomiting and nausea (Tekko et al., 2020). Drug delivery into systemic circulation via the skin is an attractive substitute for oral and subcutaneous routes of administration, but passive transdermal transport of MTX is poor (Tekko et al., 2020). This is attributed to the physicochemical properties of MTX and the stratum corneum barrier. A patch-like reservoir loaded with methotrexate (MTX-RV) was linked to hydrogel-forming microneedles. The system was used to deliver MTX into the bloodstream in a sustained fashion (Tekko et al., 2020). The authors used an aqueous blend of poly (vinyl pyrrolidone) (PVP) and poly (vinyl alcohol) (PVA) to fabricate the HFMNs. Citric acid was employed as a crosslinker. Glycerol and hydroxypropyl methylcellulose (HPMC) were utilized for the preparation of MTX-RV (Tekko et al., 2020). The MTX-RV and HFMN were used to create the integrated patch. The HFMN system was strong enough to pierce excised neonatal pig skin. A high dose of the drug (150.3 ± 5.3 μg/mg) was introduced into the MTX-RV without precipitation (Tekko et al., 2020). The integrated patch was used to transport the drug ex vivo at a steady-state flux of 506.8 ± 136.9 μg cm^2/h (Tekko et al., 2020). Remarkably, 7.6 ± 2.0 nM of the drug was detected in the blood of Sprague Dawley rats 1 h after patch administration (Tekko et al., 2020). At 24 h, the concentration of the drug in the blood increased steadily to reach its peak (C_{max}) of 35.1 ± 5.1 nM. Significantly, the authors removed the HFMNs intact from the skin with only mild erythema (Tekko et al., 2020).

Local ischemia, air/fluid leakage, and the ensuing impairment of healing are consequences of wound closure in internal organs (Jeon et al., 2019). Jeon and coworkers used a hydrogel-forming double-layered adhesive microneedle (MN) patch to surmount these challenges (Jeon et al., 2019). The inspiration for this development came from endoparasites that swell their proboscis to dock in the host's intestines (Jeon et al., 2019). The MN patch was made up of a non-swellable silk fibroin (SF)-based core and a swellable mussel adhesive protein (MAP)-based shell (Jeon et al., 2019). The authors prepared 35 μL of the prehydrogel solution made up of 15% (wt/vol) hyaluronic acid, 35% (wt/vol) MAP, 1 mM Ru(II)bpy3 (2+) and 30 mM sodium persulfate (SPS) in phosphate buffered saline and then applied same to a PDMS mold that was maintained at −85 kPa (Jeon et al., 2019). This was the approach adopted for the fabrication of the MN patch shell (Jeon et al., 2019). Twenty microliters of PBS were spread onto the surface of the residual solution/gel, and the excess sample was gently removed (Jeon et al., 2019). Next, the authors cast 35 μL of the prehydrogel solution made up of 1 mM Ru(II)bpy3 2+, 70% (wt/vol) SF, and 30 mM SPS dissolved in de-ionized water onto the mold at −85 kPa for an additional 12 h to create the microneedle core (Jeon et al., 2019). The dried microneedle patch was gently detached from the mold. The hydrogel-forming adhesive MN patch displayed superb ex vivo wound sealing capacity against luminal leaks (139.7 ± 14.1 mmHg) due to MAP-derived surface adhesion, tissue insertion capability that was 7-times greater than the force required for pig skin piercing, and selective swelling-induced physical entanglement (Jeon et al., 2019). The wound sealing property was similar to what was achieved by employing suture (151.0 ± 23.3 mmHg). In addition, there was excellent in vivo performance for dynamic and/or wet internal and external tissues (Jeon et al., 2019).

Trametinib (Tra) and doxorubicin (DOX) are commonly used anti-cancer agents (Mistarz et al., 2021; Wang et al., 2021). In addition, it has been shown that Tra can also reverse P-glycoprotein -mediated multidrug resistance, thus inhibiting the efflux of doxorubicin (Huang et al., 2020). Huang and coworkers synthesized photocrosslinkable dextran methacrylate (DexMA) (Huang et al., 2020). The authors used DexMA hydrogel to fabricate a novel type of microneedles for sustained drug administration. The authors introduced 30% DexMA gel solution into the cavity of a polydimethylsiloxane (PDMS) mold (Huang et al., 2020). The mold was subjected to vacuum at -28 MPa to remove air bubbles (Huang et al., 2020). The PDMS mold was subsequently irradiated at 365 nm UV light for 60 s, and then kept in a room temperature and dark environment for 1 h (Huang et al., 2020). The DexMA microneedle patch was then detached from the mold (Huang et al., 2020). The prepared hydrogel MNs were used for the transdermal delivery of trametinib and doxorubicin (Huang et al., 2020).

McAlister and coworkers recently fabricated hydrogel-forming microneedles and tethered them to novel, nonaqueous-based, directly compressed tablets (DCTs) (McAlister et al., 2021). Different therapeutic agents were incorporated into these microneedle systems. Poly (vinyl alcohol)-based MN systems were used for experiments (McAlister et al., 2021). In vitro transport experiments as well as animal pharmacokinetic studies were carried out in rats using microneedle patches loaded with levodopa/carbidopa (LD/CD), AMX, and levofloxacin (LVX) (McAlister et al., 2021). Therapeutically effective blood levels of LD ($\geq$0.5 μg mL-1), AMX ($\geq$2 μg mL-1), and LVX ($\geq$0.2 μg mL-1) were obtained at 2, 1 and 1 h, respectively (McAlister et al., 2021). The authors showed that DCTs can be used to widen the range of therapeutic agents transported with microneedle patches via the skin into the circulatory system (McAlister et al., 2021).

Anjani et al. used hydrogel-forming microneedles for the transdermal transport of ethambutol, isoniazid, rifampicin, and pyrazinamide (Anjani et al., 2020). The investigators incorporated the medications individually into three categories of drug reservoirs (directly compressed tablets, lyophilized tablets, and poly (ethylene glycol) tablets) (Anjani et al., 2020). The different formulations were integrated with HFMNs. In vitro transport experiments across dermatomed neonatal pig skin samples were conducted with these devices (Anjani et al., 2020). It was revealed that reservoir design and drug solubility had significant impact on the amount of drug that permeated from the hydrogel-forming MN arrays across pig skin (Anjani et al., 2020). The greatest transdermal transport of ethambutol (46.99 mg) and rifampicin (3.64 mg) were obtained via microneedle arrays that were linked to directly compressed tablet and poly (ethylene glycol) tablets respectively (Anjani et al., 2020). The highest drug permeation was attained using lyophilized reservoir in the case of pyrazinamide and isoniazid and the corresponding amount of drug delivered was about 20.08 mg and 58.45 mg respectively (Anjani et al., 2020). Significantly, the authors demonstrated that hydrogel forming microneedles can be used for the transdermal delivery of approximately 75% (rifampicin), 79% (isoniazid), 20% (pyrazinamide) and 47% (ethambutol) of the drugs incorporated into the reservoirs (Anjani et al., 2020).

Bevacizumab (BEV) is a monoclonal antibody that is used for the management of different types of cancers (Chen et al., 2020; Sato et al., 2020) as well as diabetic macular edema (DME) (Vader et al., 2020). Taking into consideration the poor oral bioavailability of monoclonal antibodies, investigators normally administer these molecules as parenteral injections (Courtenay et al., 2018). Courtenay et al. fabricated HFMNs for the percutaneous delivery of BEV (Courtenay et al., 2018). The authors used bevacizumab specific ELISA to carry out drug recovery and transdermal penetration experiments. Microneedle-patches were studied in vivo using Sprague-Dawley rats. The investigators studied BEV concentrations in the skin, spleen, axial lymph nodes, exterior lumbar and serum (Courtenay et al., 2018). HFMNs were used in this research project to deliver high doses of a monoclonal antibody. The following pharmacokinetic parameters were determined: C_{ss} 942 ± 95 ng/mL, T_{max} 48 h, AUC 44,357 ± 4540 and C_{max} 358.2 ± 100.4 ng/mL (Courtenay et al., 2018). The HFMNs delivered BEV to the lymph and the bloodstream in a sustained manner (Courtenay et al., 2018). It was shown by the authors that HFMNs can be utilized for the targeted delivery of drugs to the lymphatic system and this is particularly important for the management of metastatic cancer (Courtenay et al., 2018).

Syringes and hypodermic needles are typically used for vaccinations (Courtenay et al., 2019). Nevertheless, newer techniques of vaccine delivery are needed to reduce costs and improve coverage (Courtenay et al., 2019). Several authors have used dissolving microneedles (DMNs) as an option for the hypodermic needle. This can ultimately lead to enhanced immunogenicity through direct targeting of skin dendritic cells, as well as provide an effective means of conducting self-vaccination (Courtenay et al., 2019). Courtenay et al. carried out a comparative investigation of hydrogel-forming MNs and dissolving MNs for the delivery of ovalbumin (OVA) (Courtenay et al., 2019). The authors studied in vitro parameters a well as in vivo immunogenicity profiles (Courtenay et al., 2019). Insertion characteristics and mechanical integrity were tested using a previously validated skin insertion model. Anti-OVA specific immunoglobulin G (IgG) generated in BALB/c mice was employed as an index of immune response and delivery efficiency (Courtenay et al., 2019). The authors provided evidence which showed that vaccination with dissolving MNs resulted in substantially higher IgG titers even though the quantity of OVA incorporated into the DMNs was markedly lower compared to that loaded into HFMNs (Courtenay et al., 2019).

Yao and coworkers prepared hydrogel microneedles via high-precision digital light processing three-dimensional printing (Yao et al., 2019). The raw material of MNs was liquid initially but subsequently, the authors used blue light to crosslink the monomer dispersion (Yao et al., 2019).

Polyethylene glycol diacrylate (PEG400DA, monomer, molecular weight 508) was first transferred into a brown sealing bottle and then phenylbis (2,4,6-trimethylbenzoyl) phosphine oxide was added as a photo-initiator (Yao et al., 2019). The experiments were conducted in the dark because phenylbis (2,4,6-trimethylbenzoyl) phosphine oxide is sensitive to light. The three-dimensional printing system was then used to convert the precursor into solid microneedles (Yao et al., 2019). The fabricated hydrogel microneedles were micro-porous and sharp with the potential of being utilized as minimally invasive devices for drug delivery (Yao et al., 2019). The mechanical properties of microneedles as well as the fabrication procedures were evaluated with the aim of finding a balance between stiffness and precision (Yao et al., 2019). Optimally selected printing parameters resulted in a reasonable balance between the two mechanical properties (stiffness and precision) (Yao et al., 2019). The bio-compatible and low-cost hydrogel microneedles fabricated via 3D printing can be used for drug delivery (Yao et al., 2019).

Microneedles can be utilized for intravitreal drug delivery (Amer & Chen, 2020). The layout and number of microneedles (MNs) can be optimized to ease the administration of the MN array to the eye and to achieve sustained drug release (Amer & Chen, 2020). Amer and Chen investigated the drug release kinetics of HFMNs with disparate numbers of MNs (8, 12, and 16) over four weeks (Amer & Chen, 2020). The microneedle arrays displayed steady drug release kinetics in contrast to single injections. The 16-needle MN array fully released all the incorporated medication at the end of 4 weeks. Both 8- and 12-needle arrays showed uniform release profiles over the 4-week period (Amer & Chen, 2020). The authors developed zero-order models and used them to predict drug release profiles for the three MN arrays. It was computed that the 8-needle microneedle array can deliver the drug for up to 6 weeks (Amer & Chen, 2020). The models can also be utilized for the design of HFMNs with a targeted sustained release profile (Amer & Chen, 2020).

The poor aqueous solubility of drug molecules poses a tremendous challenge to pharmaceutical scientists (Kearney et al., 2019). Therapeutic agents with low aqueous solubility and Log P that is more than 3 do not cross the skin at therapeutically relevant rates but their solubility values can be improved by using co-solvents (Kearney et al., 2019). Kearney et al. utilized co-solvents as reservoirs in combination with hydrogel-forming microneedles to increase the transdermal transport of olanzapine, Nile red, and atorvastatin (hydrophobic molecules) (Kearney et al., 2019).

The HFMNs were fabricated from 7.5% w/w PEG 10,000, 20% w/w Gantrez® S-97, and 3% w/w anhydrous sodium carbonate (Kearney et al., 2019). Laser engineered silicone molds were used to prepare the HFMNs. The authors employed a custom-made Franz diffusion cell to study the permeation of Nile red dye from HFMNs tethered to a liquid drug reservoir (Kearney et al., 2019). The use of co-solvents was productive as 83.30% ± 9.38% of the solution of Nile red dye in poly (ethylene glycol) (PEG 400) was transported across neonatal pig skin in 24 h (Kearney et al., 2019). Propylene glycol and PEG 400 were appropriate reservoir vehicles for atorvastatin and olanzapine, with about 50% of each compound transported over 24 h (Kearney et al., 2019). The authors showed that by adjusting HFMN reservoir properties, it is possible to increase microneedle-assisted delivery of hydrophobic molecules (Kearney et al., 2019).

Hardy et al. used stimulus-responsive HFMN arrays in combination with light to improve the transport of ibuprofen (IBN) (Hardy et al., 2016). The HFMNs were fabricated from a polymer prepared from ethylene glycol dimethacrylate (EGDMA) and 2-hydroxyethyl methacrylate (HEMA) (Hardy et al., 2016). Micromolding was the fabrication method and the HFMNs displayed acceptable mechanical properties. The investigators loaded the system with up to 5% (w/w) of the drug enclosed in a light-responsive 3,5-dimethoxybenzoin conjugate (Hardy et al., 2016). Raman spectroscopy was used to detect the conjugate inside the hydrogel forming microneedles (Hardy et al., 2016). The authors also conducted experiments which showed that the microneedles can deliver up to three doses of 50 mg of ibuprofen following the application of light over a sustained period of time (up to 160 h) (Hardy et al., 2016). Stimulus-responsive HFMNs can be utilized as controlled release systems with prolonged-release properties (Hardy et al., 2016). This technology can also be used for "on-demand" delivery of different medications in diverse clinical settings (Hardy et al., 2016).

Migdadi et al. used a hydrogel-forming microneedle patch for transdermal delivery of the high-dose drug metformin hydrochloride (Migdadi et al., 2018). Patches (two layers) were fabricated from a lyophilized drug reservoir layer, with the microneedle layer prepared from aqueous blend of 20% w/w poly (methylvinylether-*co*-maleic acid) crosslinked with 7.5% w/w poly (ethylene glycol) 10,000 Da (Migdadi et al., 2018). Fig. 13.2 shows the in vitro cumulative amount of metformin HCl permeated across dermatomed neonatal porcine skin from a reservoir formulation (75 mg) in combination with hydrogel-forming MNs (Migdadi et al., 2018).

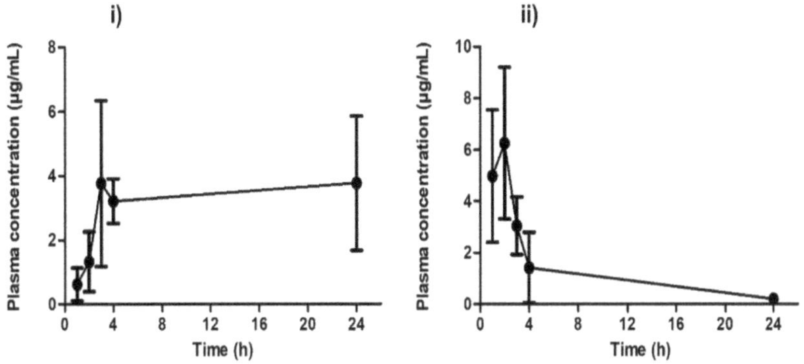

Fig. 13.2 In vivo plasma profiles of metformin hydrochloride (i) Plasma profiles following application of hydrogel-forming microneedles and solid metformin HCl-containing reservoir integrated patches at a dose of 100 mg. (ii) The in vivo plasma profiles of metformin HCl following oral gavage at a dose of 100 mg/kg of rat weight, (means ± SD, n at least = 4, for 24 h, n = 8). *(From Migdadi, E. M., Courtenay, A. J., Tekko, I. A., McCrudden, M. T. C., Kearney, M. C., McAlister, E., McCarthy, H. O. & Donnelly, R. F. (2018). Hydrogel-forming microneedles enhance transdermal delivery of metformin hydrochloride.* Journal of Controlled Release, 285, *142–151. https://doi.org/10.1016/j.jconrel.2018.07.009.)*

The hydrogel forming microneedle system (containing 75 mg or 50 mg metformin HCl, respectively) transported 9.71 ± 2.22 mg and 10.04 ± 1.92 mg at 6 h, respectively, and 28.15 ± 2.37 mg and 23.25 ± 3.58 mg at 24 h, respectively across neonatal pig skin (Migdadi et al., 2018). On the other hand, 0.34 ± 0.39 mg and 0.85 ± 0.68 mg of the drug were transported at 6 h, respectively, and 0.39 ± 0.39 mg and 1.01 ± 0.84 mg were transported at 24 h, respectively, from a control system that utilized only the drug reservoirs (Migdadi et al., 2018). The authors also carried out animal experiments to evaluate metformin transdermal bioavailability. The bioavailability of the drug following the use of microneedles was 30% (Migdadi et al., 2018).

Low molecular weight protein-based macromolecules, such as recombinant antibody fragments (10–70 kDa) have been developed with the aim of enhancing monoclonal antibody-based treatment options (Hutton et al., 2020). HFMNs can be utilized for the delivery of these macromolecules (Hutton et al., 2020). Hutton et al. employed FITC–dextran of different MWs (10, 70 and 150 kDa) as models of compounds (Hutton et al., 2020). Monoclonal antibodies (mAbs) with MWs of about 150 kDa were investigated (Hutton et al., 2020). The authors studied fluorescein sodium as a model hydrophilic, low molecular weight molecule. It was reported that fluorescein sodium had the highest cumulative permeation

(637.4 ± 42.69 μg) (Hutton et al., 2020). The quantities of FITC–dextran 150 kDa and 10 kDa which were transported across neonatal pig skin in vitro were 213.54 ± 15.19 μg and 462.17 ± 65.85 μg after 24 h, respectively (Hutton et al., 2020). The results of the experiments indicated that the transport of low molecular weight protein-based compounds, via "super swelling" hydrogel-forming MNs, can lead to greater transport across human skin, in comparison to passive transdermal delivery (Hutton et al., 2020).

Ripolin and colleagues designed, fabricated and assessed large microneedle patches fabricated from 16 separate microneedle arrays (needle height 600 μm) (Ripolin et al., 2017). Aqueous blends consisting of 7.5% w/w PEG 10,000 and 15% w/w Gantrez S-97 were used for the fabrication of the HFMNs (Ripolin et al., 2017). The authors fed the polymer matrix (0.5 g/cm^2) into silicone micromolds (600 μm, 14 × 14 array on a 0.5 cm^2 area) via a syringe (Ripolin et al., 2017). The holes were filled following the application of positive pressure (3–4 bar) to the polymeric matrix for 15 min (Ripolin et al., 2017). The HFMNs were dried for 48 h at room temperature and then crosslinked via esterification under 80C for 24 h (Ripolin et al., 2017). The HFMNs were kept under ambient temperature until they were assembled into large patches (Ripolin et al., 2017). Fourier-transform infrared spectroscopy was used to confirm the ester-based crosslinking and the HFMNs displayed acceptable mechanical strength (Ripolin et al., 2017). The depth of insertion of these needles in a validated skin model was about 500 μm. The mean insertion depth was between 300 and 450 μm. This was similar to what was obtained by the same volunteers using single unit microneedle patches (Ripolin et al., 2017). Ten human volunteers self-inserted the microneedles of these large microneedle patches into their skin, and this was validated by measuring transepidermal water loss (Ripolin et al., 2017). It was estimated by the authors that microneedle patch sizes larger than 1–2 cm^2 will be needed following the translation of this technology into clinical practice for the delivery of therapeutic agents (Ripolin et al., 2017).

References

Alexeeva, E., Dvoryakovskaya, T., Denisova, R., Sleptsova, T., Isaeva, K., Chomahidze, A., Fetisova, A., Mamutova, A., Alshevskaya, A., Gladkikh, V., & Moskalev, A. (2019). Dynamics of concomitant therapy in children with juvenile idiopathic arthritis treated with etanercept and methotrexate. *Pediatrics and Neonatology, 60*(5), 549–555. https://doi.org/10.1016/j.pedneo.2019.02.003

Amer, M., & Chen, R. K. (2020). Hydrogel-forming microneedle arrays for sustained and controlled ocular drug delivery. *Journal of Engineering and Science in Medical Diagnostics and Therapy, 3*(4). https://doi.org/10.1115/1.4048481

Anjani, Q. K., Permana, A. D., Cárcamo-Martínez, Á., Domínguez-Robles, J., Tekko, I. A., Larrañeta, E., Vora, L. K., Ramadon, D., & Donnelly, R. F. (2020). Versatility of hydrogel-forming microneedles in in vitro transdermal delivery of tuberculosis drugs. *European Journal of Pharmaceutics and Biopharmaceutics, 158*, 294–312. https://doi.org/10.1016/j.ejpb.2020.12.003

Aung, N. N., Ngawhirunpat, T., Rojanarata, T., Patrojanasophon, P., Pamornpathomkul, B., & Opanasopit, P. (2020). Fabrication, characterization and comparison of α-arbutin loaded dissolving and hydrogel forming microneedles. *International Journal of Pharmaceutics, 586*, 119508. https://doi.org/10.1016/j.ijpharm.2020.119508

Chen, Y., Guo, L., Li, X., Liu, R., Ren, C., & Du, S. (2020). Reduced-dose bevacizumab vs. standard-dose bevacizumab in recurrent high-grade glioma: Which one is better? A meta-analysis. *Clinical Neurology and Neurosurgery, 198*, 106239. https://doi.org/10.1016/j.clineuro.2020.106239

Courtenay, A. J., McAlister, E., McCrudden, M. T. C., Vora, L., Steiner, L., Levin, G., Levy-Nissenbaum, E., Shterman, N., Kearney, M.-C., & McCarthy, H. O. (2020). Hydrogel-forming microneedle arrays as a therapeutic option for transdermal esketamine delivery. *Journal of Controlled Release, 322*, 177–186.

Courtenay, Aaron J., McCrudden, M., McAvoy, K. J., McCarthy, H. O., & Donnelly, R. F. (2018). Microneedle-mediated transdermal delivery of bevacizumab. *Molecular Pharmaceutics, 15*(8), 3545–3556. https://doi.org/10.1021/acs.molpharmaceut.8b00544

Courtenay, A., Rodgers, A. M., McCrudden, M., McCarthy, H. O., & Donnelly, R. F. (2019). Novel hydrogel-forming microneedle array for intradermal vaccination in mice using ovalbumin as a model protein antigen. *Molecular Pharmaceutics, 16*(1), 118–127. https://doi.org/10.1021/acs.molpharmaceut.8b00895

Demir, Y. K., Akan, Z., & Kerimoglu, O. (2013). Characterization of polymeric microneedle arrays for transdermal drug delivery. *PloS One, 8*(10), e77289. https://doi.org/10.1371/journal.pone.0077289

Demir, Y. K., Metin, A.Ü., Satıroğlu, B., Solmaz, M. E., Kayser, V., & Mäder, K. (2017). Poly (methyl vinyl ether-co-maleic acid) – pectin based hydrogel-forming systems: Gel, film, and microneedles. *European Journal of Pharmaceutics and Biopharmaceutics, 117*, 182–194. https://doi.org/10.1016/j.ejpb.2017.04.018

Denyer, A. L., Catchpole, B., & Davison, L. J. (2021). Genetics of canine diabetes mellitus part 1: Phenotypes of disease. *The Veterinary Journal, 270*, 105611. https://doi.org/10.1016/j.tvjl.2021.105611

Hardy, J. G., Larrañeta, E., Donnelly, R. F., McGoldrick, N., Migalska, K., McCrudden, M. T. C., Irwin, N. J., Donnelly, L., & McCoy, C. P. (2016). Hydrogel-forming microneedle arrays made from light-responsive materials for on-demand transdermal drug delivery. *Molecular Pharmaceutics, 13*(3), 907–914. https://doi.org/10.1021/acs.molpharmaceut.5b00807

Huang, S., Liu, H., Huang, S., Fu, T., Xue, W., & Guo, R. (2020). Dextran methacrylate hydrogel microneedles loaded with doxorubicin and trametinib for continuous transdermal administration of melanoma. *Carbohydrate Polymers, 246*, 116650. https://doi.org/10.1016/j.carbpol.2020.116650

Hutton, A. R. J., McCrudden, M. T. C., Larrañeta, E., & Donnelly, R. F. (2020). Influence of molecular weight on transdermal delivery of model macromolecules using hydrogel-forming microneedles: Potential to enhance the administration of novel low molecular weight biotherapeutics. *Journal of Materials Chemistry B, 8*(19), 4202–4209. https://doi.org/10.1039/D0TB00021C

Ita, K., & Ita, K. (2020). *Chapter 8 - microneedles*. Academic Press. https://doi.org/10.1016/B978-0-12-822550-9.00008-9

Jeon, E. Y., Lee, J., Kim, B. J., Joo, K. I., Kim, K. H., Lim, G., & Cha, H. J. (2019). Bio-inspired swellable hydrogel-forming double-layered adhesive microneedle protein patch for regenerative internal/external surgical closure. *Biomaterials, 222*, 119439. https://doi.org/10.1016/j.biomaterials.2019.119439

Kearney, M.-C., McKenna, P. E., Quinn, H. L., Courtenay, A. J., Larrañeta, E., & Donnelly, R. F. (2019). Design and development of liquid drug reservoirs for microneedle delivery of poorly soluble drug molecules. *Pharmaceutics, 11*(11), 605. https://doi.org/10.3390/pharmaceutics11110605

Lee, K. J., Jeong, S. S., Roh, D. H., Kim, D. Y., Choi, H.-K., & Lee, E. H. (2019). A practical guide to the development of microneedle systems - in clinical trials or on the market. *International Journal of Pharmaceutics, 573*, 118778. https://doi.org/10.1016/j.ijpharm.2019.118778

Luedders, B. A., Johnson, T. M., Sayles, H., Thiele, G. M., Mikuls, T. R., O'Dell, J. R., & England, B. R. (2020). Predictive ability, validity, and responsiveness of the multi-biomarker disease activity score in patients with rheumatoid arthritis initiating methotrexate. *Seminars in Arthritis and Rheumatism, 50*(5), 1058–1063. https://doi.org/10.1016/j.semarthrit.2020.05.019

Lutton, R. E. M., Larrañeta, E., Kearney, M.-C., Boyd, P., Woolfson, A. D., & Donnelly, R. F. (2015). A novel scalable manufacturing process for the production of hydrogel-forming microneedle arrays. *International Journal of Pharmaceutics, 494*(1), 417–429. https://doi.org/10.1016/j.ijpharm.2015.08.049

McAlister, E., Dutton, B., Vora, L. K., Zhao, L., Ripolin, A., Zahari, D. S. Z. B. P. H., Quinn, H. L., Tekko, I. A., Courtenay, A. J., Kelly, S. A., Rodgers, A. M., Steiner, L., Levin, G., Levy-Nissenbaum, E., Shterman, N., McCarthy, H. O., & Donnelly, R. F. (2021). Directly compressed tablets: A novel drug-containing reservoir combined with hydrogel-forming microneedle arrays for transdermal drug delivery. *Advanced Healthcare Materials, 10*(3), 2001256. https://doi.org/10.1002/adhm.202001256

Migdadi, E. M., Courtenay, A. J., Tekko, I. A., McCrudden, M. T. C., Kearney, M.-C., McAlister, E., McCarthy, H. O., & Donnelly, R. F. (2018). Hydrogel-forming microneedles enhance transdermal delivery of metformin hydrochloride. *Journal of Controlled Release, 285*, 142–151. https://doi.org/10.1016/j.jconrel.2018.07.009

Mistarz, A., Graczyk, M., Winkler, M., Singh, P. K., Cortes, E., Miliotto, A., Liu, S., Long, M., Yan, L., Stablewski, A., O'Loughlin, K., Minderman, H., Odunsi, K., Rokita, H., McGray, A. R., Zsiros, E., & Kozbor, D. (2021). Induction of cell death in ovarian cancer cells by doxorubicin and oncolytic vaccinia virus is associated with CREB3L1 activation. *Molecular Therapy - Oncolytics*. https://doi.org/10.1016/j.omto.2021.04.014

Peng, K., Vora, L. K., Domínguez-Robles, J., Naser, Y. A., Li, M., Larrañeta, E., & Donnelly, R. F. (2021). Hydrogel-forming microneedles for rapid and efficient skin deposition of controlled release tip-implants. *Materials Science and Engineering: C, 127*, 112226. https://doi.org/10.1016/j.msec.2021.112226

Ramadon, D., McCrudden, M. T. C., Courtenay, A. J., & Donnelly, R. F. (2021). Enhancement strategies for transdermal drug delivery systems: Current trends and applications. *Drug Deliv Transl Res, 1–34*. https://doi.org/10.1007/s13346-021-00909-6

Ripolin, A., Quinn, J., Larrañeta, E., Vicente-Perez, E. M., Barry, J., & Donnelly, R. F. (2017). Successful application of large microneedle patches by human volunteers. *International Journal of Pharmaceutics, 521*(1), 92–101. https://doi.org/10.1016/j.ijpharm.2017.02.011

Sato, H., Nagashima, H., Akiyama, M., Ito, T., Hashimoto, T., Saikawa, H., Utsumi, Y., & Maemondo, M. (2020). Analysis of bevacizumab treatments and metastatic sites of lung cancer. *Cancer Treatment and Research Communications, 26*, 100290. https://doi.org/10.1016/j.ctarc.2020.100290

Shen, D., Yu, H., Wang, L., Chen, X., Feng, J., Li, C., Xiong, W., & Zhang, Q. (2021). Glucose-responsive hydrogel-based microneedles containing phenylborate ester bonds and N-isopropylacrylamide moieties and their transdermal drug delivery properties. *European Polymer Journal, 148*, 110348. https://doi.org/10.1016/j.eurpolymj.2021.110348

Tekko, I. A., Chen, G., Domínguez-Robles, J., Thakur, R. R. S., Hamdan, I. M. N., Vora, L., Larrañeta, E., McElnay, J. C., McCarthy, H. O., Rooney, M., & Donnelly, R. F. (2020). Development and characterisation of novel poly (vinyl alcohol)/poly (vinyl pyrrolidone)-based hydrogel-forming microneedle arrays for enhanced and sustained transdermal delivery of methotrexate. *International Journal of Pharmaceutics, 586*, 119580. https://doi.org/10.1016/j.ijpharm.2020.119580

Turner, J. G., White, L. R., Estrela, P., & Leese, H. S. (2021). Hydrogel-forming microneedles: Current advancements and future trends. *Macromolecular Bioscience, 21*, 2000307.

Vader, M. J. C., Schauwvlieghe, A.-S. M. E., Verbraak, F. D., Dijkman, G., Hooymans, J. M. M., Los, L. I., Zwinderman, A. H., Peto, T., Hoyng, C. B., van Leeuwen, R., Vingerling, J. R., Moll, A. C., van Lith-Verhoeven, J. J. C., Dijkgraaf, M. G. W., & Schlingemann, R. O. (2020). Comparing the efficacy of bevacizumab and Ranibizumab in patients with diabetic macular edema (BRDME): The BRDME study, a randomized trial. *Ophthalmology Retina, 4*(8), 777–788. https://doi.org/10.1016/j.oret.2020.02.008

Wang, W., Louise Smith, J., Salvatore Carlino, M., Burmeister, B., Blayne Pinkham, M., Blaise Fogarty, G., Robert Harry Christie, D., Estall, V., Shackleton, M., Clements, A., Wolfe, R., Thi Phuong Thao, L., Jane Paton, E., Steel, V., & Catherine Williams, N. (2021). Phase I/II trial of concurrent extracranial palliative radiation therapy with Dabrafenib and Trametinib in metastatic BRAF V600E/K mutation-positive cutaneous Melanoma. *Clinical and Translational Radiation Oncology*. https://doi.org/10.1016/j.ctro.2021.08.006

Yao, W., Li, D., Zhao, Y., Zhan, Z., Jin, G., Liang, H., & Yang, R. (2019). 3D printed multi-functional hydrogel microneedles based on high-precision digital light processing. *Micromachines, 11*(Issue 1). https://doi.org/10.3390/mi11010017

Index

Note: 'Page numbers followed by "f" indicate figures.'

I

J

K

L

M

N

O

P

R

S

T

U

V

W

Y

Z

Kolibri 360 GmbH
Ferdinand-Jühlke-Straße 7
99095 Erfurt, Deutschland
produktsicherheit@kolibri360.de